T0295246

CHIRAL MATTER

Proceedings of the Nobel Symposium 167

Nobel Symposium 167: Chiral Matter

Högberga Gård, Lidingö, 28 June – 2 July 2021

CHIRAL MATTER
Proceedings of the Nobel Symposium 167

Nobel Symposium 167: Chiral Matter
Högberga Gård, Lidingö, 28 June – 2 July 2021

Egor Babaev
KTH Royal Institute of Technology, Sweden

Dmitri Kharzeev
Stony Brook University, USA & Brookhaven National Laboratory, USA

Mats Larsson
Stockholm University, Sweden

Alexander Molochkov
Far Eastern Federal University, Russia

Vitali Zhaunerchyk
University of Gothenburg, Sweden

 **World Scientific**

NEW JERSEY · LONDON · SINGAPORE · BEIJING · SHANGHAI · HONG KONG · TAIPEI · CHENNAI · TOKYO

Published by

World Scientific Publishing Co. Pte. Ltd.

5 Toh Tuck Link, Singapore 596224

USA office: 27 Warren Street, Suite 401-402, Hackensack, NJ 07601

UK office: 57 Shelton Street, Covent Garden, London WC2H 9HE

Library of Congress Cataloging-in-Publication Data

Names: Nobel Symposium (167th : 2021 : Lidingö, Sweden), author. | Babaev, Egor, 1973–   editor. |
   Kharzeev, Dmitri, editor. | Larsson, Mats (Physics teacher), editor. |
   Molochkov, Alexander, editor. | Zhaunerchyk, Vitali, editor.
Title: Chiral matter : proceedings of the Nobel Symposium 167 / Egor Babaev
   (KTH Royal Institute of Technology, Sweden), Dmitri Kharzeev
   (Stony Brook University, USA & Brookhaven National Laboratory, USA), Mats Larsson
   (Stockholm University, Sweden), Alexander Molochkov (Far Eastern Federal University, Russia),
   Vitali Zhaunerchyk (University of Gothenburg, Sweden).
Other titles: Nobel Symposium 167
Description: New Jersey : World Scientific, [2023] | "Nobel Symposium 167: Chiral matter,
   Högberga Gård, Lidingö, 28 June-2 July 2021"-- title page verso. |
   Includes bibliographical references.
Identifiers: LCCN 2022052457 | ISBN 9789811265051 (hardcover) |
   ISBN 9789811265068 (ebook for institutions) | ISBN 9789811265075 (ebook for individuals)
Subjects: LCSH: Particles (Nuclear physics)--Chirality--Congresses.
Classification: LCC QC793.3.C54 N63 2023 | DDC 539.7/25--dc23/eng20230117
LC record available at https://lccn.loc.gov/2022052457

British Library Cataloguing-in-Publication Data
A catalogue record for this book is available from the British Library.

For any available supplementary material, please visit
https://www.worldscientific.com/worldscibooks/10.1142/13107#t=suppl

Typeset by Stallion Press
Email: enquiries@stallionpress.com

Preface

Nobel Symposia were established in 1965 by the Nobel Foundation. The symposia are devoted to areas of science where breakthroughs are occurring or deal with other topics of primary cultural or social significance. They should be devoted to areas or scholarly disciplines related to the Nobel Prize areas, and should have a limited number of active participants, 30 to 40 by invitation only. A Nobel Symposium must maintain an excellent international standard, and provides a possibility for Swedish scientists to have close contacts with the internationally leading scientists in many fields.

Although the administration and funding of Nobel Symposia have changed over the years, the format has remained the same. Yet, the Nobel Symposium number 167 on Chiral Matter was organized differently due to the pandemic. It was originally scheduled for late June 2020, but was postponed by one year. During late spring 2021 it seemed like the travel restrictions to Sweden would be lifted for many countries, however, this was not the case. Apart from the Nordic and EU countries, and a few exempted countries outside of EU (Israel, Switzerland), most countries were still affected by the travel restrictions imposed by the Swedish government. The organizing committee then took the decision not to postpone further, but to operate the symposium in hybrid form during June 28 to July 2, 2021. The conference venue was Högberga Gård on the island of Lidingö outside Stockholm.

The present book is the proceedings from the Nobel Symposium 167 on Chiral Matter. It provides a very good insight into the remarkable diversity of the concept of chirality. Initially, with the discovery of molecular chirality by Louis Pasteur in 1848, it is now more than 170 years later to be found in cosmology, heavy-ion collisions, quantum electrodynamic and quantum chromodynamics, condensed matter physics, molecular physics, chemistry and biology. This book covers all these topics, with an opening keynote address given by the 2004 year's Nobel laureate in physics, Frank Wilzcek.

We would like to express our gratitude to the Nobel Foundation, the Knut & Alice Wallenberg Foundation, and the Erling-Persson Family Foundation for their generous financial support of Nobel Symposium 167.

Egor Babaev
Dmitri Kharzeev
Mats Larsson
Alexander Molochkov
Vitaliy Zhaunerchyk

Contents

Chirality: A Scientific Leitmotif*

Frank Wilczek

Center for Theoretical Physics, MIT, Cambridge, MA 02139 USA;
T. D. Lee Institute and Wilczek Quantum Center,
Shanghai Jiao Tong University, Shanghai, China;
Arizona State University, Tempe, AZ, USA;
Stockholm University, Stockholm, Sweden

Handedness, or chirality, has been a continuing source of inspiration across a wide range of scientific problems. After a quick review of some important, instructive historical examples, I present three contemporary case studies involving sophisticated applications of chirality at the frontier of present-day science in the measurement of the muon magnetic moment, in topological physics, and in exploring the "chirality" of time. Finally, I briefly discuss chirality as a source of fertile questions.

1. Highlights from the history of chirality

We may never know when human — or protohuman — minds first noticed the extraordinary fact that though their two hands are precisely the same geometric form, they cannot be brought to coincide by continuous motions. The original artists and audiences at the famous "Cave of the Hands" in Santa Cruz, Argentina (now a UNESCO world heritage site), starting around 7300 B. C. according to radiocarbon dating, could hardly avoid holding their own hands up for comparison and noticing that only one of them could fit a given image palm on, while the other could only fit palm away. (Figure 1.) Perhaps a few of them made the imaginative leap to compare in a similar way the virtual "objects" reflected on a still lake to their sources, and ponder their hand-like difference. Later artists in many cultures have sensed and exploited the dynamism that chiral themes can bring to designs.

The study of magnetism, with its proliferation of "right-hand rules", brought in more systematic consideration of the role of chirality in nature. The response of magnetized compass needles to the electric currents, discovered by Oersted in 1820, on the face of it, appears to violate spatial reflection symmetry, or in physics jargon parity P (and also time reversal symmetry T). Indeed, if the "polar" structure of the needle reflected its manifest geometric form, parity really would be broken. The idea that magnetic dipoles are not true vectors but rather axial vectors suggested to Ampere that the source of a ferromagnet's magnetism might be alignment of the axes among circulating molecular electric currents. This proposal restores the parity symmetry (and also the time reversal symmetry) of the physical effect, while ascribing the apparent breaking of those symmetries to the influence of hidden structure within the compass needle. In hindsight we can see recognize that alignment of the molecular currents in a ferromagnet exemplifies spontaneous symmetry breaking, which together with symmetry itself emerged as a dominant theme in

---

*Keynote talk at Nobel Symposium 167, "Chiral Matter", Stockholm June 2021.

Fig. 1. Wall from the "Cave of the Hands", Santa Cruz, Argentina.

twentieth-century physics. When properly viewed, Oersted's experiment becomes a profoundly instructive example of spontaneously broken parity and time reversal symmetry in action. Asymmetry in the behavior, and aspiration for symmetry in the laws, guided Ampere to a bold, fruitful and essentially correct hypothesis about the origin of material magnetism, to wit that it arises from circulating electric currents at the molecular level.

Aspiring to symmetry in underlying laws, while ascribing manifest symmetry breaking to hidden structure, has been a rich source of discovery throughout the history of science.

The strange behavior of magnetic compasses also made a big impression on young Albert Einstein, as he recounts in his *Autobiographical Notes* [1]: "I encountered a wonder of such a kind as a child of 4 or 5 years when my father showed me a compass. That this needle behaved in such a determined way did not fit into the way of incidents at all which could find a place in the unconscious vocabulary of concepts (action connected with "touch"). I still remember — or I think I do — that this incident has left with me a deep impression. There must have been something behind things that was deeply hidden.". It does not seem too large a stretch to see in these early wonderings the seeds of his later obsession with problems of symmetry and apparent asymmetry in electromagnetism. Indeed, his original paper on special relativity [2] begins by referring to asymmetries in Maxwell's electrodynamics that "do not appear to be inherent in the phenomena" and in the second sentence continues "Take, for example, the reciprocal action of a magnet and a conductor ..."

Rotation of the plane of polarization of light as it propagates through a material — optical activity — is a clear indication of parity breaking. In 1811 Francois Arago observed this effect in quartz crystals, and in 1829 the renowned astronomer Herschel discovered that different individual quartz crystals, whose crystal structures formed mirror images, were optically active in opposite senses. Here we have spontaneous breaking of parity in the process of crystal formation. In 1815 Jean Baptiste Biot reported optical activity in liquids and vapors containing substances of biological origin. His results were useful in the sugar industry. In 1849 Louis Pasteur made the epochal discovery that solutions of tartaric acid derived from wine lees are optically active in a definite sense, while solutions of tartaric acid derived by conventional chemical synthesis are inactive. Crystals of synthetic tartaric acid come in two mirror-related forms, like quartz, but biology produces only one of those forms. Today we know that the different mirror forms of many molecules behave very differently in biological processes [3].

People still debate whether this biological breaking of parity is "spontaneous", in the sense that a mirror version of life would for all practical purposes work the same way as the version that actually exists, so that the choice between them is an accident of history, or whether the typically tiny effect of microscopic parity violation under normal terrestrial conditions is somehow amplified in biology. My impression is that most scientists belong to the "spontaneous" camp.

It is appropriate to remark, in this context, that the mechanism of this spontaneous breaking at work here (assuming it is valid) is rather different than what we generally consider in physics. In physics, spontaneous symmetry breaking is generally an idealization of behavior that becomes stable only in the limit of infinite volume, and arises from minimizing a system's energy (or free energy). Those concepts do not apply in a straightforward way to the biological application. In the biological context, it seems that cooperative kinetics, as opposed to cooperative energetics, is the dominant consideration: A population will outcompete and out-replicate the competition, if its members come to agree to use just one class of mirror molecules for a given purpose. (Of course, the choice of chirality for one molecule can dictate what is the favorable choice for others — such *correlations* do not involve parity violation.) The choice of one or another overall chirality would then reflect an accident of history. Note that the effect of cooperative kinetics could easily overwhelm a possible slight intrinsic advantage for growth or survival deriving from microphysical parity violation at the level of individual cells.

It would be difficult to overstate the impact of a short paper by T. D. Lee and C. N. Yang entitled "Question of Parity Conservation in Weak Interactions", written in 1956 [4]. In this paper they pointed out that while there was extensive evidence for accurate parity symmetry P in the strong nuclear and electromagnetic interactions (and, though they did not mention it, gravity), there was no such evidence in the weak interactions. Indeed, assuming the validity of parity symmetry led to the so-called $\theta - \tau$ puzzle, whereby two particles with the same charge, spin, mass and lifetime were distinguished by their opposite intrinsic parity. Within a few months C. S. Wu and collaborators [5], and then others, demonstrated experimentally that parity is very badly broken in weak interactions. The study of parity breaking triggered rapid progress, soon leading to the $V - A$ theory of weak currents, and from there to the concept of gauge symmetries that act only on left-handed particles, and ultimately to the idea that the masses of quarks and leptons (which mix their left- and right-handed forms) are not intrinsic, but arise from their interactions with a universal condensate, i.e., the Higgs condensate. This is a great but oft-told story. Here that bare indication will have to suffice.

2. Chirality at the frontier of precision

The magnetic moment of the muon correlates two axial vectors, namely the muon's spin and its dipolar magnetic field. Its magnitude is captured in the dimensionless quantity (g factor)

$$\vec{\mu} \equiv g_\mu \frac{e}{2m_\mu} \vec{S} \tag{1}$$

in units with $\hbar = c = 1$, so that the magnitude of $\vec{S}$ is $1/2$. Drilling down one more level into the definitions, the interaction of a muon with a static magnetic field is specified by the Hamiltonian $H_{\text{int.}} = -\vec{\mu} \cdot \vec{B}$.

g_μ occupies a very special place in natural philosophy [6]. It is a quantity that both the concept-world of quantum field theory and the tangible world of experimental physics can address with extraordinary precision – using *very* different tools – and support sharp comparison. In this way it epitomizes the ideal of the reductionist program, which seeks to find an exact mapping between those two worlds.

On the experimental side, the basic approach is to measure the precession of a muon's spin as it is subjected to a magnetic field. At first hearing it might seem odd to use highly unstable particles (the muon's lifetime is about 2 microseconds) for precision measurements, but actually the *chiral* decay of the muon can be leveraged to advantage. First of all, we should say that at a modern accelerator it is cheap to produce muons in great abundance, and then inject them into a storage ring and guide them in precisely controlled orbits. Furthermore 2 microseconds is a very comfortable time for modern electronics, and that in that time a relativistic muon will travel a macroscopic distance — hundreds of meters. Since the angular distribution of the decay products of a muon is correlated with the muon's spin (thanks to parity violation!), by observing the decay products experimenters can track the muon's spin precession as a function of time. This exploitation of parity violation is a lovely example of how in research physics often "yesterday's sensation is today's calibration".

On the theoretical side, magnetic moments showcase the truly *quantum* aspect of quantum field theory, namely the importance of spontaneous activity – "virtual particles" – in determining physical behavior. At the classical (i.e., tree-graph or zero-loop) level of the Dirac equation, one obtains the bare value $g = 2$. Virtual particles dress the physical muon. To isolate their effect, it is convenient to consider the anomalous moment

$$a_\mu \equiv \frac{g_\mu - 2}{2} \tag{2}$$

In reaching precise predictions for a_μ there are two big theoretical challenges. One is to account for the effects of electromagnetic and electroweak corrections. The numerically dominant contribution comes from the ordinary quantum electrodynamics of virtual photons, electrons, and muons. These must be calculated to high order in perturbation theory, which is a challenging but highly developed art (Figure 2).

The second is to include the effects of virtual strongly interacting particles. These come in indirectly, primarily through vacuum polarization but also through virtual light-by-light scattering (Figure 3). Two methods have been used to calculate these contributions. One method is to express these virtual processes in terms of related real processes, that can be measured empirically. A variety of tricks including dispersion relations and chiral perturbation theory are used to extrapolate from real to virtual. The other method is to calculate directly from the basic

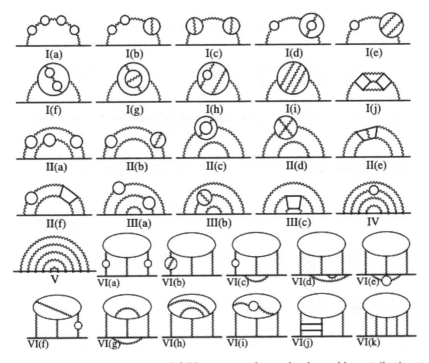

Fig. 2.   A selection of high-order virtual QED processes that make observable contributions to a_μ.

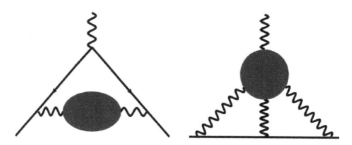

Fig. 3.   Strong interaction modifications of virtual photon "vacuum polarization" and virtual light-by-light scattering make measurable contributions to a_μ. The blobs indicate strongly interacting virtual quarks and gluons, whose behavior must be calculated nonperturbatively.

equations of quantum chromodynamics (QCD). Here perturbation theory is useless, and the work involves number-crunching that on state-of-the-art supercomputers.

The result of all this work is, at the present, both glorious and inconclusive. On the experimental side, we have [7]

$$a_\mu(\text{expt.}) = 116\,592\,061(41) \times 10^{-11} \qquad (3)$$

to be compared with the semi-empirical theoretical calculation [8]

$$a_\mu(\text{semi} - \text{emp.}) = 116\,591\,810(40) \times 10^{-11} \qquad (4)$$

and the numerical calculation [9]

$$a_\mu(\text{num.}) = 116\,591\,945(50) \times 10^{-11} \qquad (5)$$

The first and most profound thing to appreciate here is that the experimental and theoretical values agree to a few parts per ten billion. This is the glorious result. There are few if any other results in science that so convincingly demonstrate the power of mind to comprehend matter.

On the other hand, the two theoretical calculations are (at that level) significantly discrepant. One of them is comfortably compatible with the experimental result, while the other shows an $\approx 4\sigma$ deviation. Since the theoretical calculations, using either method, are extraordinarily complicated to perform and interpret, it is amazing that they agree as well as they do! In any case, this is the inconclusive state of affairs today.

Active efforts to improve the reliability and precision of all three numbers are afoot. If a discrepancy between experiment and (consensus) theory persists, it would suggest the existence of new particles, beyond the standard model, whose contribution to dressing the muon have not been taken into account.

3. Chirality empowering topology

Gauss' integral definition of linking number was one of the earliest results in topology [10]. It has a striking physical interpretation, that connects back to Oersted and Ampere – and forward to Dirac, and beyond.

The linking integral is

$$\text{link}(\gamma_1, \gamma_2) = \frac{1}{4\pi} \oint_{\gamma_1} \oint_{\gamma_2} \frac{\mathbf{r_1} - \mathbf{r_2}}{|\mathbf{r_1} - \mathbf{r_2}|^3} \cdot (d\mathbf{r_1} \times d\mathbf{r_2})$$

$$= \frac{1}{4\pi} \int_{S^1 \times S^1} \frac{\det(\dot{\gamma}_1(s), \dot{\gamma}_2(t), \gamma_1(s) - \gamma_2(t))}{|\gamma_1(s) - \gamma_2(t)|^3} \, ds dt \tag{6}$$

Here in the first form of the integral we integrate over two oriented closed paths in three-dimensional space, and in the second form we realize those paths as parameterized images of two circles. The linking integral is an integer that, as the name suggests, counts the number of times one path winds around the other.

The linking number integral has remarkable physical interpretations. Classically, is the work done by the magnetic field generated by a unit current along γ_1 on a unit magnetic charge traversing γ_2. Alternatively, in quantum mechanics the integrand is the phase of a multiplicative factor accumulated by a unit electric charge traversing γ_2 when γ_1 contains a unit magnetic flux tube, divided by 2π. The total integrated factor is $e^{2\pi i \text{link}(\gamma_1, \gamma_2)}$, which is trivial for integer values of the linking integral. This expresses the invisibility of "Dirac strings" of flux emanating from magnetic monopoles that satisfy his quantization condition. The flux-tube interpretation of the integrand plays a central role in anyon physics, as I'll discuss momentarily.

The linking number integral changes sign upon spatial inversion, and also upon reversing the orientation of either path. Evidently, it is a highly chiral object! It is easy to understand this sensitivity to chirality heuristically: we want linking to be a signed quantity, such that successive windings add up, while motion back and forth cancels. More generally, chiral structure is crucial within most of topology. At the highest of abstraction, one finds that many topological invariants are available only for "oriented" – i.e., chiral – objects; at a more tangible level, one notices the ubiquitous appearance of (P odd) tensor ϵ symbols, either explicit, or implicit in the definitions of Jacobians and exterior derivatives.

Recently those concepts, in a modified and generalized form, have energized a lively and rapidly expanding frontier of quantum theory.

The world-lines of particles in $2 + 1$ dimensions define strands that can wind around (i.e., link with) one another. Suitable states of matter, notably including essentially all states that exhibit the fractional quantum Hall effect, contain quasi-particles whose wave-functions respond to winding in a way similar to how the wave functions of charged particles respond to winding around flux tubes. For historical reasons, the response of multi-(quasi-)particle wave-functions to topological entanglement of their world-lines is called "quantum statistics". Quasi-particles whose many-body wave-functions are sensitive to the topology of world-lines are called *anyons*. Here are a few highlights from the theory of anyons:

- *fractional statistics*: The coefficient α of the linking integral is generally fractional. Thus the accumulated phase for winding of two quasiparticles is

$$e^{2\pi i \alpha \text{link}(\gamma_1, \gamma_2)} \tag{7}$$

with a fractional value of α. This is factor is generally non-trivial even for allowed, i.e., integer, values of the linking integral. Thus winding leaves a non-trivial imprint on the wave function.

- *mutual statistics*: One can have linking between different species A, B of quasi-particles, with a coefficient α_{AB} appearing in Eqn. (7).
- *non-abelian statistics*: The charges and fluxes associated with quasi-particles be chosen from a non-abelian group. In this case the quasi-particles have internal quantum numbers, and the linking integrals become path-ordered integrals giving rise to matrices.

Recent experiments have produced decisive observations of fractional statistics in the $\nu = 1/3$ fractional quantum Hall state, demonstrations of mutual statistics in engineered superconducting circuits, and suggestive evidence for non-abelian statistics in a different fractional quantum Hall state.

Braids can get very complicated, especially if we allow their strands to have different, changeable colors. Ancient South American civilizations including the Inca used knotted braids called *quipu* to encode and transmit complex information that could be shared by far-flung speakers of different languages [11].

Quantum quipu, in the form of geometrically entangled anyon world-lines, can support capacious storage and sophisticated parallel processing, up to and including universal quantum computation, once we bring in mutual and non-abelian statistics. This is the program of "topological quantum computation". By controlling the braiding process, one can perform programmed operations on the anyons' wave function. It is a technology that Microsoft is pursuing aggressively. Topology, with its discrete invariants, imparts a quasi-digital aspect to this form of processing. Indeed, since it is impossible to make small errors in discrete quantities, only errors that exceed a finite threshold can be effective.

For a longer non-technical discussion of anyons, including much more in the way of background and context, I'll happily refer you to a piece I recently contributed to *Inference* [12].

4. "Chirality" in time

Given that consideration of reflection in space has proved so fruitful, it is natural to inquire about reflection in time — temporal "chirality". Does a movie run backwards show a sequence of events that obeys the fundamental laws of physics? This is the question of T symmetry.

Prior to 1964, all empirical evidence was consistent with T symmetry. The fundamental laws of general relativity and quantum electrodynamics, and the phenomenological description of strong and weak processes, were all consistent with it. But in 1964 Cronin and Fitch observed subtle phenomena in K-meson decays that violate T symmetry. (They actually observed violation of CP symmetry, where C is charge conjugation; but since there are compelling theoretical reasons to think that CPT symmetry is very accurate, CP violation strongly suggested T violation. Later work has vindicated that inference.)

While exact T symmetry might be taken as a fundamental principle of physical law, surely "not quite exact T symmetry" cannot be so taken. It begs for deeper explanation.

In 1973 this challenge led the Japanese physicists Makoto Kobayashi and Toshihide Maskawa to a brilliant insight [13]. The basic principles of relativistic quantum field theory, together with the gauge symmetries of our theories of the fundamental interactions, severely constrain the possibilities for couplings among fundamental particles. Kobayashi and Maskawa showed that these constraints are so powerful that they ruled out, as an indirect consequence, any possibility of T violation, given the particles known at the time. But they went on to show that if one expanded the particle spectrum to include a third family – including a new charged lepton τ to accompany the electron e and muon μ and a new quark doublet t, b to accompany the up-down u, d and the charm-strange c, s doublets – then there is exactly possible one coupling that violates T. And because that coupling brings in heavy quarks and the weak interaction its manifestations are, for most practical purposes, small and subtle. The Cronin-Fitch effect, specifically, arises from exchange of virtual heavy quarks, and becomes visible only due to very special features of the K-meson system.

Soon afterward evidence for a third family began to accumulate. By now it is well established. Moreover, the specific T-violating coupling that Kobayashi and Maskawa proposed has been vindicated in detailed quantitative studies of heavy quark decays. The story of T violation has been a triumph worthy to stand beside the corresponding story of P violation [14].

The triumph is not complete, however. Deep theoretical analysis in quantum chromodynamics (QCD) not long after the KM work revealed that there is one other possible T-violating interaction that is consistent with all known general principles, besides the one KM identified. It is the so-called "θ term", described by the Lagrangian density

$$\Delta \mathcal{L} = \frac{g^2 \theta}{8\pi^2} \mathbf{E}^a \cdot \mathbf{B}^a \tag{8}$$

where $\mathbf{E}^a$ and $\mathbf{B}^a$ are the color electric and magnetic fields and g is the strong coupling constant. It changes sign under T (and also under P). The θ term does not involve heavy quarks or the weak interaction. It can be calculated to induce an *electric* dipole moment for the neutron. This makes it dangerous phenomenologically, because the experimental bounds on a neutron electron dipole moment are quite stringent. Upon comparing calculations with experiment, one deduces the bound

$$|\theta| \leq 10^{-10} \tag{9}$$

whereas dimensional analysis would suggest $\theta \sim 1$. Our understanding of approximate T symmetry cannot be considered satisfactory without an explanation for the smallness of θ. This problem is called the "strong P, T problem": Why does the strong interaction, i.e., QCD, obey P and T symmetry so accurately?

During the 40+ years that have passed since the strong P, T problem was clearly articulated several ideas have been put forward to address it, but only one has stood the test of time. It is called the Peccei-Quinn (PQ) mechanism, after Roberto Peccei and Helen Quinn, who first proposed it [15].

The PQ mechanism is best understood as a theory of evolution, applied to the θ parameter. Basically, the number θ becomes a dynamic field

$$\theta \to \theta(\mathbf{x}, t) \tag{10}$$

that minimizes its energy when $\theta(\mathbf{x}, t) \approx 0$. This can be arranged in a simple, natural way, within a modest expansion of the standard model that incorporates extra symmetry (PQ symmetry). Up to a few discrete choices, we arrive at a one-parameter theory, corresponding to the normalization F of the $\theta(\mathbf{x}, t)$ field's kinetic energy $F^2 \partial_\mu \theta \partial^\mu \theta$.

As pointed out by Steven Weinberg and me [16], the quanta of the $\theta(\mathbf{x}, t)$ field are a remarkable new kind of particle, that I christened the *axion*. For all allowed values of F – ranging from $F \sim 10^9$ GeV to $F \sim 10^{18}$ GeV – the axion is a very light, very feebly interacting spin-0 particle.

The earliest papers on axions simply took it as granted that the $\theta(\mathbf{x}, t)$ field would take on its minimum energy value. This would explain the observed smallness of T violation in the strong interaction, and complete the explanation of its approximate validity throughout fundamental physics. A few years later some of us decided to take the evolutionary part of this theory of evolution seriously, and to work out the cosmological history of the $\theta(\mathbf{x}, t)$ [17].

The result came as a stunning and wonderful surprise. The $\theta(\mathbf{x}, t)$ field does settle down very close to 0, so it does its job for fundamental physics. But the residual oscillations, though they are small numerically, carry a lot of energy, since the large value of F implies that $\theta(\mathbf{x}, t)$ is very stiff. The residual oscillations correspond to a cosmological background of axions, broadly similar in its origin to the cosmic microwave background radiation, but with significant differences:

1. Axions are predicted to interact much more feebly with ordinary matter than do photons.
2. Axions have a small but non-zero mass $m_a \sim 10^{-2} \text{GeV}^2 / F$.
3. Axions are produced cold - i.e., they are moving much slower than the speed of light when then drop out of equilibrium with the expanding, cooling big bang fireball.
4. Axions have not been observed.

These properties of axions, together with their calculated abundance, makes them a plausible candidate to supply the astronomers' "dark matter". Indeed, *if axions exist at all, it is hard to avoid the conclusion that they contribute a significant fraction of the observed dark matter* – and thus, according to William of Ockham and Thomas Bayes, plausibly most of it.

Many groups of experimenters around the world are looking for axions [18]. The weakness of their interaction with ordinary matter, and their unknown mass, makes the search challenging. But it doesn't look hopeless. There are genuine prospects that in coming years the required sensitivity will be attained for broad ranges of F.

5. Chirality as a source of questions

Let me conclude with a few more provocative chirality-related questions that I think will be fruitful going forward.

In the macroscopic form of many organisms we see strikingly accurate realizations of reflection symmetry (Figure 4). How and why is that symmetry set up and maintained in a noisy, asymmetric environment? In others we observe spiral symmetry patterns, that break parity; the same questions arise.

Is time reversal symmetry, like spatial parity, broken at the molecular level in biology?

Of course macroscopic biology, like all macroscopic processes, feels the thermodynamic arrow of time. Dissipative processes, aging, and processes that rely directly on oriented flows of energy, like photosynthetic use of incoming sunlight, manifestly differentiate between past and future. It remains a meaningful question, nevertheless, whether fundamental time-reversal symmetry T applies straightforwardly to biological materials at the molecular level or whether, like P, it is broken at that level.

For ease of discussion, let us consider a molecular species M whose geometric structure admits no non-trivial symmetry. Then we can define a body-fixed

Fig. 4. How is the symmetry of this body established, and maintained, in its complex, noisy environment?

coordinate system for M unambiguously. Let us suppose that M molecules have, in that co-ordinate system, an effective electric dipole moment $\vec{d}$ and an effective magnetic dipole moment $\vec{\mu}$. Under T, $\vec{d}$ remains invariant but $\vec{\mu}$ changes sign. Thus, the dot product $\vec{d} \cdot \vec{\mu}$, which measures the correlation of the moments, is odd under T. T symmetry at the molecular level therefore predicts that we will find that in a sample of M molecules values of $\vec{d} \cdot \vec{\mu}$ with opposite signs occur with equal probability. If we find, to the contrary, a favored sign for $\vec{d} \cdot \vec{\mu}$ in biological samples of a specified type (*e.g.*, within a specific organelle within a specific cell type within a specific individual), then we will have found biological violation of T symmetry.

In the preceding paragraph, the word "effective" alludes to an important subtlety. Thoughtful chemists and biologists might be surprised to hear, as I mentioned in passing above, that physicists test T symmetry by looking for – and, so far, not finding – non-vanishing electric dipole moments. After all, in any chemistry text you'll find extensive tabulations of non-vanishing electric dipole moments, and successful computations of them! The difference is that for, say, a neutron the only available vector degree of freedom is its spin direction $\vec{s}$, and thus a hypothetical neutron electric dipole moment needs to obey $\vec{d} \propto \vec{s}$. But $\vec{s}$ changes sign under T, while $\vec{d}$ would not. (A more rigorous and general version of this argument invokes Kramers' theorem in quantum theory.) But molecules interacting with common environments can be stabilized in states that have additional structures besides a spin direction, *e.g.*, stable average shapes, and $\vec{d}$ can be oriented relative to such structures without violating T. Those average structures define effective molecules and the textbooks tabulate their effective dipole moments.

Note that under spatial inversion P, $\vec{d}$ changes sign while $\vec{\mu}$ remains invariant, so that $\vec{d} \cdot \vec{\mu}$ is odd under P. Thus P symmetry at the molecular level separately predicts that $\vec{d} \cdot \vec{\mu}$ occurs equally with opposite signs in an invariant sample. If, however, we focus on a chiral species M, which differs from its P-transformed version M^P, then a vanishing average $\vec{d} \cdot \vec{\mu}$ within a biological sample that contains a preponderance of M over M' cannot be ascribed to P (since the sample itself violates P), but only to T.

T symmetry at the molecular level predicts equal probabilities for either sign of $\vec{\mu}$ (even without reference to $\vec{d}$). However, just as for P, we can have effective molecules M that differ significantly from their T-transformed versions M^T, say by having free spins oppositely oriented or effective currents circulating in opposite directions (in the body-fixed system). In order for an environment to be sensitive to the difference between M and M^T, the environment itself must be T-violating. Thus, as in all forms of spontaneous symmetry breaking, stabilization of a particular choice must involve a cooperative interaction among many molecules, each providing part of the environment for others.

Thus far in our discussion of T violation $\vec{d}$ has merely provided a convenient marker for the shape variables that $\vec{\mu}$ must depend on. But it is a useful one, since a non-vanishing correlation $\kappa \equiv \vec{d} \cdot \vec{\mu}$ would be significant both experimentally and functionally.

Experimentally, a non-vanishing κ would imply that application of an external electric field to a sample featuring our molecules would induce a magnetic field in response, and *vice versa*.

Functionally, a non-vanishing κ would allow, through local contact interactions, transfer between charge and spin orientations. This could be a useful ability to have, since charge orientations (charge "bits") are relatively easy to create and manipulate, while spin orientation (spin "bits") are relative easy to store stably.

More broadly and playfully, we might consider a complex, dynamic biological system as a special case of an industrial economy. It is obviously useful, within an industrial economy, to have an agreed convention about which chirality of screws to employ. Such a choice involves spontaneous breaking of P. It could also be useful to have an agreement about which way clocks run. Such a choice involves spontaneous breaking of T.

It is noteworthy that there is a known class of materials, the so-called multiferroics, that do feature both electric and magnetic polarization in bulk – i.e., that are both ferroelectric and ferromagnetic – with non-zero alignment. Of course, either sign of the alignment will occur equally often in crystals derived from symmetric mixtures. But a sample characterized by a single sign could serve as a template for cultivation of T-violating networks of molecules, as contemplated here – including, in the context of biology, transmission of a preferred molecular time orientation by heredity.

Spontaneous symmetry breaking by cooperative kinetics is an idea suggested by biology, as we've discussed. But it is certainly not restricted to biology, as a matter of logic. It should apply, for example, to crystal growth — the form that emerges from rapid (non-equilibrium) precipitation need not be the thermodynamically favored form. Does this kind of spontaneous symmetry breaking, like more conventional symmetry breaking, have universal consequences? Is it common, or at least easy to observe and characterize?

Our subject, chirality, began with ancient humans and their experience with their hands. Despite the enormous progress of science and our success in asking and answering many sophisticated questions, many naive questions about humans and their hands remain puzzling. Why are most but (by far) not all people right-handed? Why do most but not all people have their hearts on the left, their verbal centers in their left semi-brain, and other such chiral asymmetries, against a backdrop of overall bilateral symmetry? Why are those various asymmetries somewhat, but imperfectly correlated?

Chirality has been, is now, and will remain for the foreseeable future a potent source of attractive questions.

Acknowledgement

This work is supported by the U.S. Department of Energy under grant Contract Number DE-SC0012567, by the European Research Council under grant 742104, and by the Swedish Research Council under Contract No. 335-2014-7424.

References

1. A. Einstein, "Autobiographical Notes" In P. A. Schilpp (Ed.), *Albert Einstein, Philosopher-Scientist*. The Library of Living Philosophers, Open Court, La Salle IL. (1949).
2. A. Einstein, *Annalen der Physik* 17 (10): 891 (1905).
3. M. Inaki, J. Liu, and K. Matsuno, *Phil Trans. R. Soc. London B* 371(1710): 20150403 (2016).
4. T. D. Lee and C. N. Yang, *Phys. Rev.* 104: 254 (1956); E *Phys. Rev.* 106: 1371 (1957).
5. C. S. Wu, E. Ambler, R. Hayward, D. Hoppes, and R. Hudson, *Physical Review* 105 (4): 1413 (1957).
6. F. Jegerlehner, *The Anomalous Magnetic Moment of the Muon* Springer (2017).
7. B. Abi *et al.*, *Phys. Rev. Lett.* 126: 141801 (2021).
8. T. Aoyama *et al.*, *Physics Reports* 887 (2020).
9. S. Borsanyi *et al. Nature* 593: 51 (2021).
10. R. Ricca and B. Nipoti, *Journal of Knot Theory* 20: 1325 (2011).
11. See the excellent article "Quipu" in Wikipedia.
12. F. Wilczek, "Quanta of the Third Kind" in *Inference* 6(3) (2021).
13. M. Kobayashi and T. Maskawa, *Prog. Theor. Phys.* 49(2) 652 (1973).
14. A. Höcker and Z. Ligeti, *Ann. Rev.Nucl. Part. Sci.* 56: 501 (2006).
15. R. Peccei and H. Quinn, *Phys. Rev. Lett.* 38(25): 1440 (1977); *Phys. Rev.* D16 (6): 179 (1977).
16. S. Weinberg, *Phys. Rev. Lett.* 40(4) 223 (1978); F. Wilczek, *Phys. Rev. Lett.* 40(5) 279 (1978).
17. J. Preskill, M. Wise, and F. Wilczek, *Phys. Lett.* B120 (1–3): 127–132 (1983).
18. P. Sikivie, *Rev. Mod. Phys.* 93: 015005 (2021).

Chirality in Astrophysics

Axel Brandenburg

Nordita, KTH Royal Institute of Technology and Stockholm University, and
The Oskar Klein Centre, Department of Astronomy,
Stockholm University, Stockholm, Sweden
* E-mail: brandenb@nordita.org
https://www.nordita.org/~brandenb

Chirality, or handedness, enters astrophysics in three distinct ways. First, magnetic field and vortex lines tend to be helical and have a systematic twist in the northern and southern hemispheres of a star or a galaxy. Helicity is here driven by external factors. Second, chirality can also enter at the microphysical level and can then be traced back to the parity-breaking weak force. Third, chirality can arise spontaneously, but this requires not only the presence of an instability, but also the action of nonlinearity. Examples can be found both in magnetohydrodynamics and in astrobiology, where homochirality among biomolecules probably got established at the origin of life. In this review, all three types of chirality production will be explored and compared.

Keywords: Magnetic helicity; chiral magnetic effect; gravitational waves; homochirality

1. Introduction

Chirality, or handedness, plays important roles in many different fields of astrophysics, including astrobiology. There are three distinct types of chirality production: (i) driven chirality due to external factors, (ii) driven chirality due to intrinsic properties, and (iii) spontaneous chirality production due to instability and nonlinearity. The primary applications for these three types of chirality production are rather distinct, but one can find unifying circumstances under which the different types can be demonstrated and compared. One such circumstance is given by the presence of magnetic fields.

Magnetic fields can experience twisting that makes them helical, but the central question is what determines the sign of this twist—especially if it is a systematic one, always being in the same sense. When there are extrinsic or intrinsic factors such as the combination of rotation and stratification in a star, or intrinsic factors such as the presence of fermions of one of two handednesses, the answer is in principal clear. However, there can also be spontaneous helicity production, where the sign depends ultimately on chance, so both signs are possible under almost identical conditions. An example that we discuss at the end of this review is in the field of magnetohydrodynamics (MHD). Here, a magnetic field in a stratified atmosphere exhibits a magnetic buoyancy instability where, in the end, once nonlinearity plays a role, one particular handedness dominates to nearly hundred percent. It is this example that, in a broad sense, also carries over to astrobiology and the origin of life, where one particular chirality of biomolecules eventually dominates and leads, to what is known as homochirality. Magnetic fields are probably not involved in the origin of life, but there are simple mathematical analogies in both cases.

16

Table 1. Summary of three types of chirality production in astrophysics.

What	Where	How
Helicity, driven by stratification & rotation	stars, planets, galaxies	magnetic field from Zeeman effect, polarization, in situ in solar wind
Fermion asymmetry, axions	entire universe	polarization patters, photon arrival direction statistics, circularly polarized gravitational waves
Spontaneous chirality production	astrobiology, MHD	enantiospecific uptake of nutrients

In Table 1, we summarize the three types of chirality production under the "what" column or category. The "where" category lists some specific examples, and the "how" category highlights some specific techniques for measuring chirality for those three types.

We begin with some historical remarks highlighting the significance of helicity of magnetic fields (Sect. 2). In Sects. 3–5, we discuss aspects of the three types in more detail and then conclude in Sect. 6 with some additional reflections. We emphasize that we use the terms chirality, helicity, and handedness rather interchangeably, although technically this is not always accurate.

2. Historical remarks

2.1. *Helicities in fluid dynamics and MHD*

In the context of fluid dynamics, the term helicity was coined by H. Keith Moffatt [1], who identified the topological equivalence between the knottedness of vortex lines in fluid dynamics and the kinetic helicity. In fact, the term helicity was already used by Robert Betchov [2] in 1961, but Moffatt proposed this name in his 1969 paper on the grounds that this term is also used in subatomic physics to describe the alignment or anti-alignment of spin and momentum of fermions, for example. Mathematically, the mean kinetic helicity density is defined as $\langle \boldsymbol{\omega} \cdot \boldsymbol{u} \rangle$, where the vorticity $\boldsymbol{\omega} = \boldsymbol{\nabla} \times \boldsymbol{u}$ is the curl of the velocity $\boldsymbol{u}$. Kinetic helicity is a pseudoscalar, i.e., it changes its sign when the system is inspected through a mirror. Likewise, $\boldsymbol{\omega}$ is a pseudovector, so it is more meaningful to plot it with its sense of rotation (which changes in a mirror), rather than a vector with an arrow at its end.

In the magnetic context, the corresponding quantity that we now call magnetic helicity was already studied by Lodewijk Woltjer [3] in 1958 to characterize force-free magnetic fields and by John Bryan (J.B.) Taylor [4] in 1974 to describe the relaxation of a toroidal plasma.

2.2. *Magnetic helicity conservation*

The quantities that we shall focus on here are mainly the mean magnetic helicity density $\langle \boldsymbol{A} \cdot \boldsymbol{B} \rangle$, where $\boldsymbol{B} = \boldsymbol{\nabla} \times \boldsymbol{A}$ is the magnetic field expressed in terms of

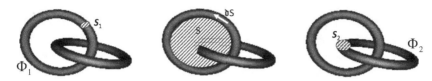

Fig. 1. Illustration showing that the line integral $\oint_{\partial S} \boldsymbol{A} \cdot \mathrm{d}\boldsymbol{\ell}$ along flux tube Φ_1, can be written as a surface integral over the enclosed surface S, $\int_S (\boldsymbol{\nabla} \times \boldsymbol{A}) \cdot \mathrm{d}\boldsymbol{S}$, where ∂S is the line along flux tube Φ_1. However, the only nonvanishing contribution comes from flux tube Φ_2, i.e., $\int_{S_2} \boldsymbol{B} \cdot \mathrm{d}\boldsymbol{S}$.

its vector potential $\boldsymbol{A}$, and the magnetic helicity over a volume containing two interlocked flux loops, $\int_{V_1+V_2} \boldsymbol{A} \cdot \boldsymbol{B} \, \mathrm{d}V = \int_{V_1} \boldsymbol{A} \cdot \boldsymbol{B} \, \mathrm{d}V + \int_{V_2} \boldsymbol{A} \cdot \boldsymbol{B} \, \mathrm{d}V$. The volumes V_1 and V_2 are those of the two tubes. For V_1, we can split the volume integral into a line integral $\oint_{\partial S} \boldsymbol{A} \cdot \mathrm{d}\boldsymbol{\ell}$ along the flux tube and a surface integral $\int_{S_1} \boldsymbol{B} \cdot \mathrm{d}\boldsymbol{S}$ across the tube, i.e.,

$$\int_{V_1} \boldsymbol{A} \cdot \boldsymbol{B} \, \mathrm{d}V = \left(\int_{S_1} \boldsymbol{B} \cdot \mathrm{d}\boldsymbol{S} \right) \left(\oint_{\partial S} \boldsymbol{A} \cdot \mathrm{d}\boldsymbol{\ell} \right); \tag{1}$$

see Figure 1. Using Stokes' theorem, the line integral along tube 1 can be rewritten as a surface integral over the magnetic field going through the flux ring, but its only nonvanishing contribution comes from the other intersecting tube with surface S_2. Thus, we have $\int_{V_1} \boldsymbol{A} \cdot \boldsymbol{B} \, \mathrm{d}V = \pm \Phi_1 \Phi_2$, and likewise for V_2, so we get a factor 2. The sign depends on the relative orientation of the field vectors in the two tubes, so we write here $\int \boldsymbol{A} \cdot \boldsymbol{B} \, \mathrm{d}V = \pm 2\Phi_1 \Phi_2$.

The magnetic field is a pseudovector. Its evolution is governed by the homogeneous Maxwell equation $\partial \boldsymbol{B}/\partial t = -\nabla \times \boldsymbol{E}$, where t is time and $\boldsymbol{E}$ is the electric field. Since $\boldsymbol{\nabla} \cdot \boldsymbol{B} = 0$, it is convenient to consider the uncurled induction equation

$$\partial \boldsymbol{A}/\partial t = -\boldsymbol{E} - \boldsymbol{\nabla}\phi, \tag{2}$$

where ϕ is the electric (or scalar) potential. To obtain an evolution equation for $\boldsymbol{A} \cdot \boldsymbol{B}$, we compute $\boldsymbol{A} \cdot \dot{\boldsymbol{B}} + \dot{\boldsymbol{A}} \cdot \boldsymbol{B}$, where dots denote partial time differentiation, and find

$$\frac{\partial}{\partial t} (\boldsymbol{A} \cdot \boldsymbol{B}) = -2\boldsymbol{E} \cdot \boldsymbol{B} - \boldsymbol{\nabla} \cdot \underbrace{(\phi \boldsymbol{B} + \boldsymbol{E} \times \boldsymbol{A})}_{\text{helicity flux}}. \tag{3}$$

We see that the magnetic helicity production depends on $\boldsymbol{E} \cdot \boldsymbol{B}$ and on the presence of magnetic helicity fluxes, $\mathcal{F}_H = \phi\boldsymbol{B} + \boldsymbol{E} \times \boldsymbol{A}$. The $\boldsymbol{E} \cdot \boldsymbol{B}$ term plays important roles in electrically non-conducting environments, for example during inflation in the early universe when space-time became extremely diluted. By contrast, in the contemporary universe, and even in the space between galaxy clusters, there is still sufficient conductivity so that the laws of MHD apply, and not the vacuum equations for electromagnetic waves, which apply during inflation. In MHD, the electric field is given by $\boldsymbol{E} = -\boldsymbol{u} \times \boldsymbol{B} + \boldsymbol{J}/\sigma$, where $\boldsymbol{J}$ is the current density and σ the electric conductivity. We see that the first term does not contribute to $\boldsymbol{E} \cdot \boldsymbol{B}$. Thus, the only contribution comes from $\boldsymbol{J} \cdot \boldsymbol{B}/\sigma$, so we have

$$\frac{\partial}{\partial t}\langle \boldsymbol{A} \cdot \boldsymbol{B} \rangle = -2\eta \langle \boldsymbol{J} \cdot \boldsymbol{B} \rangle, \tag{4}$$

where $\eta = (\mu_0 \sigma)^{-1}$ is the magnetic diffusivity, and angle brackets denote volume averaging. When the conductivity is large, η is small, and $\eta \langle \boldsymbol{J} \cdot \boldsymbol{B} \rangle$ converges to zero like $\eta^{1/2}$ as $\eta \to 0$.

The smallness of η in many astrophysical settings makes $\langle \boldsymbol{A} \cdot \boldsymbol{B} \rangle$ nearly perfectly conserved [3], except for the presence of magnetic helicity fluxes. Those vanish in homogeneous systems (e.g., in homogeneous helical turbulence), but astrophysical dynamos are usually not homogeneous and magnetic helicity fluxes, for example out of the star or between its northern and southern hemispheres, are believed to play important roles in astrophysical dynamos to alleviate some serious constraints [5] arising from the magnetic helicity conservation otherwise.

2.3. Helicity-driven large-scale dynamos

To understand the aforementioned constraint, we need to emphasize that magnetic helicity is usually connected with the presence of kinetic helicity, $\langle \boldsymbol{\omega} \cdot \boldsymbol{u} \rangle$. It can lead to an electromotive force along the mean magnetic field, $\overline{\boldsymbol{B}}$, of the form

$$\overline{\boldsymbol{u} \times \boldsymbol{b}} = \alpha \overline{\boldsymbol{B}} - \eta_{\mathrm{t}} \mu_0 \overline{\boldsymbol{J}}, \tag{5}$$

where $\boldsymbol{u} = \boldsymbol{U} - \overline{\boldsymbol{U}}$ and $\boldsymbol{b} = \boldsymbol{B} - \overline{\boldsymbol{B}}$ are fluctuations of velocity and magnetic fields, and $\overline{\boldsymbol{J}} = \boldsymbol{\nabla} \times \overline{\boldsymbol{B}}/\mu_0$ is the mean current density. Here, overbars denote averaging (for example planar or xy averaging), but the type of averaging depends on the particular problem. Under the assumption of isotropy and high conductivity [6], we have $\alpha = -\langle \boldsymbol{\omega} \cdot \boldsymbol{u} \rangle \tau/3$ and $\eta_{\mathrm{t}} = \langle \boldsymbol{u}^2 \rangle \tau/3$, and τ is the correlation time. The $\alpha \overline{\boldsymbol{B}}$ term can lead to an exponential growth of $\overline{\boldsymbol{B}}$. The second term just leads to an enhancement of the microphysical magnetic diffusion, $\eta \mu_0 \overline{\boldsymbol{J}}$.

The α effect leads to the generation of magnetic helicity of the mean field with $\overline{\boldsymbol{A}} \cdot \overline{\boldsymbol{B}} = O(\overline{\boldsymbol{B}}^2/k_{\mathrm{m}}) \neq 0$, where k_{m} is the typical wavenumber of the magnetic field. However, since the total magnetic helicity is conserved (and vanishing if the field was very small initially), we must generate small-scale magnetic helicity, $\langle \boldsymbol{a} \cdot \boldsymbol{b} \rangle$, of opposite sign. The resulting current helicity, $\langle \boldsymbol{j} \cdot \boldsymbol{b} \rangle \approx k_{\mathrm{f}}^2 \langle \boldsymbol{a} \cdot \boldsymbol{b} \rangle$, with some typical wavenumber k_{f} characterizing the fluctuation scales, quenches the α effect and leads to slow dynamo saturation [7].

We will not go into further details here, but refer the reader to reviews on the subject [5, 8]. This field of research remains very active and there are still important questions regarding magnetic helicity fluxes that remain controversial.

3. Chirality driven by external factors

3.1. *Examples of helicities: even and odd in B*

We discuss here three different helicities: kinetic helicity $\langle \boldsymbol{\omega} \cdot \boldsymbol{u} \rangle$, magnetic helicity $\langle \boldsymbol{A} \cdot \boldsymbol{B} \rangle$, and cross helicity $\langle \boldsymbol{u} \cdot \boldsymbol{B} \rangle$. The latter reflects the linkage between an $\boldsymbol{\omega}$ tube (vortex tube) and a $\boldsymbol{B}$ tube (magnetic flux tube). To understand the production of kinetic helicity, one has to realize that it is a pseudoscalar. This means, that it is the product of a polar vector and an axial vector. Rotation, for example, is an axial vector, but that alone cannot produce magnetic or kinetic helicity. However, in the presence of both rotation $\boldsymbol{\Omega}$ and gravitational stratification characterized by the gravitational acceleration $\boldsymbol{g}$, which is a polar vector, we can produce the pseudoscalar $\boldsymbol{g} \cdot \boldsymbol{\Omega}$. Thus, kinetic helicity can be produced if one can identify external factors such as the combined presence of $\boldsymbol{g}$ and $\boldsymbol{\Omega}$ that could explain the presence of a non-vanishing helicity. However, these external factors must also be even in the magnetic field, so a nonvanishing $\langle \boldsymbol{u} \cdot \boldsymbol{B} \rangle$ with a systematic sign cannot be explained in that way. Finite cross helicity can, however, be explained if rotation was replaced by an imposed magnetic field $\boldsymbol{B}_0$, so that $\boldsymbol{g} \cdot \boldsymbol{B}_0$ would be finite.

We thus see that $\langle \boldsymbol{\omega} \cdot \boldsymbol{u} \rangle$ may be linked to $\boldsymbol{g} \cdot \boldsymbol{\Omega}$, although this needs to be (and has been) verified using a detailed calculation [6]. In spherical coordinates (r, θ, ϕ), where θ is colatitude and the unit vectors of $\boldsymbol{g}$ and $\boldsymbol{\Omega}$ are $\hat{\boldsymbol{g}} = (-1, 0, 0)$ and $\hat{\boldsymbol{\Omega}} = (\cos\theta, -\sin\theta, 0)$, respectively, we have $\boldsymbol{g} \cdot \boldsymbol{\Omega} = -\cos\theta$, which is negative in the north and positive in the south. This is indeed consistent with the observed sign of $\langle \boldsymbol{\omega} \cdot \boldsymbol{u} \rangle$, and it is also found to govern the sign of the magnetic helicity, but only that at small and moderate length scales, i.e., $\langle \boldsymbol{a} \cdot \boldsymbol{b} \rangle$.

Regarding the cross helicity, there is indeed a systematic large-scale magnetic field $\boldsymbol{B}_0$ at the solar surface, although there can be several sign changes in each hemisphere. (The radius of the Sun is 700 Mm, and 1 Mm = 1000 km, a useful unit in solar physics!) On smaller scales of ≥ 20 Mm, it also changes between the two sides of a sunspot pair, which is consistent with observations [9]. It turns out that $\langle \boldsymbol{u} \cdot \boldsymbol{B} \rangle \approx -(\eta_t/c_s^2)\, \boldsymbol{g} \cdot \boldsymbol{B}_0$, where η_t is the turbulent magnetic diffusivity and c_s is the sound speed [10]. This production mechanism of cross helicity may play a role in theories of shallow sunspot formation [11, 12].

3.2. *Observing helicity*

3.2.1. *Measuring magnetic helicity from solar magnetograms*

Above the surface of the Sun, one often sees twisted structures in extreme ultraviolet and x-ray images, which are suggestive of helicity [13–15]. Even space observations of the surface of the Earth reveal cyclonic cloud patterns that have opposite orientation in the northern and southern hemispheres. However, to draw a connection with magnetic helicity, one must first detect the magnetic field. This is possible through the Zeeman effect, which causes circular polarization proportional to the line-of-sight magnetic field, and linear polarization related to the perpendicular magnetic field component – except for a π ambiguity, which means that polarization measurements are never able to tell where the tip of the magnetic field vector is, so there is an uncertainty with respect to $180°$. For strong enough magnetic fields, a "disambiguation procedure" based on a minimal magnetic energy assumption allows one to determine the full $\boldsymbol{B}$ vector at the solar surface [16, 17]. However, there is still not enough information about the changes of polarization parameters along the line of sight, i.e., below and above the surface of the Sun.

To make progress, one has to make some extra assumptions. One possibility is to determine just the components normal to the surface, $B_\| \equiv B_z$ and $J_\| \equiv J_z = \partial_x B_y - \partial_y B_x$, where local Cartesian coordinates (x, y, z) have been employed, and compute $J_\| B_\|$. This was first done by Seehafer [18], who found that $J_\| B_\|$ is negative in most of the active regions in the northern hemisphere of the Sun, and positive in most of the active regions in the southern one. From $J_\|$ and $B_\|$, one can also determine a proxy of the magnetic helicity spectra, $\tilde{A}_\|(\boldsymbol{k}_\perp)\tilde{B}_\|^*(\boldsymbol{k}_\perp)$, where tildes denote Fourier transformation, $\boldsymbol{k}_\perp$ is the wavevector in the horizontal (xy surface) plane, and the asterisk denotes the complex conjugate. Here, $\tilde{A}_\| = \tilde{J}_\|/\boldsymbol{k}_\perp^2$. This has been done [19, 20] and magnetic energy and helicity spectra are shown in Figure 2(a) for active region AR 11158. Subsequent work [9] also presented cross helicity spectra.

3.2.2. *Magnetic helicity spectra from a time series*

A completely different approach, due to Matthaeus and *et al.* [21], which also makes use of Fourier transformation, is to use *in situ* observations of time series of the $\boldsymbol{B}$ vector in the solar wind at one point in space. One can then make use of what is known as the Taylor hypothesis to associate the temporal changes with different positions through $\boldsymbol{r} = \boldsymbol{r}_0 - \boldsymbol{v}t$, where $\boldsymbol{v}$ is the velocity vector of the solar wind, and $\boldsymbol{r}_0$ is a reference position. Assuming homogeneity, i.e., that the statistical properties are independent of position, one can write the magnetic two-point correlation tensor in Fourier space as

$$4\pi\langle \tilde{B}_i(\boldsymbol{k})\tilde{B}_j(\boldsymbol{k})\rangle = (\delta_{ij} - \hat{k}_i \hat{k}_j)2\mu_0 E_M(k) - i\epsilon_{ijk}\hat{k}_k H_M(k), \tag{6}$$

where $E_M(k)$ is the magnetic energy spectrum, normalized such that $\int E_M(k)\,dk = \langle \boldsymbol{B}^2 \rangle/2\mu_0$, and $H_M(k)$ is the magnetic helicity spectrum with $\int H_M(k)\,dk = \langle \boldsymbol{A}\cdot\boldsymbol{B}\rangle$. This procedure revealed a clear hemispheric antisymmetry with respect to the solar equator [22].

Comparing Figures 2(a) and (b), which are here both for the southern hemisphere, we see that at the solar surface, $H_M(k)$ has the expected sign at small and intermediate scales ($k > 0.1\,\mathrm{Mm}^{-1}$). In the solar wind,[a] however, a positive sign is only seen at very large scales ($k < 30\,\mathrm{AU}^{-1} \approx 0.0002\,\mathrm{Mm}^{-1}$). The typical wavenumber at which the spectral helicity changes sign at the solar surface (radius

---

[a]The mean distance between the Sun and the Earth is one astronomical unit ($1\,\mathrm{AU} \approx 149,600\,\mathrm{Mm}$).

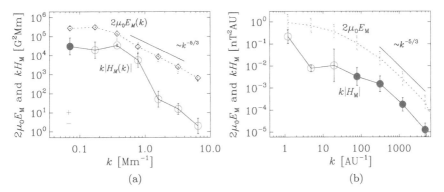

Fig. 2. Magnetic energy and magnetic helicity spectra for southern latitudes (a) at the solar surface in active region AR 11158, and (b) in the solar wind at ~ 1 AU distance ($1\,\mathrm{AU} \approx 149{,}600\,\mathrm{Mm}$). Positive (negative) signs are shown as red open (blue filled) symbols. Positive signs are the solar surface at intermediate and large k correspond to positive values in the solar wind at small k. Note that $1\,\mathrm{G} = 10^{-4}\,\mathrm{T} = 10^{5}\,\mathrm{nT}$.

$700\,\mathrm{Mm}$) is $k \approx 0.1\,\mathrm{Mm}^{-1}$. Expanding this linearly to a distance of 1 AU yields a corresponding wavenumber of $k = 70\,\mathrm{AU}^{-1}$, which is indeed where the spectrum in the solar wind changes sign; see Figure 2(b). However, the change is here the other way around: from positive to negative at large k. The reason for this sign mismatch is not yet fully understood, but it is worth noting that a sign reversal has also been seen in idealized numerical simulations of stellar and galactic dynamos embedded in a turbulent exterior [23–25]. Chiral solar wind MHD turbulence has also been studied in the equatorial plane, but then both signs of helicity are possible [26, 27]. Both signs of magnetic helicity have also been found from multi-spacecraft measurements in close proximity of the equatorial plane [28].

3.2.3. Canceling Faraday depolarization with helicity

There is an intriguing possibility to determine magnetic helicity from the cancelation of Faraday depolarization [29, 30]. It might be particularly suitable for calculating the magnetic helicity in the outskirts of edge-on galaxies, where one might see the sign of magnetic helicity being reversed.

Normally, in the absence of magnetic helicity, Faraday rotation causes Faraday depolarization. This is caused by the superposition of different polarization planes from different depths along the line of sight. At the same time, however, the perpendicular component of the magnetic field itself can rotate about the line of sign if it is helical. These two effects can then either enhance each other (and make the Faraday depolarization more complete), or they can offset each other and lead to increased transmission or reduced depolarization.

Mathematically, this can be seen by considering the observable polarization, written in complex form as

$$P(x, z, \lambda^2) \equiv Q + iU = p_0 \int_{-\infty}^{\infty} \epsilon \, e^{2i(\psi_P + \phi\lambda^2)} \, dy, \qquad (7)$$

where $\psi_P = \psi_B + \pi/2$ is the electric field angle, $\psi_B = \mathrm{atan}(B_y, B_x)$ is the magnetic field angle, $\epsilon(x, y, z)$ is the emissivity, p_0 is the degree of polarization, λ is the wavelength,

$$\phi(x, y, z) = -K \int_{-\infty}^{y} n_e(x, y', z) \, B_\parallel(x, y', z) \, \mathrm{d}y' \tag{8}$$

is the Faraday depth, with n_e being the electron density and $K = 0.81\,\mathrm{m}^{-2}\,\mathrm{cm}^3$ $\mu\mathrm{G}^{-1}\,\mathrm{pc}^{-1} = 2.6 \times 10^{-17}\,\mathrm{G}^{-1}$ being a constant. Evidently, Faraday depolarization is canceled if $\psi_P + \phi\lambda^2 = 0$.

In essence, for edge-on galaxies, we expect maximum polarized emission in diagonally opposite quadrants of a galaxy; see Fig. 19.12 of Ref. 31. This technique has also been applied to synthetic data of the solar corona [32].

3.2.4. *Magnetic helicity proxy*

In the context of cosmology, a proxy of parity breaking and finite helicity of the cosmic microwave background and the Galactic foreground is obtained by decomposing the observed linear polarization in the sky into parity-even and parity-odd contributions. This is done by expanding the complex polarization $P \equiv Q + iU$ with Stokes parameters Q and U into a spin-2 spherical harmonics as [33, 34]

$$\tilde{R}_{\ell m} = \int_{4\pi} (Q + iU) \, _2Y_{\ell m}^*(\theta, \phi) \, \sin\theta \, \mathrm{d}\theta \, \mathrm{d}\phi. \tag{9}$$

The parity-even (E) and parity-odd (B) contributions are obtained as the real and imaginary parts of the return transformation as [35, 36]

$$E + iB \equiv R = \sum_{\ell=2}^{N_\ell} \sum_{m=-\ell}^{\ell} \tilde{R}_{\ell m} Y_{\ell m}(\theta, \phi). \tag{10}$$

In cosmology, one usually considers correlations between E and B, as well as temperature and B, for example. However, it is also useful to consider B on its own, as was done in the context of the solar magnetic field [37] and in the context of the Galactic magnetic field [38]. The result for our Galaxy is shown in Figure 3.

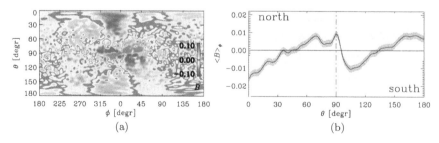

Fig. 3. Left: Galactic B mode polarization. Right: longitudinally averaged B mode polarization. Here, θ and ϕ are Galactic colatitude ($= 90° - $ latitude) and longitude.

We see that the B polarization is locally antisymmetric about the equatorial plane, but there are canceling contributions from different longitudes, so the longitudinal average is much smaller. Nevertheless, even the longitudinal average does show a hemispheric antisymmetry.

Comparing with synthetic observations using the magnetic field from idealized dynamo models [39] shows that this hemispheric dependence does not originate from the different signs of the magnetic helicity expected in the northern and southern hemispheres, but from the spiral pattern of our Galaxy, where the views from the north and south correspond to mirror images of each other.

4. Magnetic helicity throughout the whole Universe

There is at present no definitive observation of finite helicity throughout all of the Universe, but the possibility certainly exists [40]. In this section, we first discuss two quite different mechanisms. One is related to the chiral magnetic effect (CME) and the other to inflationary magnetogenesis. Both generate helical magnetic fields, and the electromagnetic stress can generate relic gravitational waves (GWs) that are circularly polarized. Such waves, once generated, would not dissipate and would only dilute under the cosmic expansion. They could still be observable with space interferometers [41–43] and with pulsar timing arrays [44]. Measuring circular polarization could provide a clean mechanism for determining the sign of helicity in the Universe.

In the following, we describe the two generation mechanisms and compare in Figure 4 the growth of the resulting magnetic field and the circular polarization of GWs that could be observed in future using space interferometers in the millihertz range [41–43].

4.1. The CME

The CME is a quantum effect associated with the systematic alignment of the spin s of electrically charged fermions (electrons or positrons, for example) with the momentum p. A nonvanishing net helicity $s \cdot p$ originates from the parity-breaking weak force and manifests itself in the β decay, for example, where spin and momentum of electrons are antialigned, i.e., $s \cdot p < 0$. The sign would be opposite for antimatter, i.e., for positrons, for example. However, at low energies, spin flipping occurs [45], making this effect important only for highly relativistic plasmas at high enough temperatures.

In the presence of a magnetic field, the spin of chiral fermions aligns itself with the magnetic field, and, owing to the finite momentum and charge of the fermions, they produce a net current along the magnetic field [46–50],

$$J = 24\alpha_{\mathrm{em}} \left(n_{\mathrm{L}} - n_{\mathrm{R}}\right) \left(\frac{k_{\mathrm{B}}T}{\hbar c}\right)^2 B, \tag{11}$$

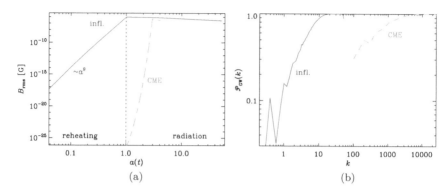

(a)   (b)

Fig. 4. (a) Growth of magnetic field from the CME (orange) and inflationary magnetogenesis (red), where time is here expressed in terms of the scale factor of the Universe $a(t)$, which increases monotonically and is set to unity at the beginning of the radiation-dominated era. Note the algebraic increase $B_{rms} \propto a^9$. (b) Circular polarization spectra of GWs produced from the CME (orange) and inflationary magnetogenesis (red).

where $\alpha_{em} \approx 1/137$ is the fine structure constant, n_L and n_R are the number densities of left- and right-handed fermions, respectively, k_B is the Boltzmann constant, T is the temperature of the Universe today, $\hbar$ is the reduced Planck constant, and c is the speed of light.

We see the analogy with Eq. (5), where we had a mean electromotive force $\overline{\mathcal{E}} \equiv \overline{\boldsymbol{u} \times \boldsymbol{b}}$ with a mean current $\overline{\boldsymbol{J}} = (\alpha/\eta_t \mu_0)\,\overline{\boldsymbol{B}}$ produced along the mean magnetic field for turbulent for helical turbulence. Similar to the dynamo effect in helical turbulence, there is a dynamo effect associated with chiral plasmas. This was discussed in the context of cosmology [47] as a mechanism for producing magnetic fields in the early Universe. This effect produces helical magnetic fields, and the associated helicity reduces the net chirality of the fermions such that the total chirality is conserved, i.e.,

$$(n_L - n_R) + \frac{4\alpha_{em}}{\hbar c}\langle \boldsymbol{A} \cdot \boldsymbol{B}\rangle = \text{const.} \tag{12}$$

The length scales are small compared with the Hubble radius at any given time and the resulting field strengths are limited to [51]

$$|\langle \boldsymbol{A} \cdot \boldsymbol{B}\rangle| \lesssim \xi_M \langle \boldsymbol{B}^2\rangle \lesssim (0.5 \times 10^{-18}\,\text{G})^2\,\text{Mpc}, \tag{13}$$

where ξ_M is the magnetic correlation length, measured here in megaparsec (1 Mpc $\approx$ 3×10^{24} cm). This value of $\xi_M \langle \boldsymbol{B}^2\rangle$ is below the lower limits derived from the non-observation of secondary GeV photons that are expected from inverse Compton scattering of TeV photons from energetic blazers. One can therefore not be very hopeful that those lower limits on the magnetic field can be explained as a result of the CME. However, stronger fields could still be produced during inflation, as will be discussed next.

4.2. Inflationary magnetogenesis

Models of inflationary magnetogenesis tend to invoke conformal invariance-breaking to generate magnetic fields as a result of stretching without diluting the magnetic energy during the inflationary expansion and the subsequent reheating phase when most of the relevant particles and photons were produced. Conformal invariance-breaking means that the term $F_{\mu\nu}F^{\mu\nu}$ in the Lagrangian density is replaced by $f^2 F_{\mu\nu}F^{\mu\nu}$ with $f \neq 1$ during inflation [52] and reheating [53]. In the presence of such a term, the vacuum evolution equation for the vector potential changes from a standard wave equation $(\partial^2/\partial t^2 + k^2)\tilde{\mathbf{A}} = 0$ to $(\partial^2/\partial t^2 + k^2 - f''/f)(f\tilde{\mathbf{A}}) = 0$, where t is now conformal time and the primes on f denote derivatives with respect to conformal time.

Commonly adopted forms of f include $f \propto a^\alpha$ during inflation and $f \propto a^{-\beta}$ during reheating [54–57]. In the presence of pseudoscalars γ, such as axions, one also expects terms of the form $\gamma f^2 F_{\mu\nu}\tilde{F}^{\mu\nu}$ in the Lagrangian density. The evolution equation for $\tilde{\mathbf{A}}$ takes then the form

$$\left(\frac{\partial^2}{\partial t^2} + k^2 \pm 2\gamma k \frac{f'}{f} - \frac{f''}{f}\right)(f\tilde{A}_\pm) = 0, \tag{14}$$

where $\tilde{\mathbf{A}} = \tilde{A}_+\hat{\mathbf{e}}_+ + \tilde{A}_-\hat{\mathbf{e}}_-$ has been expressed in terms of the polarization basis $\tilde{\mathbf{e}}_\pm(\mathbf{k}) = [\tilde{\mathbf{e}}_1(\mathbf{k}) \pm i\tilde{\mathbf{e}}_2(\mathbf{k})]/\sqrt{2}\,i$ with $i\mathbf{k} \times \tilde{\mathbf{e}}_\pm = \pm k\tilde{\mathbf{e}}_\pm$, and $\tilde{\mathbf{e}}_1(\mathbf{k})$, $\tilde{\mathbf{e}}_2(\mathbf{k})$ represent units vectors orthogonal to $\mathbf{k}$ and orthogonal to each other. Figure 4(a) shows the resulting algebraic growth of the magnetic field in comparison with the exponential growth from the CME, and panel (b) shows the circular polarization spectrum $\mathcal{P}_{\mathrm{GW}}(k)$. It is defined as [58]

$$\mathcal{P}_{\mathrm{GW}}(k) = \int 2\,\mathrm{Im}\,\tilde{h}_+\tilde{h}_\times^*\,k^2\mathrm{d}\Omega_k \Big/ \int \left(|\tilde{h}_+|^2 + |\tilde{h}_\times|^2\right)k^2\mathrm{d}\Omega_k. \tag{15}$$

We see that the degree of polarization reaches nearly 100% in a certain range [59, 60].

4.3. Detecting handedness from unit vectors in the sky

In addition to measuring the polarization of GWs as an indicator of the helicity of the underlying magnetic field, there is yet another interesting method that we describe here briefly. Suppose we observed energetic photons from a particular astrophysical source from three slightly different directions, $\hat{\mathbf{n}}_1$, $\hat{\mathbf{n}}_2$, and $\hat{\mathbf{n}}_3$, and that their energies E_i are ordered such that $E_1 < E_2 < E_3$, then we can construct a pseudoscalar

$$Q = (\hat{\mathbf{n}}_1 \times \hat{\mathbf{n}}_2) \cdot \hat{\mathbf{n}}_3. \tag{16}$$

The quantity Q would change sign in a mirror image of our Universe. The value of Q may well be zero within error bars, but if it is not, its sign must mean something.

Detailed calculations have shown the TeV photons from blazars (i.e., the accretion disk around supermassive black holes) can upscatter on the cosmic background light and produce GeV photons [61]. Using about 10,000 photons observed with Fermi Large Area Telescope data over a period of about five years, Tashiro *et al.* [62] found $Q < 0$ for all possible photon triples in certain energy ranges. They interpreted this as evidence in favor of a baryogenesis scenario that proceeds through changes in the Chern-Simons number, which implies the generation of magnetic fields of negative helicity [63]; see also Ref. [64] for a review.

In all the studies of Q done since then [65–68], one introduced a cutoff toward low Galactic latitudes so as to avoid excess contamination from our Galaxy. Using synthetic data, it has been found [69] that it is this procedure that leads to the occurrence of large statistical errors in the estimate of Q. Updated observations covering 11 years turned out to be no longer compatible with a detection of a negative value of Q [69]. This has been confirmed in subsequent work [70]. Thus, although this method could work in principle, it would require much better statistics.

In this connection, it is interesting to note that an equivalent quantity Q can be determined for a variety of different observations. Suppose there are three sunspots of different strengths on the surface of the Sun, this again implies a finite handedness. We can then ask whether this handedness could be linked to the helicity of the underlying magnetic field. Idealized model calculations have shown that this is indeed the case [71].

5. Spontaneous chirality production

5.1. *Biological homochirality*

In astrobiology, an important question concerns the origin of biological homochirality [72]. In solution, many organic molecules tend to rotate the polarization plane of linearly polarized light. One refers to the substances as either levorotatory (L for left-handed) or dextrorotatory (D for right-handed). Almost all amino acids of terrestrial life are of the L form, and almost all sugars are of the D form, for example the sugars in the phosphorus backbones of deoxyribose nucleic acid (DNA). The origin of this homochirality can be explained in terms of two essential processes: autocatalysis and mutual antagonism – an old idea that goes back to a paper by F. C. Frank [73] of 1953. Interestingly, this is the same year when Watson and Crick [74] discovered the helix structure of DNA.

Autocatalysis produces "more of itself", i.e., it can catalyze the formation of chiral molecules of the L form from an achiral substrate A in the presence of L and, conversely, it can catalyze the formation of molecules of the D form in the presence of D. The corresponding reactions

$$L + A \to 2L \quad \text{and} \quad D + A \to 2D, \tag{17}$$

with rate coefficient k_C, imply that the associated concentrations $[L]$ and $[D]$ obey

$$\frac{\mathrm{d}}{\mathrm{d}t}[L] = k_C[A][L] - \dots \quad \text{and} \quad \frac{\mathrm{d}}{\mathrm{d}t}[D] = k_C[A][D] - \dots, \tag{18}$$

which leads to exponential growth with time t of both $[L] = [L]_0 e^{k_C[A]t}$ and $[D] = [D]_0 e^{k_C[A]t}$, with initial values $[L]_0$ and $[D]_0$. However, this process alone does not change the enantiomeric excess (e.e.), e.e. $= ([L]-[D])/([L]+[D])$, which will always be equal to the initial value. This is because we still need mutual antagonism, which will be explained next.

For a long time, it remained unclear what would correspond to Frank's mutual antagonism. The relevant understanding was put forward by Sandars [75]. At that time, it was thought that homochirality was a prerequisite to the origin of life. This idea was based on an experimental result by Joyce *et al.* [76], which showed that in template-directed polymerization of oligomers of one chirality, polymers of the opposite chirality terminate further polymerization. This was called enantiomeric cross-inhibition, and was regarded as a serious problem for the origin of life, and that life could only emerge in a fully homochiral environment [77]. Sandars realized that enantiomeric cross-inhibition could just be the crucial mechanism that corresponds to Frank's mutual antagonism that would lead to the emergence of homochirality. The corresponding reaction and rate equations, with rate coefficient k_I, are

$$L + D \rightarrow 2A \quad \text{and} \quad \frac{\mathrm{d}}{\mathrm{d}t}[A] = 2k_I[L][D] - \dots . \tag{19}$$

Multiple extensions of Sanders' model have been produced [78–81] and there are also other variants that are not based on nucleotides, but on peptides [82–85]. If one regards these first polymerization reactions as the first steps toward life, one could then say that homochirality emerges as a consequence of life, and not as a prerequisite [86, 87].

The lessons learnt from astrobiology may well be applicable to other fields of physics, and in particular to MHD. Examples were found in the context of the magnetobuoyancy instability [88] and the Tayler instability [89, 90]. In those cases, there are two unstable eigenfunctions that are helical and have positive and negative helicities, respectively, but their growth rates are equal. This process corresponds to autocatalysis. The nonlinearity in the MHD equations, associated with the Lorentz force, corresponds to mutual antagonism. The resulting amplitude equations in MHD are the same as in the production of homochirality [90]. Another more recent example of this type has been found in studies of the CME when the chiral chemical potential is fluctuating around zero [91, 92]. Again, one chirality becomes eventually dominant, and this choice depends on details in the initial conditions.

5.2. *Spatially distinct domains of chirality*

In spatially extended domains, the evolution equations for $[L]$ and $[D]$ attain additional spatial diffusion terms and become then similar to the Fisher equation [93] which describes front propagation. In the present case, once fronts of L and D polymers come into contact, the front between them comes to a halt and cannot propagate any further, unless the front is curved. If it is curved, the front continues to propagate in the direction of largest curvature [94]. This leads in the end to small

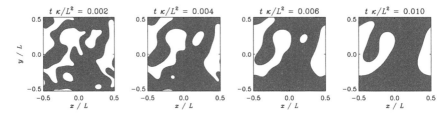

Fig. 5. Gradual shrinking of isolated islands on one handedness (white), leading to the eventual dominance of the other (gray).

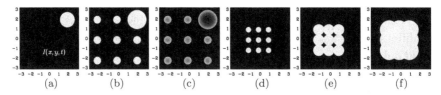

Fig. 6. The number of infected $I(x, y)$ in a two-dimensional Cartesian plane (x, y), as obtained from model calculations [95], showing (a) a circular spreading center in the upper right corner, (b) the subsequent emergence of eight additional spreading centers that (c) continue to grow, but with decreasing $I(x, y)$ in their centers due to recovery or death. Note that the total length of the periphery increases when the number of spreading centers increases. Panels (d)–(f) show a case where spreading centers merge, so the growth in the length of the periphery declines.

near-circular islands that shrink until they disappear; see Figure 5. This means that the enantiomeric excess changes in a piecewise linear fashion.

5.3. *Analogous mechanisms in other systems*

Given that closed patches of one handedness always shrink and eventually disappear, it is clear that the dominant chirality must in the end be that of the outside of the last surviving patch. Thus, it is not necessarily the one that was initially the most dominant one.

A piecewise linear evolution is common to many spatially extended systems, including those describing the spread of SARS-CoV-2 over the past three years. Here, however, it is not the number N itself, but its square root, $N^{1/2}$. In Figure 6, we show the spatial geometry of a hypothetic spreading center in the upper right corner. The speed of growth depends just on the length of the periphery. At a later time, there will be new spreading centers, so the total length of the periphery increases. This happened in the middle of March 2020; see Figure 7. During the first part of the epidemic (denoted by a red A), the evolution was comparatively slow and the disease was essentially confined just to China. During the subsequent phase (denoted by a blue B) it spread all over the world. This led to an increase in the total length of the periphery. Later, different spreading centers began to merge, so the total periphery has now decreased again, and the growth has slowed down (denoted by the green and orange segments C and D, respectively), but it always

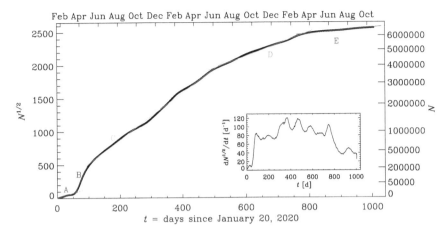

Fig. 7. Square root of the number N of deaths, which is regarded as a proxy of the number of infected that is more reliable than the reported number of SARS-CoV-2. Note the piecewise linear growth in $N^{1/2}$, corresponding to a piecewise quadratic growth. The line segments A–D are described in the text.

remained piecewise quadratic and was never exponential [95]. The spreading of SARS-CoV-2 is obviously no longer directly related to the topic of chirality in astrophysics, but it is interesting to see that the mathematics of front propagation in spatially extended domains is similar to that of left and right handed life forms invading the early Earth [94].

6. Conclusions

In this work, we have sketched three rather different ways of achieving chirality in astrophysics and astrobiology: externally driven, intrinsically driven, and spontaneous chirality production. The first two mechanisms are particularly relevant to fluid dynamics and magnetic fields, while the last one of spontaneous chirality production is mainly relevant to astrobiology and to the origin of life, but it remains hypothetic until one is able to find an example of another genesis of life independent of that on Earth, for example on Mars or on some of the icy moons in our solar system [72].

The idea of propagating fronts of life forms of opposite handedness is intriguing and it would be useful to reproduce this in the lab. This may not be easy because such fronts propagate relatively rapidly under laboratory conditions. In fact, their propagation resembles the propagation of epidemiological fronts, such as the black death [96] and perhaps even SARS-CoV-2 [95]. However, the application to the early Earth implies much larger spatial scales. Earlier work quoted half a billion years as a relevant time scale [94]. However, if one thinks of the deep biosphere of Mars, it may not be impossible to explain a possible detection of opposite chiralities of DNA, if such should ever be observed on Mars or in its permafrost.

Alternatively, it is possible that the chirality of biomolecules is determined through an external influence [97]. Such an influence would then, similarly to the intrinsically driven chirality discussed in Sect. 4, be related to the parity-breaking weak force. This possibility cannot easily be dismissed. In particular, a 2% enantiomeric excess in favor of the L form has been found for amino acids in the Murchison meteorite [98]. This was a pristine meteorite rich in organics, as already evidenced by the smell reported by initial eyewitnesses. On the other hand, those molecules are also susceptible to contamination, while those not susceptible to contamination did not show any enantiomeric excess.

Determining a global chirality that is the same throughout the entire Universe would be a major discovery. Measuring the chirality through circular polarization of GWs would likely be the most definitive proof of parity violation in the Universe.

Acknowledgments

I thank Tanmay Vachaspati, Jérémy Vachier, and Jian-Zhou Zhu for comments and noticing some mistakes. I am particularly grateful to Mats Larsson for having provided us with an early opportunity for an in-person meeting in June 2021, right when the quadratic growth of SARS-CoV-2 showed a break; see Figure 7. This work was supported in part through the Swedish Research Council, grant 2019-04234. I acknowledge the allocation of computing resources provided by the Swedish National Allocations Committee at the Center for Parallel Computers at the Royal Institute of Technology in Stockholm.

References

1. H. K. Moffatt, The degree of knottedness of tangled vortex lines, *J. Fluid Mech.* **35**, 117 (1969).
2. R. Betchov, Semi-Isotropic Turbulence and Helicoidal Flows, *Phys. Fluids* **4**, 925 (1961).
3. L. Woltjer, A theorem on force-free magnetic fields, *Proc. Nat. Acad. Sci.* **44**, 489 (1958).
4. J. B. Taylor, Relaxation of Toroidal Plasma and Generation of Reverse Magnetic Fields, *Phys. Rev. Lett.* **33**, 1139 (1974).
5. A. Brandenburg and K. Subramanian, Astrophysical magnetic fields and nonlinear dynamo theory, *Phys. Rep.* **417**, 1 (2005).
6. F. Krause and K.-H. Rädler, *Mean-Field Magnetohydrodynamics and Dynamo Theory* (Pergamon Press (also Akademie-Verlag: Berlin), Oxford, 1980).
7. A. Brandenburg, The inverse cascade and nonlinear alpha-effect in simulations of isotropic helical hydromagnetic turbulence, *Astrophys. J.* **550**, 824 (2001).
8. F. Rincon, Dynamo theories, *J. Plasma Phys.* **85**, 205850401 (2019).
9. H. Zhang and A. Brandenburg, Solar kinetic energy and cross helicity spectra, *Astrophys. J. Lett.* **862**, L17 (2018).
10. G. Rüdiger, L. L. Kitchatinov, and A. Brandenburg, Cross helicity and turbulent magnetic diffusivity in the solar convection zone, *Sol. Phys.* **269**, 3 (2011).
11. A. Brandenburg, N. Kleeorin and I. Rogachevskii, Self-assembly of shallow magnetic spots through strongly stratified turbulence, *Astrophys. J. Lett.* **776**, L23 (2013).

12. A. Brandenburg, O. Gressel, S. Jabbari, N. Kleeorin and I. Rogachevskii, Mean-field and direct numerical simulations of magnetic flux concentrations from vertical field, *Astron. Astrophys.* **562**, A53 (2014).
13. P. Demoulin and E. R. Priest, A twisted flux model for solar prominences. II - Formation of a dip in a magnetic structure before the formation of a solar prominence, *Astron. Astrophys.* **214**, 360 (1989).
14. S. E. Gibson, L. Fletcher, G. Del Zanna, C. D. Pike, H. E. Mason, C. H. Mandrini, P. Démoulin, H. Gilbert, J. Burkepile, T. Holzer, D. Alexander, Y. Liu, N. Nitta, J. Qiu, B. Schmieder and B. J. Thompson, The structure and evolution of a sigmoidal active region, *Astrophys. J.* **574**, 1021 (2002).
15. Y. Guo, M. D. Ding, X. Cheng, J. Zhao and E. Pariat, Twist accumulation and topology structure of a solar magnetic flux rope, *Astrophys. J.* **779**, 157 (2013).
16. M. K. Georgoulis, A new technique for a routine azimuth disambiguation of solar vector magnetograms, *Astrophys. J. Lett.* **629**, L69 (2005).
17. G. V. Rudenko and S. A. Anfinogentov, Very fast and accurate azimuth disambiguation of vector magnetograms, *Sol. Phys.* **289**, 1499 (2014).
18. N. Seehafer, Electric current helicity in the solar atmosphere, *Sol. Phys.* **125**, 219 (1990).
19. H. Zhang, A. Brandenburg and D. D. Sokoloff, Magnetic helicity and energy spectra of a solar active region, *Astrophys. J. Lett.* **784**, L45 (2014).
20. H. Zhang, A. Brandenburg and D. D. Sokoloff, Evolution of magnetic helicity and energy spectra of solar active regions, *Astrophys. J.* **819**, 146 (2016).
21. W. H. Matthaeus, M. L. Goldstein and C. Smith, Evaluation of magnetic helicity in homogeneous turbulence, *Phys. Rev. Lett.* **48**, 1256 (1982).
22. A. Brandenburg, K. Subramanian, A. Balogh and M. L. Goldstein, Scale dependence of magnetic helicity in the solar wind, *Astrophys. J.* **734**, 9 (2011).
23. A. Brandenburg, S. Candelaresi and P. Chatterjee, Small-scale magnetic helicity losses from a mean-field dynamo, *Mon. Not. Roy. Astron. Soc.* **398**, 1414 (2009).
24. J. Warnecke, A. Brandenburg and D. Mitra, Dynamo-driven plasmoid ejections above a spherical surface, *Astron. Astrophys.* **534**, A11 (2011).
25. J. Warnecke, A. Brandenburg and D. Mitra, Magnetic twist: a source and property of space weather, *J. Space Weather Space Climate* **2**, A260000 (2012).
26. J.-Z. Zhu, W. Yang and G.-Y. Zhu, Purely helical absolute equilibria and chirality of (magneto)fluid turbulence, *J. Fluid Mech.* **739**, 479 (2014).
27. J.-Z. Zhu, Chirality, extended magnetohydrodynamics statistics and topological constraints for solar wind turbulence, *Mon. Not. Roy. Astron. Soc.* **470**, L87 (2017).
28. Y. Narita, G. Kleindienst and K. H. Glassmeier, Evaluation of magnetic helicity density in the wave number domain using multi-point measurements in space, *Annales Geophysicae* **27**, 3967 (2009).
29. A. Brandenburg and R. Stepanov, Faraday signature of magnetic helicity from reduced depolarization, *Astrophys. J.* **786**, 91 (2014).
30. C. Horellou and A. Fletcher, Magnetic field tomography, helical magnetic fields and Faraday depolarization, *Mon. Not. Roy. Astron. Soc.* **441**, 2049 (2014).
31. A. Brandenburg, Simulations of galactic dynamos, *Astrophys. Spa. Sci. Lib.* **407**, 529 (2015).
32. A. Brandenburg, M. B. Ashurova and S. Jabbari, Compensating Faraday depolarization by magnetic helicity in the solar corona, *Astrophys. J.* **845**, L15 (2017).
33. M. Kamionkowski, A. Kosowsky and A. Stebbins, A probe of primordial gravity waves and vorticity, *Phys. Rev. Lett.* **78**, 2058 (1997).
34. U. Seljak and M. Zaldarriaga, Signature of gravity waves in the polarization of the microwave background, *Phys. Rev. Lett.* **78**, 2054 (1997).

35. R. Durrer, *The Cosmic Microwave Background* (Cambridge University Press, Cambridge, 2008).
36. M. Kamionkowski and E. D. Kovetz, The quest for B modes from inflationary gravitational waves, *Ann. Rev. Astron. Astrophys.* **54**, 227 (2016).
37. A. Brandenburg, A global two-scale helicity proxy from π-ambiguous solar magnetic fields, *Astrophys. J.* **883**, 119 (2019).
38. A. Brandenburg and M. Brüggen, Hemispheric handedness in the Galactic synchrotron polarization foreground, *Astrophys. J. Lett.* **896**, L14 (2020).
39. A. Brandenburg and R. S. Furuya, Application of a helicity proxy to edge-on galaxies, *Mon. Not. Roy. Astron. Soc.* **496**, 4749 (2020).
40. T. Vachaspati, Estimate of the primordial magnetic field helicity, *Phys. Rev. Lett.* **87**, 251302 (2001).
41. C. Caprini, M. Hindmarsh, S. Huber, T. Konstandin, J. Kozaczuk, G. Nardini, J. M. No, A. Petiteau, P. Schwaller, G. Servant and D. J. Weir, Science with the space-based interferometer eLISA. II: gravitational waves from cosmological phase transitions, *J. Cosm. Astropart. Phys.* **2016**, 001 (2016).
42. P. Amaro-Seoane, H. Audley, S. Babak, J. Baker, E. Barausse, P. Bender, E. Berti, P. Binetruy, M. Born, D. Bortoluzzi, J. Camp, C. Caprini, V. Cardoso, M. Colpi, J. Conklin, N. Cornish, C. Cutler, K. Danzmann, R. Dolesi, L. Ferraioli, V. Ferroni, E. Fitzsimons, J. Gair, L. Gesa Bote, D. Giardini, F. Gibert, C. Grimani, H. Halloin, G. Heinzel, T. Hertog, M. Hewitson, K. Holley-Bockelmann, D. Hollington, M. Hueller, H. Inchauspe, P. Jetzer, N. Karnesis, C. Killow, A. Klein, B. Klipstein, N. Korsakova, S. L. Larson, J. Livas, I. Lloro, N. Man, D. Mance, J. Martino, I. Mateos, K. McKenzie, S. T. McWilliams, C. Miller, G. Mueller, G. Nardini, G. Nelemans, M. Nofrarias, A. Petiteau, P. Pivato, E. Plagnol, E. Porter, J. Reiche, D. Robertson, N. Robertson, E. Rossi, G. Russano, B. Schutz, A. Sesana, D. Shoemaker, J. Slutsky, C. F. Sopuerta, T. Sumner, N. Tamanini, I. Thorpe, M. Troebs, M. Vallisneri, A. Vecchio, D. Vetrugno, S. Vitale, M. Volonteri, G. Wanner, H. Ward, P. Wass, W. Weber, J. Ziemer and P. Zweifel, Laser Interferometer Space Antenna, arXiv:1702.00786 (2017).
43. Y.-L. Taiji Scientific Collaboration, Wu, Z.-R. Luo, J.-Y. Wang, M. Bai, W. Bian, R.-G. Cai, Z.-M. Cai, J. Cao, D.-J. Chen, L. Chen, L.-S. Chen, M.-W. Chen, W.-B. Chen, Z.-Y. Chen, L.-X. Cong, J.-F. Deng, X.-L. Dong, L. Duan, S.-Q. Fan, S.-S. Fan, C. Fang, Y. Fang, K. Feng, P. Feng, Z. Feng, R.-H. Gao, R.-L. Gao, Z.-K. Guo, J.-W. He, J.-B. He, X. Hou, L. Hu, W.-R. Hu, Z.-Q. Hu, M.-J. Huang, J.-J. Jia, K.-L. Jiang, G. Jin, H.-B. Jin, Q. Kang, J.-G. Lei, B.-Q. Li, D.-J. Li, F. Li, H.-S. Li, H.-W. Li, L.-F. Li, W. Li, X.-K. Li, Y.-M. Li, Y.-G. Li, Y.-P. Li, Y.-P. Li, Z. Li, Z.-Y. Lin, C. Liu, D.-B. Liu, H.-S. Liu, H. Liu, P. Liu, Y.-R. Liu, Z.-Y. Lu, H.-W. Luo, F.-L. Ma, L.-F. Ma, X.-S. Ma, X. Ma, Y.-C. Man, J. Min, Y. Niu, J.-K. Peng, X.-D. Peng, K.-Q. Qi, L.-É. Qiang, C.-F. Qiao, Y.-X. Qu, W.-H. Ruan, W. Sha, J. Shen, X.-J. Shi, R. Shu, J. Su, Y.-L. Sui, G.-W. Sun, W.-L. Tang, H.-J. Tao, W.-Z. Tao, Z. Tian, L.-F. Wan, C.-Y. Wang, J. Wang, J. Wang, L.-L. Wang, S.-X. Wang, X.-P. Wang, Y.-K. Wang, Z. Wang, Z.-L. Wang, Y.-X. Wei, L.-M. Di Wu, Wu, P.-Z. Wu, Z.-H. Wu, D.-X. Xi, Y.-F. Xie, G.-F. Xin, L.-X. Xu, P. Xu, S.-Y. Xu, Y. Xu, S.-W. Xue, Z.-B. Xue, C. Yang, R. Yang, S.-J. Yang, S. Yang, Y. Yang, Z.-G. Yang, Y.-L. Yin, J.-P. Yu, T. Yu, À.-B. Zhang, C. Zhang, M. Zhang, X.-Q. Zhang, Y.-Z. Zhang, J. Zhao, W.-W. Zhao, Y. Zhao, J.-H. Zheng, C.-Y. Zhou, Z.-C. Zhu, X.-B. Zou and Z.-M. Zou, China's first step towards probing the expanding universe and the nature of gravity using a space borne gravitational wave antenna, *CmPhy* **4**, 34 (2021).
44. Z. Arzoumanian, P. T. Baker, H. Blumer, B. Bécsy, A. Brazier, P. R. Brook, S. Burke-Spolaor, S. Chatterjee, S. Chen, J. M. Cordes, N. J. Cornish, F. Crawford, H. T. Cromartie, M. E. Decesar, P. B. Demorest, T. Dolch, J. A. Ellis, E. C. Ferrara,

W. Fiore, E. Fonseca, N. Garver-Daniels, P. A. Gentile, D. C. Good, J. S. Hazboun, A. M. Holgado, K. Islo, R. J. Jennings, M. L. Jones, A. R. Kaiser, D. L. Kaplan, L. Z. Kelley, J. S. Key, N. Laal, M. T. Lam, T. J. W. Lazio, D. R. Lorimer, J. Luo, R. S. Lynch, D. R. Madison, M. A. McLaughlin, C. M. F. Mingarelli, C. Ng, D. J. Nice, T. T. Pennucci, N. S. Pol, S. M. Ransom, P. S. Ray, B. J. Shapiro-Albert, X. Siemens, J. Simon, R. Spiewak, I. H. Stairs, D. R. Stinebring, K. Stovall, J. P. Sun, J. K. Swiggum, S. R. Taylor, J. E. Turner, M. Vallisneri, S. J. Vigeland, C. A. Witt and Nanograv Collaboration, The NANOGrav 12.5 yr data set: Search for an isotropic stochastic gravitational-wave background, *Astrophys. J. Lett.* **905**, L34 (2020).
45. A. Boyarsky, V. Cheianov, O. Ruchayskiy and O. Sobol, Evolution of the primordial axial charge across cosmic times, *Phys. Rev. Lett.* **126**, 021801 (2021).
46. A. Vilenkin, Equilibrium parity-violating current in a magnetic field, *Phys. Rev. D* **22**, 3080 (1980).
47. M. Joyce and M. Shaposhnikov, Primordial magnetic fields, right electrons, and the Abelian anomaly, *Phys. Rev. Lett.* **79**, 1193 (1997).
48. A. Boyarsky, J. Fröhlich and O. Ruchayskiy, Self-consistent evolution of magnetic fields and chiral asymmetry in the early Universe, *Phys. Rev. Lett.* **108**, 031301 (2012).
49. A. Boyarsky, J. Fröhlich and O. Ruchayskiy, Magnetohydrodynamics of chiral relativistic fluids, *Phys. Rev. D* **92**, 043004 (2015).
50. I. Rogachevskii, O. Ruchayskiy, A. Boyarsky, J. Fröhlich, N. Kleeorin, A. Brandenburg and J. Schober, Laminar and turbulent dynamos in chiral magnetohydrodynamics. I. Theory, *Astrophys. J.* **846**, 153 (2017).
51. A. Brandenburg, J. Schober, I. Rogachevskii, T. Kahniashvili, A. Boyarsky, J. Fröhlich, O. Ruchayskiy and N. Kleeorin, The turbulent chiral magnetic cascade in the early Universe, *Astrophys. J.* **845**, L21 (2017).
52. B. Ratra, Cosmological "Seed" Magnetic Field from Inflation, *Astrophys. J. Lett.* **391**, L1 (1992).
53. V. Demozzi and C. Ringeval, Reheating constraints in inflationary magnetogenesis, *J. Cosm. Astropart. Phys.* **2012**, 009 (2012).
54. R. J. Z. Ferreira, R. K. Jain and M. S. Sloth, Inflationary magnetogenesis without the strong coupling problem, *J. Cosm. Astropart. Phys.* **2013**, 004 (2013).
55. R. Sharma, S. Jagannathan, T. R. Seshadri and K. Subramanian, Challenges in inflationary magnetogenesis: Constraints from strong coupling, backreaction, and the Schwinger effect, *Phys. Rev. D* **96**, 083511 (2017).
56. R. Sharma, K. Subramanian and T. R. Seshadri, Generation of helical magnetic field in a viable scenario of inflationary magnetogenesis, *Phys. Rev. D* **97**, 083503 (2018).
57. A. Brandenburg and R. Sharma, Simulating relic gravitational waves from inflationary magnetogenesis, *Astrophys. J.* **920**, 26 (2021).
58. A. Roper Pol, A. Brandenburg, T. Kahniashvili, A. Kosowsky and S. Mandal, The timestep constraint in solving the gravitational wave equations sourced by hydromagnetic turbulence, *Geophys. Astrophys. Fluid Dynam.* **114**, 130 (2020).
59. A. Brandenburg, Y. He, T. Kahniashvili, M. Rheinhardt and J. Schober, Relic Gravitational Waves from the Chiral Magnetic Effect, *Astrophys. J.* **911**, 110 (2021).
60. A. Brandenburg, Y. He and R. Sharma, Simulations of Helical Inflationary Magnetogenesis and Gravitational Waves, *Astrophys. J.* **922**, p. 192 (2021).
61. H. Tashiro and T. Vachaspati, Cosmological magnetic field correlators from blazar induced cascade, *Phys. Rev. D* **87**, 123527 (2013).
62. H. Tashiro, W. Chen, F. Ferrer and T. Vachaspati, Search for CP violating signature of intergalactic magnetic helicity in the gamma-ray sky., *Mon. Not. Roy. Astron. Soc.* **445**, L41 (2014).

63. T. Vachaspati, Estimate of the primordial magnetic field helicity, *Phys. Rev. Lett.* **87**, 251302 (2001).
64. T. Vachaspati, Progress on cosmological magnetic fields, *Rep. Prog. Phys.* (2021).
65. W. Chen, J. H. Buckley and F. Ferrer, Search for GeV γ-ray pair halos around low redshift blazars, *Phys. Rev. Lett.* **115**, 211103 (2015).
66. W. Chen, B. D. Chowdhury, F. Ferrer, H. Tashiro and T. Vachaspati, Intergalactic magnetic field spectra from diffuse gamma-rays, *Mon. Not. Roy. Astron. Soc.* **450**, 3371 (2015).
67. A. J. Long and T. Vachaspati, Morphology of blazar-induced gamma ray halos due to a helical intergalactic magnetic field, *J. Cosm. Astropart. Phys.* **2015**, 065 (2015).
68. A. J. Long and T. Vachaspati, Implications of a primordial magnetic field for magnetic monopoles, axions, and Dirac neutrinos, *Phys. Rev. D* **91**, 103522 (2015).
69. J. Asplund, G. Jóhannesson and A. Brandenburg, On the measurement of handedness in Fermi Large Area Telescope data, *Astrophys. J.* **898**, 124 (2020).
70. M. Kachelrieß and B. C. Martinez, Searching for primordial helical magnetic fields, *Phys. Rev. D* **102**, 083001 (2020).
71. P.-A. Bourdin and A. Brandenburg, Magnetic helicity from multipolar regions on the solar surface, *Astrophys. J.* **869**, 3 (2018).
72. D. Rothery, I. Gilmour and M. Sephton, *An Introduction to Astrobiology* (Cambridge University Press, 2011).
73. F. C. Frank, On spontaneous asymmetric synthesis, *Biochim. Biophys. Acta* **11**, 459 (1953).
74. J. D. Watson and F. H. C. Crick, Molecular structure of nucleic acids: a structure for deoxyribose nucleic acid, *Nature* **171**, 737 (1953).
75. P. G. H. Sandars, A Toy Model for the Generation of Homochirality during Polymerization, *Orig. Life Evol. Biosph.* **33**, 575 (2003).
76. G. F. Joyce, G. M. Visser, C. A. A. van Boeckel, J. H. van Boom, L. E. Orgel and J. van Westrenen, Chiral selection in poly(C)-directed synthesis of oligo(G), *Nature* **310**, 602 (1984).
77. V. I. Gol'danskiĭ and V. V. Kuz'min, Spontaneous breaking of mirror symmetry in nature and the origin of life, *Soviet Physics Uspekhi* **32**, 1 (1989).
78. Y. Saito and H. Hyuga, Chirality selection in open flow systems and in polymerization, *J. Phys. Soc. Jpn.* **74**, 1629 (2005).
79. A. Brandenburg, A. C. Andersen, S. Höfner and M. Nilsson, Homochiral growth through enantiomeric cross-inhibition, *Orig. Life Evol. Biosph.* **35**, 225 (2005).
80. J. A. D. Wattis and P. V. Coveney, Symmetry-breaking in chiral polymerisation, *Orig. Life Evol. Biosph.* **35**, 243 (2005).
81. M. Gleiser and S. I. Walker, An extended model for the evolution of prebiotic homochirality: a bottom-up approach to the origin of life, *Orig. Life Evol. Biosph.* **38**, 293 (2008).
82. R. Plasson, H. Bersini and A. Commeyras, Recycling Frank: Spontaneous emergence of homochirality in noncatalytic systems, *Proc. Nat. Acad. Sci.* **101**, 16733 (2004).
83. A. Brandenburg, H. J. Lehto and K. M. Lehto, Homochirality in an early peptide world, *Astrobiology* **7**, 725 (2007).
84. M. Gleiser and S. I. Walker, Toward homochiral protocells in noncatalytic peptide systems, *Orig. Life Evol. Biosph.* **39**, 479 (2009).
85. K. K. Konstantinov and A. F. Konstantinova, Chiral symmetry breaking in large peptide systems, *Orig. Life Evol. Biosph.* **50**, 99 (2020).
86. P. G. H. Sandars, Chirality in the RNA world and beyond, *Int. J. Astrobiol.* **4**, 49 (2005).

87. A. Brandenburg, Homochirality: A Prerequisite or Consequence of Life?, in *Prebiotic Chemistry and the Origin of Life*, eds. A. Neubeck and S. McMahon 2021 pp. 87–115.

88. P. Chatterjee, D. Mitra, A. Brandenburg and M. Rheinhardt, Spontaneous chiral symmetry breaking by hydromagnetic buoyancy, *Phys. Rev. E* **84**, 025403 (2011).

89. M. Gellert, G. Rüdiger and R. Hollerbach, Helicity and α-effect by current-driven instabilities of helical magnetic fields, *Mon. Not. Roy. Astron. Soc.* **414**, 2696 (2011).

90. A. Bonanno, A. Brandenburg, F. Del Sordo and D. Mitra, Breakdown of chiral symmetry during saturation of the Tayler instability, *Phys. Rev. E* **86**, 016313 (2012).

91. J. Schober, I. Rogachevskii and A. Brandenburg, Production of a Chiral Magnetic Anomaly with Emerging Turbulence and Mean-Field Dynamo Action, *Phys. Rev. Lett.* **128**, p. 065002 (2022).

92. J. Schober, I. Rogachevskii and A. Brandenburg, Dynamo instabilities in plasmas with inhomogeneous chiral chemical potential, *Phys. Rev. D* **105**, p. 043507 (2022).

93. R. A. Fisher, The wave of advance of advantageous genes, *Ann. Eugenics* **7**, 353 (1937).

94. A. Brandenburg and T. Multamäki, How long can left and right handed life forms coexist?, *Int. J. Astrobiology* **3**, 209 (2004).

95. A. Brandenburg, Piecewise quadratic growth during the 2019 novel coronavirus epidemic, *Infectious Disease Modelling* **5**, 681 (2020).

96. J. V. Noble, Geographic and temporal development of plagues, *Nature* **250**, 726 (1974).

97. N. Globus and R. D. Blandford, The chiral puzzle of life, *Astrophys. J. Lett.* **895**, L11 (2020).

98. S. Pizzarello and J. R. Cronin, Non-racemic amino acids in the Murray and Murchison meteorites, *Geochim. Cosmochim. Acta* **64**, 329 (2000).

Using Ultra-Relativistic Heavy-Ion Collisions to Search for the Chiral Magnetic Effect and Local Parity Violation

Helen Caines*

Wright Lab, Physics Department, Yale University,
New Haven, CT, U.S.A.
* E-mail: helen.caines@yale.edu

This article describes the current status of the search for evidence for the Chiral Magnetic Effect (CME) using ultra-relativistic heavy-ion collisions at the Relativistic Heavy-Ion Collider (RHIC) at BNL, New York, USA, and the Large Hadron Collider (LHC), at CERN, Geneva, Switzerland.

Keywords: Quark-Gluon Plasma, Chiral magnetic effect, CME, RHIC, STAR, Heavy-ions.

1. Introduction

Experimental and theoretical studies over the past few decades have provided strong evidence that collisions of ultra-relativistic heavy-ions create a hot and dense medium where the relevant degrees of freedom are quarks and gluons, the so-called Quark-Gluon Plasma (QGP) [1–5]. Recent Lattice Quantum Chromodynamics (QCD) calculations performed at zero baryon chemical potential indicate that a cross-over transition to the QGP occurs at a temperature, $T = 156.5 \pm 1.5\,\text{MeV}$ [6].

1.1. *Some nomenclature*

Events are typically classified experimentally via their centrality; with peripheral collisions having large impact parameters and resulting in few produced particles, and central collisions having small impact parameters and large particle production. We then relate, via the Glauber Model [7], the centrality of the collision to the number of participating nucleon, N_{part}, and the number of equivalent inelastic nucleon-nucleon collisions, N_{bin}.

Of importance to the topic of this report are the definition of three planes. The reaction plane of the event is defined as that which contains the beamline, z, and the impact parameter vector. Then there is the participant plane, which is defined by the beam direction and the average transverse position of the participant nucleons. This does not coincide with the reaction plane because of large event-by-event fluctuations. Lastly there is the third order event plane which is at an angle, Ψ_3, which is defined by the 3rd harmonic number of the Fourier expansion in the transverse plane of the azimuthal angle of emitted particles.

1.2. *Local parity violation and heavy-ion collisions*

In QCD, chiral symmetry breaking is fundamental and due to nontrivial topological solutions. Event-by-event local strong parity violation would be among the best evidence for this physics phenomenon.

While the strong interaction is believed to globally conserve parity, local strong parity violation may be observable in metastable domains within the QGP created in heavy-ion collisions [8–12]. Configurations, local in space and time, can be created within the dense gluonic fields with topological configurations which cause this CP violating effect. Each region is spontaneously produced with a random sign of P-violation, therefore averaged over many events they would not yield a finite expectation value for a P-odd observable.

The non-zero topological charge of each meta-stable region within the QGP results in a difference in the number of quarks with positive and negative chiralities. In non-central heavy-ion collisions an extremely high magnetic field, aligned with the reaction plane and angular momentum vector of the collision, is produced which generates an electric current perpendicular to the event plane resulting in a charge separation in the final-state particles, a phenomenon known as the "Chiral Magnetic Effect" (CME). Simple estimates, and more quantitative modeling, suggest that the magnetic field created in heavy-ion collisions is of the order 10^{18} Gauss [13, 14], much stronger than that on the surface of a magnetar.

In each event charge separation along the angular momentum vector can be described by the Fourier decomposition of the charged particle azimuthal distribution

$$\frac{dN_\pm}{d\phi} \propto 1 + 2v_1 cos(\phi - \Psi_{RP}) + 2v_2 cos(2(\phi - \Psi_{RP})) + \cdots + 2a_\pm sin(\phi - \Psi_{RP}) + \cdots \quad (1)$$

where Ψ_{RP} is the azimuthal angle of the reaction plane, and v_1 and v_2 are coefficients accounting for the so-called directed and elliptic flow [15]. The a parameters, $a_- = -a_+$, describe the P-violating effect. Due to the random nature of the effect $\langle a_+ \rangle = \langle a_- \rangle = 0$, where the angle brackets denote an average of many events. However, $\langle a_\alpha a_\beta \rangle$, where α and β represent the electric charge of a particle, should be non-zero. This led to the proposal [16] for experimentalists to measure

$$\gamma = \langle (\phi_\alpha + \phi_\beta - 2\Psi_{RP}) \rangle = (\langle cos\Delta\phi_\alpha cos\Delta\phi_\beta \rangle - \langle sin\Delta\phi_\alpha sin\Delta\phi_\beta \rangle) \quad (2)$$

where $\Delta\phi = (\phi - \Psi_{RP})$ and the averaging is done over all particles in an event and over all events. γ is sensitive to $-\langle a_\alpha a_\beta \rangle$ and expected to scale with the magnitude of the B-field and inversely with the number of produced charged particles. It is expected that $\gamma_{++} = \gamma_{--} < \gamma_{\pm\mp}$ and experiments often report $\Delta\gamma = \gamma os - \gamma ss$ where os and ss refer to opposite sign and same sign respectively. Care has to be taken when interpreting the results since γ is P-even so may be contaminated by other effects such as resonance decays and jets.

2. Experimental status

2.1. *First evidence*

Experimentally the first potential evidence for the CME was provided by the STAR experiment at RHIC in Au+Au collisions at center-of-mass collision energies per

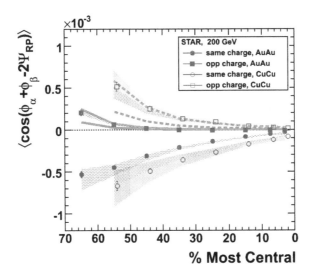

Fig. 1. γ in Au+Au and Cu+Cu collisions at $\sqrt{s_{NN}} = 200\,\mathrm{GeV}$. Figure from [17].

nucleon of 200 GeV [17]. These were followed by measurements in Cu+Cu collisions at 200 GeV and both species at 62.4 GeV [18]. All four measurements resulted in same-sign (ss) and opposite-sign (os) correlations that had the expected sign for a CME signal. In addition, at the same centrality the Cu+Cu data are larger than those from the Au+Au collisions, which is as expected for a signal from the CME due to the predicted $1/N_{ch}$ dependence. The results for Au+Au and Cu+Cu at 200 GeV are shown in Fig. 1.

The signal was shown to be largely independent of the transverse momentum difference, ruling out potential background sources such as femtoscopic correlations and Coulomb attraction/repulsion. Models including effects such as elliptic flow, resonance decays and jets but no CME also did not reproduce the observed ss and os correlations.

2.2. *Collision energy dependence*

Another important question was how the potential signal depended on the collision energy. Since the CME is believed to depend on the formation of a quark-gluon plasma and chiral symmetry restoration, the signal could be greatly suppressed, or even absent, at low collision energies where the QGP may not be formed, or its lifetime may be significantly shortened.

To explore this question STAR utilized their Beam Energy Scan (BES) data to analyze events from $\sqrt{s_{NN}} = 7.7 - 200\,\mathrm{GeV}$ [19], while ALICE studied Pb+Pb collisions at 2.76 TeV at the LHC [20]. As shown in Fig. 2, clear signals compatible with that of charge separation along the magnetic field were observed from $\sqrt{s_{NN}} = 20\,\mathrm{GeV}$ to 2.76 TeV, with little-to-no energy dependence. The potential signal then falls steeply at lower energies. This trend could be consistent with the CME, since

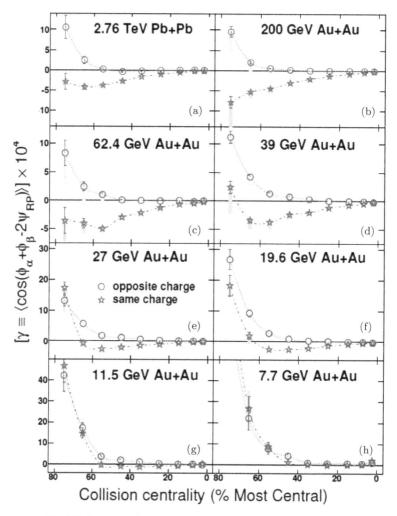

Fig. 2. γ from Pb+Pb data at $\sqrt{s_{NN}} = 2.76\,\mathrm{TeV}$ from ALICE and Au+Au from $\sqrt{s_{NN}} = 200 - 7.7\,\mathrm{GeV}$ from STAR.

at lower energies the hadronic phase has been shown to play a dominant role in the medium's evolution.

2.3. Background contributions

CMS then performed a study on Pb+Pb and p+Pb collisions at $\sqrt{s_{NN}} = 5.02\,\mathrm{TeV}$. They observed differences between the γ_{ss} and γ_{os} correlations that were of similar magnitude in both the p+Pb and Pb+Pb data [21]. These results pose a significant challenge for the CME interpretation of the charge separation. The magnetic field, and hence the CME contribution to γ, in the p+Pb collisions, is expected to be significantly smaller than that in peripheral Pb+Pb collisions with the same multiplicity. In addition, monte-carlo studies indicate that unlike in A+A events, there is little correlation between the B-field and the event plane of p+Pb collisions.

To further probe the strength of the background contributions many efforts have been made to design and compare variables known to be background driven to γ which might contain the CME signal. One example is $\gamma_{123} = \langle (\phi_\alpha - 2\phi_\beta - 3\Psi_3) \rangle$ which measures the charge correlation with respect to the third order plane, Ψ_3, as opposed to the reaction, or second order, plane. Ψ_3 is not expected to be correlated to the B-field and hence no CME signal should manifest.

With H defined as the magnitude of the CME signal, F the background, κ a scaling factor of order unity, and $\delta = \langle cos(\phi_\alpha - \phi_\beta) \rangle$ we have $\gamma = \kappa v_2 F - H$, $\delta = F + H$ and $\gamma_{123} = \kappa v_3 F$. If, as expected, the p+A data contain only background $\Delta\gamma/(v_2\Delta\delta) = \Delta\gamma_{123}/(v_3\Delta\delta)$. The similarity of the two ratios has been demonstrated in p+Pb at $\sqrt{s_{NN}} = 8.16\,\text{TeV}$ [22] and similar studies have reached similar conclusions from analyses of p+Au and d+Au data at $\sqrt{s_{NN}} = 200\,\text{GeV}$ [23]. The ratios $\Delta\gamma/(v_2\Delta\delta)$ and $\Delta\gamma_{123}/(v_3\Delta\delta)$ were also compared in the Pb+Pb data at $\sqrt{s_{NN}} = 5.02\,\text{TeV}$, see Fig. 3. From these measurements CMS concluded that the fraction of CME contributing to $\Delta\gamma$ in Pb+Pb collisions is $<7\%$ at a 95% confidence level [22].

Another attempt to disentangle the possible CME signal from the background involved comparing γ determined using Ψ_{RP} to γ_{PP} where Ψ_{PP} was used in the calculation. The v_2 dominated background component should be maximal when measured with respect to Ψ_{PP}, whereas the CME part is maximal when Ψ_{RP} is used, due to its better alignment with the B-field. $\Delta\gamma$ and $\Delta\gamma_{PP}$ therefore contain different fractional amounts of the CME and background signals. This offers the opportunity to determine these two contributions uniquely. With some thought you can see that the CME fraction, $f_{CME} = \Delta\gamma_{PP}^{CME}/\Delta\gamma_{PP} = [A/(a-1)]/[1/a^2 - 1]$ where $A = \Delta\gamma/\Delta\gamma_{PP}$ and $a = v_2/v_{2,PP} = \langle cos[2(\Psi_{PP} - \Psi_{RP})] \rangle$.

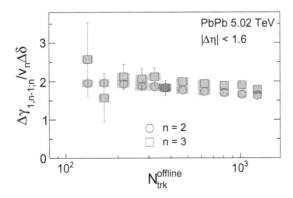

Fig. 3. The ratio of $\Delta\gamma/(v_2\Delta\delta)$ and $\Delta\gamma_{123}/(v_3\Delta\delta)$ for Pb+Pb collisions at $\sqrt{s_{NN}} = 5.02\,\text{TeV}$ from CMS. Statistical and systematic uncertainties are indicated by the error bars and shaded regions, respectively. Figure from [22].

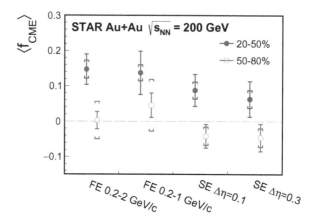

Fig. 4. f_{CME} from Au+Au data at $\sqrt{s_{NN}} = 200\,\mathrm{GeV}$ from STAR using the reaction (or spectator) plane versus the participant plane technique. Figure from [24].

Figure 4 shows the recently reported measurement by STAR using this technique [24]. Although these results make clear that the background dominates the measurements, a finite CME signal is suggested with a significance of 1–3 standard deviations in mid-central collisions. It is possible for there to be a measurable signal at RHIC energies but not at those of the LHC due to differences in the expected magnitudes of the integrated magnetic fields.

2.4. Isobar studies

Most recently, isobar collisions were proposed as a "clean" nearly model-independent method to probe for the CME even while the backgrounds remained not completely understood [25]. The idea was to keep the backgrounds, particularly the v_2 component, constant while varying the magnetic field, which drives the signal. STAR proposed to study $^{96}_{44}Ru +^{96}_{44} Ru$ and $^{96}_{40}Zr +^{96}_{40} Zr$ collisions. Due to the four extra protons in Ru the magnetic field is expected to be 10–20% higher while the expected eccentricities, which drive v_2, and multiplicities of the mid-central collisions should be similar [26]. It was determined that assuming $\Delta\gamma$ contained 80% non-CME background, a clear signal, $\sim5\sigma$, could be obtained if more than 1B events for each isobar species could be collected.

The isobar run was approved at RHIC and took place in 2018; more than 2 billion events were collected for each isobar species. The need to keep the systematic uncertainties at a few percent, and less than the statistical precision of the measurement, led to some special running conditions being implemented. The species in the beam were switched each store and long stores with level low luminosities were recorded. This minimized variations in the detector responses between the data collected for the two isobar species.

In addition it was decided to perform the CME analyses on the isobar data in a blinded fashion. A three-step analysis procedure was developed. In Step 1 analysts were provided a "reference sample" of data, comprised of a mix of events from the two species, the order of which respected time-dependent changes in run conditions. This sample was used to tune analysis codes and perform time-dependent quality assurance. In Step 2 a species-blind sample suitable for calculating efficiencies and corrections for individual data-taking runs was provided. Each data file only contained data from a single species and were limited to ~30 mins of data taking, all species-specific information was disguised. At this stage, only run-by-run corrections and code alterations subsequent to these corrections were allowed. The code was then "frozen" and Step 3 passed over the fully un-blinded data. It was agreed by the collaboration that the results from the frozen codes used during Step 3 would be released to the community without any post-blinding adjustments. More details of the blinding procedure can be found in [27].

When I gave this presentation Step 3 was still underway and we were eagerly awaiting the results. Since then the results have been announced and published [25]. A precision down to 0.4% was achieved in the relative magnitudes of the pertinent observables between the two isobar systems. This is understood to be the most precise analysis reported from heavy-ion collisions to date. The analyses revealed indications that the magnitude of the CME background is different between the two species as differences in the multiplicity and flow harmonics at matching centralities were observed. Prior to the blinded analyses starting the proposed CME signatures were defined as a significant excess of the CME-sensitive observables in Ru+Ru collisions over those in Zr+Zr collisions. No CME signature that satisfied the predefined criteria has been observed in isobar collisions via these blind analyses. However, the cause(s) of the observed multiplicity and flow differences between the isobars requires further studies. Enhanced understanding of these issues may result in a refining of the baselines used to define an observation of a CME signal.

3. Summary

Both experimentally and theoretically our understanding of a potential CME in heavy-ion collisions continues to evolve. New experimental methods to potentially tease out the signal from the large background are being explored, and there has been significant recent improvement of our understanding of the sensitivities of the established techniques.

While no strong conclusions can yet be drawn, several results are consistent with CME expectations. However a clear interpretation remains elusive as the backgrounds are large and not yet fully understood. The recent release of the isobar results by STAR made evident that significant effort remains to be done to understand these background effects to the level of precision needed to finally report if the CME is present, or not, in the medium created in ultra-relativistic heavy-ion collisions.

References

1. I. Arsene *et al.*, Quark gluon plasma and color glass condensate at RHIC? The Perspective from the BRAHMS experiment, *Nucl. Phys. A* **757**, 1 (2005).
2. K. Adcox *et al.*, Formation of dense partonic matter in relativistic nucleus-nucleus collisions at RHIC: Experimental evaluation by the PHENIX collaboration, *Nucl. Phys. A* **757**, 184 (2005).
3. B. B. Back *et al.*, The PHOBOS perspective on discoveries at RHIC, *Nucl. Phys. A* **757**, 28 (2005).
4. J. Adams *et al.*, Experimental and theoretical challenges in the search for the quark gluon plasma: The STAR Collaboration's critical assessment of the evidence from RHIC collisions, *Nucl. Phys. A* **757**, 102 (2005).
5. F. Antinori *et al.*, Proceedings of the 7th international conference on ultrarelativistic nucleus-nucleus collisions: Quark matter 2018, *Nuclear Physics A* **982**, 1 (2019).
6. A. Bazavov *et al.*, Chiral crossover in QCD at zero and non-zero chemical potentials, *Phys. Lett. B* **795**, 15 (2019).
7. M. L. Miller, K. Reygers, S. J. Sanders and P. Steinberg, Glauber modeling in high energy nuclear collisions, *Ann. Rev. Nucl. Part. Sci.* **57**, 205 (2007).
8. T. D. Lee, A theory of spontaneous t violation, *Phys. Rev. D* **8**, 1226 (Aug 1973).
9. T. D. Lee and G. C. Wick, Vacuum stability and vacuum excitation in a spin-0 field theory, *Phys. Rev. D* **9**, 2291 (Apr 1974).
10. P. D. Morley and I. A. Schmidt, Strong P, CP, T violations in heavy-ion collisions, *Zeitschrift für Physik C Particles and Fields* **26**, 627 (1985).
11. D. Kharzeev, R. D. Pisarski and M. H. G. Tytgat, Possibility of spontaneous parity violation in hot QCD, *Phys. Rev. Lett.* **81**, 512 (1998).
12. D. Kharzeev and R. D. Pisarski, Pionic measures of parity and CP violation in high-energy nuclear collisions, *Phys. Rev. D* **61**, p. 111901 (2000).
13. D. E. Kharzeev, L. D. McLerran and H. J. Warringa, The Effects of topological charge change in heavy ion collisions: 'Event by event P and CP violation', *Nucl. Phys. A* **803**, 227 (2008).
14. V. Skokov, A. Y. Illarionov and V. Toneev, Estimate of the magnetic field strength in heavy-ion collisions, *Int. J. Mod. Phys. A* **24**, 5925 (2009).
15. A. M. Poskanzer and S. A. Voloshin, Methods for analyzing anisotropic flow in relativistic nuclear collisions, *Phys. Rev. C* **58**, 1671 (1998).
16. S. A. Voloshin, Parity violation in hot QCD: How to detect it, *Phys. Rev. C* **70**, p. 057901 (2004).
17. B. I. Abelev *et al.*, Azimuthal charged-particle correlations and possible local strong parity violation, *Phys. Rev. Lett.* **103**, p. 251601 (2009).
18. B. I. Abelev *et al.*, Observation of charge-dependent azimuthal correlations and possible local strong parity violation in heavy ion collisions, *Phys. Rev. C* **81**, p. 054908 (2010).
19. L. Adamczyk *et al.*, Beam-energy dependence of charge separation along the magnetic field in Au+Au collisions at RHIC, *Phys. Rev. Lett.* **113**, p. 052302 (2014).
20. B. Abelev *et al.*, Charge separation relative to the reaction plane in Pb-Pb collisions at $\sqrt{s_{NN}} = 2.76$ TeV, *Phys. Rev. Lett.* **110**, p. 012301 (2013).
21. V. Khachatryan *et al.*, Observation of charge-dependent azimuthal correlations in p-Pb collisions and its implication for the search for the chiral magnetic effect, *Phys. Rev. Lett.* **118**, p. 122301 (2017).
22. A. M. Sirunyan *et al.*, Constraints on the chiral magnetic effect using charge-dependent azimuthal correlations in pPb and PbPb collisions at the CERN Large Hadron Collider, *Phys. Rev. C* **97**, p. 044912 (2018).

23. J. Adam *et al.*, Charge-dependent pair correlations relative to a third particle in p+Au and d+Au collisions at RHIC, *Phys. Lett. B* **798**, p. 134975 (2019).
24. M. Abdallah *et al.*, Search for the chiral magnetic effect via charge-dependent azimuthal correlations relative to spectator and participant planes in Au+Au collisions at $\sqrt{s_{NN}} = 200$ GeV (6 2021).
25. V. Koch, S. Schlichting, V. Skokov, P. Sorensen, J. Thomas, S. Voloshin, G. Wang and H.-U. Yee, Status of the chiral magnetic effect and collisions of isobars, *Chin. Phys. C* **41**, p. 072001 (2017).
26. W.-T. Deng, X.-G. Huang, G.-L. Ma and G. Wang, Testing the chiral magnetic effect with isobaric collisions, *Phys. Rev. C* **94**, p. 041901 (Oct 2016).
27. J. Adam *et al.*, Methods for a blind analysis of isobar data collected by the STAR collaboration, *Nucl. Sci. Tech.* **32**, p. 48 (2021).

Chiral Magnetic Effect in Heavy Ion Collisions and Beyond

Dmitri E. Kharzeev*

Center for Nuclear Theory,
Department of Physics and Astronomy, Stony Brook University
New York 11794-3800, USA
Department of Physics, Brookhaven National Laboratory
Upton, New York 11973-5000, USA
* E-mail: dmitri.kharzeev@stonybrook.edu
www.physics.sunysb.edu/kharzeev

Chirality is a ubiquitous concept in modern science, from particle physics to biology. In quantum physics, chirality of fermions is linked to topology of gauge fields by the chiral anomaly. While the chiral anomaly is usually associated with the short-distance behavior in field theory, in recent years it has been realized that it also affects the macroscopic behavior of systems with chiral fermions. In particular, the local imbalance between left- and right-handed fermions in the presence of a magnetic field induces non-dissipative transport of electric charge ("the Chiral Magnetic Effect", CME). In heavy ion collisions, there is an ongoing search for this effect at Relativistic Heavy Ion Collider, with results from a dedicated isobar run presented very recently. An observation of CME in heavy ion collisions could shed light on the mechanism of baryon asymmetry generation in the Early Universe. Recently, the CME has been discovered in Dirac and Weyl semimetals possessing chiral quasi-particles. This observation opens a path towards quantum sensors, and potentially a new kind of quantum computers.

Keywords: Chiral magnetic effect; chiral matter; quantum chromodynamics; heavy ion collisions.

1. Chirality in subatomic world

In 1874, Louis Pasteur wrote [1]: *"The universe is asymmetric and I am persuaded that life, as it is known to us, is a direct result of the asymmetry of the universe or of its indirect consequences."* Life *is* asymmetric, and the concept of chirality – the distinction between the left and right, or between an object and its reflection in the mirror – is quite literally ingrained in our DNA. The DNA double helix is chiral; moreover, DNA molecules also tend to form knots [2] – and most of the complex knots are chiral as well.

Pasteur's prophecy is that the chiral asymmetry of life originates from the asymmetry of the Universe. At present we still do not understand how this connection operates – but we do know already that the Universe is asymmetric. As predicted in 1956 by T.D. Lee and C.N. Yang [3], and established shortly afterwards through the observation of angular asymmetry in β decay by C.S. Wu [4], weak interactions involve only left-handed fermions. This discovery brought to focus the crucial role of chirality in subatomic world.

For fermions, chirality is defined as the projection of spin on momentum, with a minus sign for antifermions. In Dirac theory, it is an eigenvalue of γ^5 matrix, with projectors on the right and left chiral states given by

$$\mathcal{P}^+ = \frac{1}{2}(1 + \gamma^5); \quad \mathcal{P}^- = \frac{1}{2}(1 - \gamma^5).$$

A Dirac spinor ψ can thus be decomposed into right- and left-handed chiral states by $\psi^+ = \mathcal{P}^+\psi$ and $\psi^- = \mathcal{P}^-\psi$. Dirac equation for massless fermions is $i\hat{D}\psi \equiv i\gamma^\mu\partial_\mu\psi = 0$, and Dirac operator can also be decomposed into right- and left-handed components:

$$D = i\hat{D}\mathcal{P}^+; \quad D^\dagger = i\hat{D}\mathcal{P}^-. \tag{1}$$

For massless fermions, chirality is conserved in the interactions with gauge fields (e.g. photons and gluons) on the classical level - so a left-handed fermion is expected to always stay left-handed. However quantum chiral anomaly [5, 6] allows for a chirality transmutation – a left-handed fermion can become right-handed if it interacts with a configuration of gauge field that can change its own chirality. While the chirality of a fermion is a familiar concept, how does one characterize chirality of a gauge field?

2. Chirality of gauge fields

It appears that the key to understanding the chirality of gauge fields is offered by the example of DNA knots mentioned above. Indeed, let us consider a knot of magnetic flux. The chiral invariant of this gauge field configuration is known as *magnetic helicity*:

$$\text{CS}[A] = \int d^3x \, \mathbf{A} \cdot \mathbf{B}, \tag{2}$$

where $\mathbf{A}$ is the vector gauge potential, and $\mathbf{B} = \nabla \times \mathbf{A}$ is magnetic field. Using analogy between the gauge potential and velocity $\mathbf{v}$, we can easily understand that (2) detects chirality – indeed, when $\mathbf{A} \to \mathbf{v}$, magnetic field gets replaced by vorticity $\mathbf{B} \to \mathbf{\Omega} = \frac{1}{2}\nabla \times \mathbf{v}$, and a non-zero scalar product $\mathbf{v} \cdot \mathbf{\Omega}$ implies a helical motion.

For an Abelian gauge theory (such as electrodynamics), magnetic helicity coincides with a more general 3-form derived by Chern and Simons [7] in differential geometry to characterize the global topology of a manifold with a Lie algebra valued 1-form $\mathbf{A}$ over it:

$$\text{CS}[A] = \text{Tr}\left[\mathbf{F} \wedge \mathbf{A} - \frac{1}{3}\mathbf{A} \wedge \mathbf{A} \wedge \mathbf{A}\right], \tag{3}$$

where the curvature (field strength tensor) is defined as $\mathbf{F} = d\mathbf{A} + \mathbf{A} \wedge \mathbf{A}$. In physical terms, the invariant (2) corresponds to the chirality of the knot made of magnetic flux - for example, a torus would be characterized by $\text{CS} = 0$, a right-handed trefoil knot by $\text{CS} = +1$, and a left-handed one by $\text{CS} = -1$.

Chern-Simons p-form can be defined for any odd space-time dimension p. We happen to live in 3+1 dimensional space-time with even $p = 4$ – does this mean that Chern-Simons invariant is irrelevant? Of course not – what this means is that in our four-dimensional space-time Chern-Simons invariant can change in time, with non-vanishing exterior derivative

$$dCS[A] = \text{Tr}\left[\mathbf{F} \wedge \mathbf{F}\right],\tag{4}$$

which is known as the Chern-Pontryagin invariant. In a more familiar for physicists notation, this relation implies that

$$\frac{\partial\, CS[A]}{\partial t} = -2 \int d^3x\, \mathbf{E} \cdot \mathbf{B},\tag{5}$$

which can be readily obtained from (2) using the Coulomb gauge where the electric field $\mathbf{E}$ is related to the gauge potential by $\mathbf{E} = -\dot{\mathbf{A}}$. Therefore, a change of topology of the gauge field in time quantified by (5) gives rise to an electric field parallel to the magnetic one. What is the effect of parallel electric and magnetic fields on chirality of chiral fermions? Can this electric field drive a current?

3. Chiral anomaly

Charged particles experience a Lorentz force $\mathbf{F}_m = e\, \mathbf{v} \times \mathbf{B}$ from an external magnetic field $\mathbf{B}$. If the projection of their velocities on the direction of magnetic field is equal to zero, this leads to the motion along closed cyclotron orbits with $\mathbf{\Omega} = \nabla \times \mathbf{v} \neq 0$, but $\mathbf{v} \cdot \mathbf{\Omega} = 0$. Even if the charge has a non-zero velocity component along $\mathbf{\Omega}$, but there is no external force directed along $\mathbf{B}$, then we can always choose a frame in which $\mathbf{v} \cdot \mathbf{B} = \mathbf{v} \cdot \mathbf{\Omega} = 0$, i.e. the motion of the charge is not helical. This is not true if there is a force applied along the direction of $\mathbf{B}$, e.g. a Lorentz force $\mathbf{F}_e = e\, \mathbf{E}$ resulting from an electric field $\mathbf{E}$ parallel to $\mathbf{B}$ - then the motion is helical in any inertial frame.

   In quantum theory, charged particles occupy quantized Landau levels, and for massless fermions the lowest Landau level (LLL) is chiral and has a zero energy – qualitatively, this appears due to a cancellation between a positive kinetic energy and negative Zeeman energy of the interaction between magnetic field and spin. Therefore, on the LLL the spins of positive (negative) fermions are aligned along (against) the direction of magnetic field. All excited levels are degenerate in spin, and are thus not chiral.

   More formally, this can be seen as a consequence of Atiyah-Singer index theorem [8] that relates the analytical index of Dirac operator to its topological index. In other words, it relates the number of zero modes of Dirac operator acting on a manifold M to the topology of this manifold. The analytical index of Dirac operator (1) is given by the number of zero energy modes with right (ν_+) and left (ν_+) chirality:

$$\text{ind}\, D = \dim \ker D - \dim \ker D^\dagger = \nu_+ - \nu_-,\tag{6}$$

where $\ker D$ is the subspace spanned by the kernel of the operator D, i.e. the subspace of states that obey $D\psi = 0$, see (1).

For two dimensional manifold M, the topological index of this operator is equal to $\frac{1}{2\pi} \int_M \mathrm{tr}\, \mathbf{F}$, and Atiyah-Singer index theorem thus states that

$$\nu_+ - \nu_- = \frac{1}{2\pi} \int_M \mathrm{Tr}\, \mathbf{F}. \tag{7}$$

Performing analytical continuation to Euclidean (x, y) space (with $\mathbf{B}$ along the $\mathbf{z}$ axis), we thus find that the number of chiral zero modes is given by the total magnetic flux through the system [9]. For positive fermions with charge $e > 0$, we have $\nu_- = 0$ and the number of right-handed chiral modes from (7) is given by

$$\nu_+ = \frac{e\Phi}{2\pi}, \tag{8}$$

which is just the number of LLLs in the transverse plane; we have included an explicit dependence on electric charge e. For negative fermions $\nu_+ = 0$ and $\nu_- = \frac{e\Phi}{2\pi}$.

Let us assume for simplicity that the charge chemical potential is equal to zero, $\mu = 0$. It is clear that $\nu_+ = \nu_-$, and the system possesses zero chirality. Let us now turn on an external electric field $\mathbf{E} \parallel \mathbf{B}$. Dynamics of fermions on the LLL is $(1+1)$ dimensional along the direction $\mathbf{B}$, and we can apply the index theorem (7) to the (z, t) manifold; for positive fermions of "+" chirality

$$\nu_+ = \frac{1}{2\pi} \int dz dt\, eE, \quad \nu_- = 0, \tag{9}$$

and for negative fermions of "−" chirality

$$\nu_+ = 0, \quad \nu_- = -\frac{1}{2\pi} \int dz dt\, eE. \tag{10}$$

These relations can be understood from a *seemingly* classical argument [10]: the positive charges are accelerated by the Lorentz force along the electric field $\mathbf{E}$, and acquire Fermi-momentum $p_F^+ = eEt$. The density of states in one spatial dimension is $p_F/(2\pi)$, and so the total number of positive fermions with positive chirality is $\nu_+ = 1/(2\pi) \int dz dt\, eE$, in accord with (9). The same argument applied to negative fermions explains (10). While the notion of acceleration by Lorentz force is classical, in assuming that it increases the Fermi momentum, we have made an assumption that there is an infinite tower of states that are accelerated by the Lorentz force. This tower of states does not exist in classical theory; however it is a crucial ingredient of the (quantum) theory of Dirac.

Multiplying the density of states in longitudinal direction $p_F/(2\pi)$ by the density of states $eB/(2\pi)$ in the transverse direction, we find from (9) and (10) that in $(3+1)$ dimensions

$$\nu_+ - \nu_- = 2 \times \frac{e^2}{4\pi^2} \int d^2x\, dz\, dt\, \mathbf{E} \cdot \mathbf{B} = \frac{e^2}{2\pi^2} \int d^2x\, dz\, dt\, \mathbf{E} \cdot \mathbf{B}, \tag{11}$$

where the factor of 2 is due to the contributions of positive and negative fermions. This relation represents Atiyah-Singer theorem for $U(1)$ in $(3+1)$ dimensions, so we could use it directly instead of relying on dimensional reduction of the LLL dynamics. Note that the quantity on the r.h.s. of (11) is nothing but the derivative of Chern-Simons three-form, see (4), (5).

The relation (11) can also be written in differential form in terms of the axial current

$$J_\mu^5 = \bar{\psi}\gamma_\mu\gamma^5\psi = J_\mu^+ - J_\mu^- \tag{12}$$

as [5, 6]

$$\partial^\mu J_\mu^5 = -\frac{e^2}{8\pi^2} \mathbf{E} \cdot \mathbf{B}. \tag{13}$$

The chiral anomaly equation (13) is an operator relation. In particular, we can use it to evaluate the matrix element of transition from a pseudoscalar excitation of Dirac vacuum (a neutral pion) into two photons. This can be done by using on the l.h.s. of (13) the PCAC relation that replaces the divergence of axial current by the interpolating pion field φ

$$\partial^\mu J_\mu^5 \simeq F_\pi M_\pi^2 \, \varphi, \tag{14}$$

where F_π and M_π are the pion decay constant and mass. Taking the matrix element of (13) between the vacuum and the two-photon states $\langle 0|\partial^\mu J_\mu^5|\gamma\gamma\rangle$ then yields the decay width of $\pi^0 \to \gamma\gamma$ decay which is a hallmark of chiral anomaly. However, the chiral anomaly has much broader implications when the classical gauge fields are involved, as we will now discuss.

4. Chiral magnetic effect: Non-dissipative quantum transport induced by chiral anomaly

As a first step, let us observe that the derivation of chiral anomaly presented above implies the existence of non-dissipative electric current in parallel electric and magnetic fields. Indeed, the (vector) electric current

$$J_\mu = \bar{\psi}\gamma_\mu\psi = J_\mu^+ + J_\mu^- \tag{15}$$

contains equal contributions from positive charge, positive chirality fermions flowing along the direction of $\mathbf{E}$ (which we assume to be parallel to $\mathbf{B}$), and negative charge, negative chirality fermions flowing in the direction opposite to $\mathbf{E}$:

$$J_z = 2 \times \frac{e^2}{4\pi^2}\mathbf{E} \cdot \mathbf{B}\, t = \frac{e^2}{2\pi^2}\mathbf{E} \cdot \mathbf{B}\, t. \tag{16}$$

In constant electric and magnetic fields, this current grows linearly in time – this means that the conductivity σ defined by $J = \sigma E$ becomes divergent, and resistivity $\rho = 1/\sigma$ vanishes. This means that the current (16) is non-dissipative, similarly to what happens in superconductors!

We can also write down the relation (16) in terms of the chemical potentials $\mu_+ = p_F^+$ and $\mu_- = p_F^-$ for right- and left-handed fermions, which for massless dispersion relation are given by the corresponding Fermi momenta $p_F^+ = eEt$ and $p_F^- = -eEt$. It is convenient to define the *chiral chemical potential*

$$\mu_5 \equiv \frac{1}{2} \left(\mu_+ - \mu_- \right); \tag{17}$$

the relation (16) then becomes [11]

$$\mathbf{J} = \frac{e^2}{2\pi^2} \mu_5 \mathbf{B}. \tag{18}$$

It is important to realize that unlike a usual chemical potential, the chiral chemical potential μ_5 does not correspond to a conserved quantity – on the contrary, the non-conservation of chiral charge due to chiral anomaly is necessary for the Chiral Magnetic Effect (CME) described by (18) to exist. Indeed, a static magnetic field cannot perform work, so the current (18) can be powered only by a change in the chiral chemical potential. Another way to see this is to consider a power [12] of the current (18): $P = \int d^3x \, \mathbf{E} \, \mathbf{J} \sim \mu_5 \int d^3x \, \mathbf{E} \, \mathbf{B}$. For a constant μ_5, it can be both positive or negative, in clear contradiction with energy conservation. On the other hand, if μ_5 is dynamically generated through the chiral anomaly, it has the same sign as $\int d^3x \, \mathbf{E} \, \mathbf{B}$, and the electric power is always positive, as it should be.

For the case of parallel $\mathbf{E}$ and $\mathbf{B}$, the CME relation (18) is a direct consequence of the Abelian chiral anomaly for the case of classical background fields. However it is valid also when the chiral chemical potential is sourced by non-Abelian anomalies [11], coupling to time-dependent axion field [13], or is just a consequence of a non-equilibrium dynamics [14].

The relation (18) was first introduced in 1980 in a pioneering paper by Vilenkin [15] who however considered the case of a constant μ_5, motivated by parity violation in weak interactions. In this case, as we have discussed above, the system is in equilibrium and the electric current cannot exist [16] – so while the relation (18) was known for some time, turning it into a real physical effect required understanding the role of anomaly, see review [17] for a detailed discussion and references. The same is true in condensed matter applications when the nodes of dispersion relations of left- and right-handed fermions are located at different energies, $E_+(p = 0) \neq E_-(p = 0)$. In this case, contrary to some early claims, the CME current does not appear [12]. For the CME to emerge, the chiral asymmetry has to appear in the occupancy of the left- and right-handed states.

5. Chiral magnetic effect as a probe of vacuum topology: Interplay of Abelian and non-Abelian chiral anomalies

The CME relation (18) can be proven also for the case of non-Abelian plasma containing chiral fermions in an external Abelian magnetic field $\mathbf{B}$ [11]. In this case, the chiral charge (and the corresponding chiral chemical potential) is created by non-Abelian anomaly due to transitions [18] between different topological sectors marked by different Chern-Simons numbers (4), and the electric current flows along (or against) the direction of **B**. This presents a unique opportunity to get a direct experimental access to the study of non-Abelian topological fluctuations [19–21]. Topological transitions in non-Abelian electroweak plasma violate the baryon number conservation and may be at the origin of baryon asymmetry of our Universe [22, 23].

The only non-Abelian theory where topological transitions are accessible to experiment is Quantum ChromoDynamics (QCD). Its chiral fermions are quarks, that are confined into hadrons. According to the Atiyah-Singer index theorem [8], topological transitions in QCD should be accompanied by the change in chirality of quarks. They can thus be responsible for spontaneous breaking of $U_+(3) \times U_-(3)$ chiral symmetry in QCD, emergence of the corresponding Goldstone bosons (pions, kaons, and η meson), and a massive flavor-singlet η' meson resulting from an explicit breaking of the flavor-singlet part of the chiral symmetry.

While there is a significant evidence for the prominent role of topological transitions (instantons, sphalerons, ...) in the structure of QCD vacuum and the properties of hadrons (see [24] for a review), such transitions have never been detected in experiment. The CME, with its directly detectable electric current, makes such an observation possible. Indeed, consider a QCD plasma is created in a heavy ion collision. Because the colliding ions possess positive electric charges, the produced plasma, at least during its early moments, is embedded into a very strong magnetic field which is on the order of a typical QCD scale [20], $eB \sim \Lambda_{QCD}^2$. In such a strong magnetic field (possibly the strongest in the present Universe), electromagnetic interactions of quarks are comparable in strength to the strong ones. In addition, the rate of topological "sphaleron" [25] transitions in hot QCD plasma is high, and this should result in the creation of chirally imbalanced domains ("$\mathcal{P}$-odd bubbles" [26]) characterized by non-zero value of the chiral chemical potential. This means that all conditions for the CME are met, and there should be an electric current (18) propagating through the QCD plasma.

6. Chiral magnetic effect in heavy ion collisions and the RHIC isobar run

The magnetic field produced by the colliding ions is directed perpendicular to the reaction plane of the collision, therefore the CME current (18) is directed perpendicular to the reaction plane as well. The magnetic field is strongest during the early moments of the collision; the sphaleron transitions are thus expected to induce the electric charge separation relative to the reaction plane early in the evolution of the system [19]. Because the QCD plasma is rapidly expanding (similarly to the Early Universe), this initial electric charge separation cannot be fully scrambled by final state interactions, and survives till the moment when the plasma cools down and transforms into hadrons. Since the electric charge is conserved throughout the hadronization, the produced hadrons should possess charge asymmetry relative to the reaction plane that can be directly detected in experiment.

Of course, QCD does not violate parity $\mathcal{P}$ symmetry globally, and the sign of chiral imbalance fluctuates event-by-event – we are thus dealing with a "local $\mathcal{P}$-violation". The experimental signature of CME is thus a dynamical enhancement of out-of-plane fluctuations of charge asymmetry, relative to the in-plane fluctuations that are not affected by CME. The corresponding experimental observable was proposed by Voloshin [27] almost immediately after the idea to search for CME in heavy ion collisions was formulated [19]. In terms of the azimuthal angles of positive ϕ^+ and negative ϕ^- hadrons, and the azimuthal angle of the reaction plane Ψ_{RP} the proposed "γ correlator" observable is [27]

$$\gamma^{+-} \equiv \left\langle \cos(\phi^+ + \phi^- - 2\Psi_{RP}) \right\rangle =$$

$$= \left\langle \cos(\phi^+ - \Psi_{RP}) \cos(\phi^- - \Psi_{RP}) \right\rangle - \left\langle \sin(\phi^+ - \Psi_{RP}) \sin(\phi^- - \Psi_{RP}) \right\rangle, \quad (19)$$

where the sum over hadrons in a given event, and then an average over many events, are assumed.

The CME would result in the electric charge separation relative to the reaction plane, and should produce, in a given event, either $\sin(\phi^+ - \Psi_{RP}) > 0$, $\sin(\phi^+ - \Psi_{RP}) < 0$, or $\sin(\phi^+ - \Psi_{RP}) < 0$, $\sin(\phi^+ - \Psi_{RP}) > 0$, depending on the sign of the chiral imbalance. In both cases however $\left\langle \sin(\phi^+ - \Psi_{RP}) \sin(\phi^- - \Psi_{RP}) \right\rangle < 0$, and the correlator (19) should be positive. Of course, background fluctuations would also give contributions to the γ correlator, but (19) is the difference of in-plane and out-of-plane fluctuations, so the backgrounds that do not depend on the reaction plane cancel out. Therefore the backgrounds that survive in (19) should depend on the reaction plane, and are thus expected to be proportional to the "elliptic flow" that describes the elliptic deformation of the event and is defined as the second Fourier harmonic of the hadron azimuthal angle distribution [27]. In hydrodynamical description of heavy ion collisions, the elliptic flow results from the pressure anisotropy generated by the geometry of an off-central collision. It is also important to note that (19) is expected to scale with inverse hadron multiplicity, both for the signal and background contributions [19, 27], see [28] for a review and detailed discussion.

STAR Collaboration performed the measurements of γ correlators, and other CME observables, over a range of collision energies and for different ions [29–31]. In addition, ALICE [32] and CMS [33] Collaborations at the LHC extended these studies to higher energies, see [28, 34] for reviews and compilations of published results. For all studied heavy ion colllisions, non-zero and positive γ^{+-} was measured. The correlators γ^{++} and γ^{--} were also non-zero and negative, as expected for CME. However, background contributions proportional to elliptic flow have also also identified, and the problem of separating the possible signal from background has assumed the center of the stage.

This is why the proposal of using isobar collisions has been made [35]. The idea is that since the isobars have the same mass number, they have *approximately* the same size and shape, and thus the elliptic flow, and the backgrounds driven by it, should be nearly identical. On the other hand, the difference in the electric charge should create a difference in the produced magnetic field, and thus in the CME which is proportional to it. The measurement of γ correlators, and other CME observables, in isobar collisions would thus allow to isolate the CME signal from the background.

A dedicated high statistics measurement of CME observables in $^{96}_{44}$Ru $^{96}_{44}$Ru and $^{96}_{40}$Zr $^{96}_{40}$Zr collisions was performed by STAR Collaboration at RHIC in 2018 [36]. The data from 3.8 billion collision events were recorded, and a blind analysis of the data (unprecedented in heavy ion physics) was performed. The expectation has been that since the electric charge of Ru is higher than that of Zr, the ratio of CME observables, e.g. γ^{Ru}/γ^{Zr} would exceed one if CME were present, and be equal to one if CME were to give a negligible contribution. In any scenario, one expects to find $\gamma^{Ru}/\gamma^{Zr} \geq 1$.

The STAR analysis however revealed that the ratio γ^{Ru}/γ^{Zr} was significantly below one! This certainly did not fit the CME expectations – and in fact *any* theory expectation, CME-based or not. Basing on these findings, and the predefined criteria assuming the identical backgrounds in $^{96}_{44}$Ru $^{96}_{44}$Ru and $^{96}_{40}$Zr $^{96}_{40}$Zr collisions, STAR Collaboration concluded that "no CME signature that satisfies the predefined criteria has been observed" [36].

While the original CME expectations for the isobar run have certainly not been met, the puzzling observation of $\gamma^{Ru}/\gamma^{Zr} < 1$ begs for an explanation. A detailed investigation is still ongoing, but the post-blinding analysis [36] has already revealed that the key to the solution is a very significant difference in hadron multiplicity in $^{96}_{44}$Ru $^{96}_{44}$Ru and $^{96}_{40}$Zr $^{96}_{40}$Zr collisions observed by STAR in centrality cuts relevant for the CME analysis. As mentioned above, the γ correlator scales, in the good first approximation, with inverse multiplicity, and the multiplicity measured in $^{96}_{44}$Ru $^{96}_{44}$Ru is significantly higher than in $^{96}_{40}$Zr $^{96}_{40}$Zr collisions in the same centrality cut. This explains the surprisingly low ratio $\gamma^{Ru}/\gamma^{Zr} < 1$. If one establishes a new baseline given by the measured ratio of inverse multiplicities, the observed γ^{Ru}/γ^{Zr} ratio in fact exceeds the baseline by $(1-4)$ σ, depending on the details of the analysis. This supports the CME interpretation, albeit with an insufficient statistical significance.

More detailed analysis of the isobar data taking account of the difference in the shape of Ru and Zr nuclei is still ongoing, but it has already become clear that the isobar run data represent an important milestone in the hunt for CME in heavy ion collisions. Nevertheless, the final conclusions will have to be based on the joint analysis of both isobar and symmetric heavy ion collisions. It will also be very important to extend the CME studies to lower collision energies, where topological fluctuations can be strongly enhanced [37] due to proximity to the critical point in the QCD phase diagram. This program is planned for the beam energy scan at RHIC, and can also be performed at future heavy ion facilities, such as NICA and FAIR.

7. Broader implications

CME is a macroscopic quantum phenomenon driven by the chiral anomaly; it is a direct probe of gauge field topology. As a result, it is induces a variety of phenomena involving chiral fermions in quark-gluon plasma, the Early Universe [38], astrophysics [39], and condensed matter physics. In the latter case, the CME has been firmly established through the studies of magnetotransport in Dirac and Weyl semimetals [40, 41], and is finding new applications. The emerging new frontier in condensed matter physics is chiral photonics, including the studies of chiral magnetic photocurrents [42] and other chiral phenomena. Among potential future directions, let us mention the use of CME for controlling "chiral qubits" [43] in quantum processors. It is clear that the chiral asymmetry of the Universe has many far-reaching consequences that we are just beginning to uncover.

This work was supported by the U.S. Department of Energy, Office of Science grants No. DE-FG88ER40388 and DE-SC0012704, and Office of Science, National Quantum Information Science Research Centers, Co-design Center for Quantum Advantage (C^2QA) under contract number DE-SC0012704.

References

1. L. Pasteur, Comptes Rendus de l'Académie des Sciences, 1874.
2. D. W. Sumners, Untangling DNA, The Mathematical Intelligencer **12**, 71 (1990).
3. T. D. Lee and C. N. Yang, Question of parity conservation in weak interactions, Phys. Rev. **104**, 254–258 (1956).
4. C. S. Wu, E. Ambler, R. W. Hayward, D. D. Hoppes and R. P. Hudson, Phys. Rev. **105**, 1413–1414 (1957).
5. S. L. Adler, Axial vector vertex in spinor electrodynamics, Phys. Rev. **177**, 2426–2438 (1969).
6. J. S. Bell and R. Jackiw, A PCAC puzzle: $\pi^0 \to \gamma\gamma$ in the σ model, Nuovo Cim. A **60**, 47–61 (1969).
7. S. S. Chern and J. Simons, Characteristic forms and geometric invariants, Annals Math. **99**, 48–69 (1974).
8. M. F. Atiyah and I. M. Singer, The Index of elliptic operators. 1,' Annals Math. **87**, 484–530 (1968).
9. Y. Aharonov and A. Casher, The Ground State of a Spin 1/2 Charged Particle in a Two-dimensional Magnetic Field, Phys. Rev. A **19**, 2461–2462 (1979) doi:10.1103/PhysRevA.19.2461
10. H. B. Nielsen and M. Ninomiya, Adler-Bell-Jackiw anomaly and Weyl fermions in crystal, Phys. Lett. B **130**, 389–396 (1983).
11. K. Fukushima, D. E. Kharzeev and H. J. Warringa, The Chiral Magnetic Effect, Phys. Rev. D **78**, 074033 (2008).
12. G. Basar, D. E. Kharzeev and H. U. Yee, Triangle anomaly in Weyl semimetals, Phys. Rev. B **89**, no. 3, 035142 (2014).
13. F. Wilczek, Two applications of axion electrodynamics, Phys. Rev. Lett. **58**, 1799 (1987).
14. D. Kharzeev, Y. Kikuchi and R. Meyer, Chiral magnetic effect without chirality source in asymmetric Weyl semimetals, Eur. Phys. J. B **91**, no. 5, 83 (2018).

15. A. Vilenkin, Equilibrium parity violating current in a magnetic field, Phys. Rev. D **22**, 3080–3084 (1980).
16. A. Vilenkin, Cancellation of equilibrium parity violating currents, Phys. Rev. D **22**, 3067–3079 (1980).
17. D. E. Kharzeev, Prog. Part. Nucl. Phys. **75**, 133–151 (2014).
18. A. A. Belavin, A. M. Polyakov, A. S. Schwartz and Y. S. Tyupkin, Pseudoparticle Solutions of the Yang-Mills Equations, Phys. Lett. B **59**, 85–87 (1975).
19. D. Kharzeev, Parity violation in hot QCD: Why it can happen, and how to look for it, Phys. Lett. B **633**, 260–264 (2006).
20. D. E. Kharzeev, L. D. McLerran and H. J. Warringa, The Effects of topological charge change in heavy ion collisions: 'Event by event P and CP violation', Nucl. Phys. A **803**, 227–253 (2008).
21. D. Kharzeev and A. Zhitnitsky, Charge separation induced by P-odd bubbles in QCD matter, Nucl. Phys. A **797**, 67–79 (2007).
22. V. A. Kuzmin, V. A. Rubakov and M. E. Shaposhnikov, On the anomalous electroweak baryon number nonconservation in the early universe, Phys. Lett. B **155**, 36 (1985).
23. V. A. Rubakov and M. E. Shaposhnikov, Electroweak baryon number nonconservation in the early universe and in high-energy collisions, Usp. Fiz. Nauk **166**, 493–537 (1996).
24. T. Schäfer and E. V. Shuryak, Instantons in QCD, Rev. Mod. Phys. **70**, 323–426 (1998).
25. F. R. Klinkhamer and N. S. Manton, Phys. Rev. D **30**, 2212 (1984).
26. D. Kharzeev, R. D. Pisarski and M. H. G. Tytgat, Possibility of spontaneous parity violation in hot QCD, Phys. Rev. Lett. **81**, 512–515 (1998).
27. S. A. Voloshin, Parity violation in hot QCD: How to detect it, Phys. Rev. C **70**, 057901 (2004).
28. D. E. Kharzeev, J. Liao, S. A. Voloshin and G. Wang, Chiral magnetic and vortical effects in high-energy nuclear collisions—A status report, Prog. Part. Nucl. Phys. **88**, 1–28 (2016).
29. B. I. Abelev *et al.* [STAR], Azimuthal charged-particle correlations and possible local strong parity violation, Phys. Rev. Lett. **103**, 251601 (2009).
30. B. I. Abelev *et al.* [STAR], Observation of charge-dependent azimuthal correlations and possible local strong parity violation in heavy ion collisions, Phys. Rev. C **81**, 054908 (2010).
31. L. Adamczyk *et al.* [STAR], Fluctuations of charge separation perpendicular to the event plane and local parity violation in $\sqrt{s_{NN}} = 200$ GeV Au+Au collisions at the BNL Relativistic Heavy Ion Collider, Phys. Rev. C **88**, no. 6, 064911 (2013).
32. B. Abelev *et al.* [ALICE], Charge separation relative to the reaction plane in Pb-Pb collisions at $\sqrt{s_{NN}} = 2.76$ TeV, Phys. Rev. Lett. **110**, no. 1, 012301 (2013).
33. A. M. Sirunyan *et al.* [CMS], Constraints on the chiral magnetic effect using charge-dependent azimuthal correlations in pPb and PbPb collisions at the CERN Large Hadron Collider, Phys. Rev. C **97**, no. 4, 044912 (2018).
34. D. E. Kharzeev and J. Liao, Chiral magnetic effect reveals the topology of gauge fields in heavy-ion collisions, Nature Rev. Phys. **3**, no. 1, 55–63 (2021).
35. S. A. Voloshin, Phys. Rev. Lett. **105**, 172301 (2010).
36. M. Abdallah *et al.* [STAR], Search for the Chiral Magnetic Effect with Isobar Collisions at $\sqrt{s_{NN}} = 200$ GeV by the STAR Collaboration at RHIC, [arXiv:2109.00131 [nucl-ex]].
37. K. Ikeda, D. E. Kharzeev and Y. Kikuchi, Real-time dynamics of Chern-Simons fluctuations near a critical point, Phys. Rev. D **103**, no. 7, L071502 (2021).

38. A. Brandenburg, J. Schober, I. Rogachevskii, T. Kahniashvili, A. Boyarsky, J. Frohlich, O. Ruchayskiy and N. Kleeorin, The turbulent chiral-magnetic cascade in the early universe, Astrophys. J. Lett. **845**, no. 2, L21 (2017) doi:10.3847/2041-8213/aa855d [arXiv:1707.03385 [astro-ph.CO]].
39. E. V. Gorbar and I. A. Shovkovy, [arXiv:2110.11380 [astro-ph.HE]].
40. Q. Li, D. E. Kharzeev, C. Zhang, Y. Huang, I. Pletikosic, A. V. Fedorov, R. D. Zhong, J. A. Schneeloch, G. D. Gu and T. Valla, Observation of the chiral magnetic effect in ZrTe5, Nature Phys. **12**, 550–554 (2016) [arXiv:1412.6543 [cond-mat.str-el]].
41. J. Xiong, S. K. Kushwaha, T. Liang, J. W. Krizan, M. Hirschberger, W. Wang, R. J. Cava and N. P. Ong, Evidence for the chiral anomaly in the Dirac semimetal Na3Bi. Science 350 (6259), 413–416 (2015).
42. S. Kaushik, D. E. Kharzeev and E. J. Philip, Phys. Rev. B **99**, no. 7, 075150 (2019) doi:10.1103/PhysRevB.99.075150 [arXiv:1810.02399 [cond-mat.mes-hall]].
43. D. E. Kharzeev and Q. Li, [arXiv:1903.07133 [quant-ph]].

Nonperturbative Casimir Effects: Vacuum Structure, Confinement, and Chiral Symmetry Breaking

A.V.Molochkov

Pacific Quantum Center, Far Eastern Federal University, Street,
Vladivostok, 690001, Russian Federation
E-mail: molochkov.alexander@gmail.com

The review of vacuum and matter restructuring in space-time with boundaries is presented. We consider phase properties of confining gauge theories and strongly interacting fermion systems. In particular, the chiral and deconfinement phase transitions properties in the presence of Casimir plates. We also discuss mass scale shifts in such systems and their possible dynamical and geometrical nature.

Keywords: Casimir effect, chiral phase transition, relativistic bound states

1. Introduction

One of the fundamental problems of quantum field theory is the consistency of non-trivial geometry and quantization. One of the brightest examples in this area is quantum fields in the space with boundaries. In the case of a pure vacuum, the certain boundaries give rise to Casimir forces [1]. Recent theoretical works have shown the possibility of vacuum restructuring effects occurring in the systems with boundaries, leading to a change in the gauge theory vacuum's phase properties and critical behaviour. In particular, it can lead to the low temperature deconfinement phase transition [2–4]. One can also assume that the effects of space-time boundaries should manifest themselves in systems with matter fields, particularly in fermionic systems [5–8]. The fundamental question is - what is the nature of these effects? Is it a consequence of the dynamics of the restructured vacuum, or is it a consequence of the interaction with the Casimir boundaries, or is it a result of a finite space-time volume?

Since its discovery, the Casimir effect has been intensively studied experimentally and theoretically [9, 10]. The experimental studies [11, 12] confirmed this effect in plate-sphere geometries in agreement with theory.

In his original work, Hendrik Casimir assumed that the vacuum gauge field's states spectrum of a system with boundaries is limited in the infrared region due to finite size, which leads to the observable difference in energy densities of the unbound vacuum and the vacuum with boundaries. As a result, the attracting forces between the boundaries arise. However, several theoretical results indicate a non-trivial change in the vacuum structure, which cannot be explained by an infrared restriction of the vacuum spectrum.

In particular, a spherical geometry leads to the repulsive Casimir force, which is acting outwards [13]

$$< E_{Sphere} >= +\frac{0.0461765}{R} \tag{1}$$

Another interesting theoretical result was obtained by Scharnhorst who found that in the Casimir vacuum in the low-frequency region $\omega \ll m$, where m is the electron mass, light propagation modes have phase velocity exceeding c [14].

In the paper [15], Scharnhorst and Barton assumed that the Casimir vacuum behaves like a passive medium $(Im\,(n_\perp(\omega)) \geq 0)$. This assumption led them to the conclusion that the front velocity of light $c/n_\perp(\infty)$ in the Casimir vacuum is greater than the speed of the light front c in the unbound vacuum. At the same time, the author emphasizes that this conclusion does not contain any severe conceptual dangers and, in particular, does not contradict the special theory of relativity.

These examples show two different sources of the Casimir vacuum change. The first example shows that the geometry of the Casimir boundaries significantly affects the properties of the vacuum. Second shows fundamental physical scales change due to the non-perturbative Casimir dynamics and the final size of the system.

In the first part of the paper, we will review the non-perturbative properties and phase structure of field theories with plain Casimir boundaries. In the second part, we will discuss the effects of the four-dimensional geometry of bounded space-time to understand its relations with mass-scale shifts in finite volume systems.

2. Phase structure of field theories in space with boundaries

This section considers three examples of the scale shift in systems with Casimir boundaries - strongly interacting fermionic system, compact electrodynamics, and SU(2)theory.

Strongly interacting fermionic system. The critical phenomenon of strong interactions is the spontaneous breaking of chiral symmetry, which occurs in the fermionic QCD sector. Chiral symmetry breaking in systems with Casimir boundaries was studied in the model of interacting fermions with the Lagrangian [5, 6]:

$$\mathcal{L} = i\bar{\Psi}\,\partial\!\!\!/\Psi + \frac{g}{2}(\bar{\Psi}\Psi)^2 \tag{2}$$

This theory is invariant under discrete Z_2 chiral transformations $\Psi \rightarrow \gamma_5 \Psi$ of a fermionic field with N-flavors.

In the unbound space, the vacuum of this model forms a dynamic chiral condensate $\langle\bar{\Psi}\Psi\rangle$ that breaks the chiral symmetry. Symmetry is restored at high temperatures through a second-order phase transition. In the presence of Casimir plates, the critical temperature decreases, and the phase transition becomes first order. The chiral symmetry is restored at a sufficiently small distance between the plates, even at zero temperature. Figure 1 shows the phase diagram of this model. The restoration of chiral symmetry due to the Casimir geometry agrees well with the observation that boundary effects restore chiral symmetry in the broken phase [7, 8].

Thus, the presence of the Casimir boundaries strongly affects the critical behaviour and symmetry properties of field theories.

Let us consider the non-perturbative analysis of the symmetry breaking in confining theories - compact electrodynamics and SU(2) gauge theory.

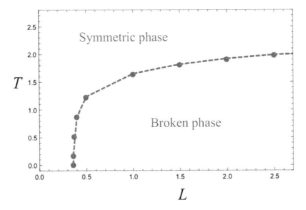

Fig. 1. The phase diagram of the $(3 + 1)$ dimensional model of interacting fermions (2) for the inter-plate distance L and temperature T, from Ref. [5]. The dimensional quantities are expressed in units of the coupling g.

Compact eletrodynamics Compact electrodynamics is another toy model with interesting nonperturbative properties similar to QCD. It has linear confinement of electric charges and the presence of non-trivial topology in physically significant cases of two and three spatial dimensions.

Below, we will briefly consider the vacuum structure transformation of the compact QED due to the presence of the Casimir boundaries. The presented analysis was performed within the first-principles simulations of lattice field theory. The technical details and definitions can be found in the papers [2–4].

The important feature of the compact QED is the presence of monopole singularities. In two spatial dimensions, the monopole is an instanton-like topological object that arises due to the compactness of the gauge group.

The presence of monopoles generates the mass gap

$$m = \frac{2\pi\sqrt{\rho}}{g} \tag{3}$$

and a finite-temperature phase transition at a certain critical temperature $T = T_c$. Here ρ is monopole density, and g is the lattice coupling.

In two spatial dimensions, the standard Casimir boundary conditions are formulated for one-dimensional objects ("wires"). A static and infinitely thin wire, made of a perfect metal, forces the tangential component of the electric field $\mathbf{E}$ to vanish at every point x of the wire, $E_\parallel(x) = 0$. The wire does not affect the pseudoscalar magnetic field B.

The Casimir energy density corresponds to the component of the canonical energy-momentum tensor $T^{00} = (E^2 + B^2)/(2g^2)$. Numerical calculations show that the presence of Abelian monopoles has a nonperturbative effect on the Casimir effect [2]. At large distances between the wires, the mass gap (3) screens the Casimir energy. At small distances, it is the wires that act on the monopoles. As the wires

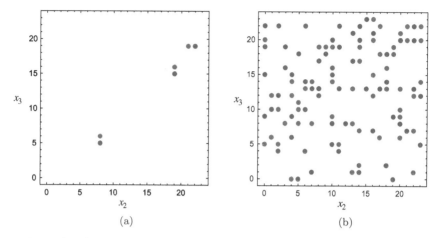

Fig. 2. Examples of typical configurations of monopoles (blue) and antimonopoles (red) in (a) a slice in between closely spaced plates and (b) in a space outside the plates (from [2]).

approach, the relatively dense monopole gas between them is continuously transformed into a dilute gas of monopole-antimonopoly pairs, as shown in Fig. 2(a) and (b) [3]. The geometry-induced binding transition is similar to the Berezinsky–Kosterlitz–Thouless (BKT)-type infinite-order phase transition [16] that occurs in the same model at a finite temperature [17].

The BKT transition is associated with a loss of the confinement property in between the metallic plates because the weak fields of the magnetic dipoles cannot lead to a disorder of the Polyakov-line deconfinement order parameter. This conclusion agrees well with expectation with a direct evaluation of the Polyakov line in between the plates [3]. Figure 3(a) shows the phase structure of the vacuum of compact electrodynamics in the space between long parallel Casimir wires at finite temperature T. The deconfinement temperature T_c is a monotonically rising function of the interwire distance R. Formally, the charge confinement disappears completely when the separation between the plates becomes smaller than the critical distance $R = R_c$ determined by the condition $T_c(R_c) = 0$. According to the numerical estimates [3], $R_c = 0.72(1)/g^2$.

SU(2) theory. Let us consider the Casimir effect for a non-Abelian gauge theory which possesses an inherently nonperturbative vacuum structure. It is instructive to consider a zero-temperature Yang-Mills theory in $(2+1)$ spacetime dimensions. The model in $(2+1)$ dimensions exhibits mass gap generation and colour confinement similarly to its $3+1$ dimensional counterpart.

The Casimir energy of gluon fluctuations per unit length of the wire is shown in Fig. 4(a). The lattice results – which exhibit excellent scaling with respect to a variation of the lattice cutoff – can be described very well by the following function:

$$V_{Cas}(R) = 3\frac{\zeta(3)}{16\pi}\frac{1}{R^2}\frac{1}{(\sqrt{\sigma}R)^\nu}e^{-M_{Cas}R}, \qquad (4)$$

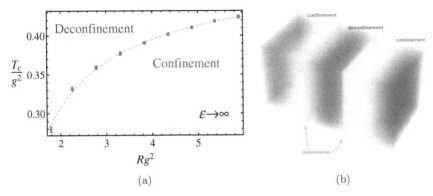

(a) (b)

Fig. 3. (a) The phase in-between the plates: the critical temperature T_c of the deconfinement transition as the function of the inter-plate distance R in units of the electric charge g in the ideal-metal limit ($\varepsilon \to \infty$). (b) An illustration of the deconfinement in the space between the plates (from Ref. [3]).

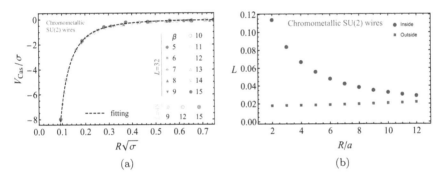

(a) (b)

Fig. 4. (a) The Casimir potential V_{Cas} for a chromometallic wire as the function of the distance R between the wires (in units of the string tension σ) at various ultraviolet lattice cutoffs controlled by the lattice spacing β. The line is the best fit (4). (b) A typical expectation value of the absolute value of the mean Polyakov line in the spaces in between and outside the wires vs. the interwire separation R [4].

where the anomalous power ν (which controls the short-distance behaviour) and the "Casimir mass" M_{Cas} (which is responsible for the screening at large inter-wire separations). The values $\nu = 0$ and $M_{Cas} = 0$ correspond to the Casimir energy of three non-interacting vector particles. In the SU(2) Yang-Mills theory, one gets:

$$M_{Cas} = 1.38(3)\sqrt{\sigma}, \qquad \nu_\infty = 0.05(2). \qquad (5)$$

Surprisingly, the Casimir mass M_{Cas} turns out to be substantially smaller than the mass of the lowest colorless excitation, the 0^{++} glueball, $M_{0^{++}} \approx 4.7\sqrt{\sigma}$ (the latter quantity has been calculated numerically in Ref. [18]). In Ref. [19] it was shown that the $(2+1)$ Casimir mass might be related to the magnetic gluon mass of the Yang-Mills theory in $(3+1)$ dimensions.

In Fig. 4(b), we show the expectation value of the order parameter of the deconfining transition, the Polyakov loop L, in the space inside and outside the wires. Similarly to the compact Abelian case, the gluons in between the wires experience a smooth deconfining transition as the wires approach each other, thus confirming the qualitative picture shown in Fig. 3(b).

3. Effects of the four-dimensional boundaries

We considered systems with time-like space boundaries that lead to non-trivial properties and vacuum structure change. Another essential configuration of the boundaries is space-like time hyper-surfaces that lead to finite time-size systems. By the analogy of a space box that as a whole has space-translation symmetry, we can consider a finite time-box that as a whole has continuous time-translational symmetry. It corresponds to the space-time box, which is at the rest frame. Here we will show how the relativistic treatment of the bound state allows us to consider it as a space-time box. We will also discuss the physical outcomes of the existence of the time-boundaries.

Let us consider a relativistic bound state of two particles. The bound particles are distributed in the bound state's internal space-time (space-time bag). Instead of the traditional formulation of the Casimir effect, where borders are externally set boundary conditions, the border here is of the dynamical origin. The boundaries result from the bound particles' dynamics, which naturally leads to the vanishing of the current's radial component at the bound state boundaries. This condition is equivalent to the MIT bag condition.

The bound state of particles can be defined as the constituents scattering amplitude pole in the momentum space:

$$G(p_1, p_2, p_1', p_2') = \frac{\Gamma(p_1, p_2)\bar{\Gamma}(p_1', p_2')}{(p_1 + p_2)^2 - M^2} + \cdots,$$ (6)

where M is the mass of the bound state, p_1, p_2-4-momenta of the constituents.

Bellow, we will discuss the relation between the appearance of the pole and the finite time size of the bound state, which in its order is related to the shifting of the particles from the mass shell.

For this purpose, let us consider the 4D kinematics of the two-particle system. To separate centre of mass constituents kinematics, instead of individual 4-momenta of the particles (p_1^μ, p_2^μ), we will use the set of 4-momenta of the centre of mass and the relative momenta of the particles (P^μ, k^μ):

$$P^\mu = p_1^\mu + p_2^\mu, \quad 2k^\mu = p_1^\mu - p_2^\mu$$ (7)

Let us consider for simplicity the center of mass rest frame. In this case the momenta p_1^μ and p_2^μ can be written in the form:

$$p_1^0 = \frac{M}{2} + k_0, \quad p_2^0 = \frac{M}{2} - k_0$$

$$\mathbf{p}_1 = -\mathbf{p}_2 = \mathbf{k}$$ (8)

The corresponding space-time variables have the following form:

$$x_1^0 = T + \frac{\tau}{2}, \quad x_2^0 = T - \frac{\tau}{2}$$

$$\mathbf{x}_1 = \mathbf{X} + \frac{\chi}{2}, \quad \mathbf{x}_2 = \mathbf{X} - \frac{\chi}{2}, \tag{9}$$

where $(T, \mathbf{X})$-center of mass space-time, (τ, χ)-space-time of constituents relative positions.

The relative energy $2k_0 \neq 0$ for the off-shell particles system only. Indeed, the on-shell conditions in terms of the variables (9) are following:

$$\frac{M}{2} + k_0 = E(\mathbf{k}) \quad \frac{M}{2} - k_0 = E(\mathbf{k}) \quad E(\mathbf{k}) = \sqrt{\mathbf{k}^2 + m^2} \tag{10}$$

Solution of the system gives the following mass-shell condition:

$$k_0 = 0, \quad M = 2E(\mathbf{k}) \geq 2m \tag{11}$$

Thus, k_0 determines the degree of the nucleon's off-mass-shell shift. Bellow, we will show that the finite time size of the bound state leads to $k_0 \neq 0$.

Let us consider the Green function of a bound state of two particles:

$$G(P) = \int \frac{d^4 k}{(2\pi)^4} \frac{\bar{\Gamma}(P, k) \Lambda_1(P, k) \otimes \Lambda_2(P, k) \Gamma(P, k)}{\left[\left(\frac{P}{2} + k \right)^2 - m^2 \right] \left[\left(\frac{P}{2} - k \right)^2 - m^2 \right]}, \tag{12}$$

where

$$\Lambda_1(P, k) = \left(\frac{P}{2} + k \right) \gamma + m, \quad \Lambda_2(P, k) = \left(\frac{P}{2} - k \right) \gamma + m.$$

Figure 5 illustrates this Green function by the corresponding Feinman diagram. It contains contributions of positive and negative energy states. Let us consider positive energy contributions. The negative energy terms can be analysed in the same way.

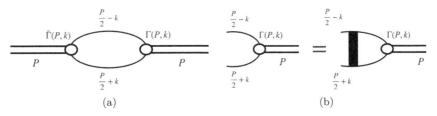

Fig. 5. (a) The diagramatic illustration of the bound state propagator. (b) The diagramatic illustration of the Bethe-Salpeter equation.

The positive term contribution to the Eq. (12) has the following form:

$$G^{++}(P) = \int \frac{d^4k}{(2\pi)^4} \frac{\bar{\Gamma}^{++}(P,k)[u(\mathbf{k})\bar{u}(\mathbf{k}) \otimes u(-\mathbf{k})\bar{u}(-\mathbf{k})]\Gamma^{++}(P,k)}{4E^2 \left[\frac{M}{2} + k_0 - E\right] \left[\frac{M}{2} - k_0 - E\right]} \qquad (13)$$

The initial and final total 4-momenta P, P' are equal to each other due to momentum conservation. This term takes into account any intermediate interaction corrections due to the Bethe-Salpeter equation (see Fig 5b). Thus, the relative 4-momenta k are conserved too.

The integrand of Eq.(13) represents the momentum distribution of the particles within the bound state. The total momentum P dependence is trivial in the case of $\mathbf{P} = 0$. The corresponding internal space-time distribution can be obtained as a Fourie transformation of the integrand.

$$g(\tau, \chi) = \int \frac{d^4k}{(2\pi)^4} \frac{-f(k)e^{ik_0\tau - i\mathbf{k}\cdot\chi}}{4E^2(k_0 - (E - \frac{M}{2}))(k_0 + (E - \frac{M}{2}))}, \qquad (14)$$

where

$$f(k) = \bar{\Gamma}^{++}(M,k)[u(\mathbf{k})\bar{u}(\mathbf{k}) \otimes u(-\mathbf{k})\bar{u}(-\mathbf{k})]\Gamma^{++}(M,k).$$

Let's integrate (14) with respect to k_0 taking into account the poles $k_0 = \pm(E - \frac{M}{2})$:

$$g(\tau,\chi) = \int \frac{d^3\mathbf{k}}{(2\pi)^3} f\left(E - \frac{M}{2}, \mathbf{k}\right) e^{-i\mathbf{k}\cdot\chi} \frac{1}{i2E^2} \left[\frac{e^{i(E-\frac{M}{2})\tau}}{2(2E - M)} - \frac{e^{-i(E-\frac{M}{2})\tau}}{2(2E - M)}\right]$$

$$= \int \frac{d^3\mathbf{k}}{(2\pi)^3} f\left(E - \frac{M}{2}, \mathbf{k}\right) e^{-i\mathbf{k}\cdot\chi} \frac{\sin((2E - M)\tau)}{2E^2(2E - M)}$$

We calculate the Fourie image with respect $\mathbf{k}$ as a Fourie transform of the product of the functions:

$$g(\tau,\chi) = \int d^3\chi' g_1(\chi - \chi')g_2(\tau, \chi') \qquad (15)$$

where

$$g_1(\chi - \chi') = \int \frac{d^3\mathbf{k}}{(2\pi)^3} f(2E - M, \mathbf{k})e^{-i\mathbf{k}\cdot(\chi-\chi')} \qquad (16)$$

and

$$g_2(\tau, \chi') = \int \frac{d^3\mathbf{k}}{(2\pi)^3} \frac{\sin((E - \frac{M}{2})\tau)}{2E^2(2E - M)} e^{i\mathbf{k}\cdot\chi'}. \qquad (17)$$

The function g_2 fully desribes the τ-dependence of the distribution function $g(\tau, \chi)$. It does not depend explicitly on the constituents' interaction and has a pure geometrical nature. According the expression (15) it plays role of function limiting the three dimensional space distribution g_1. Thus, one can consider it as a kind of integral boundary condition.

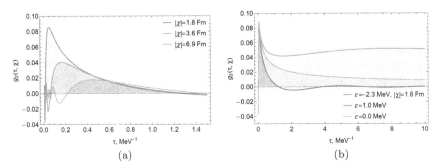

Fig. 6. (a) The time distribution $g_1(\tau, \chi)$ for the deuteron at three different values of space distance $\chi = 1.8Fm$ (blue line), $\chi = 3.6Fm$ (orange line), $\chi = 6.9Fm$ (green line). (b) The time distribution $g_2(\tau, \chi)$ for the deuteron-like bound state with different values of the binding energy at $|\chi| = 1.8Fm$.

Calculation of the integral in (17) with respect $\mathbf{k}$ gives time distribution of the constituents within the bound state:

$$g_2(\tau, \chi') = \int\limits_0^\infty |\mathbf{k}|^2 d|\mathbf{k}| \frac{sin((E - \frac{M}{2})\tau)}{E^2(2E - M)} \frac{sin(|\mathbf{k}||\chi'|)}{|\mathbf{k}||\chi'|} \qquad (18)$$

To illustrate a relation between time structure and energy scales, we choose values of m and M of a physical bound state of two nucleons - deuteron. The result of the calculation is presented in the figure 6a.

The Fig. 6a shows that time distribution $(g_2(\tau, \chi))$ of the bound state with mass-defect exhibits final time size behaviour. In other words, the final time-size of the system leads to the observable shift from the mass-shell. To define the time-size of the system we take positions of the first zero of the main peak. The following relation defines positions of the zero:

$$(2E - M)\tau_0 = \pi. \qquad (19)$$

The averge value of the τ_0 is about $1.4 \, MeV^{-1}$, what corresponds $276 \, Fm$ or $9.2 \cdot 10^{-22}$ seconds.

The Eq.(19) shows a relation between the off-mass-shell shift and the final time-size of the system. The Fig. 6b presents the comparison of the calculations with negative binding energy, zero and small positive binding energy. From the figure we see that at zero binding energy the time-size of the bound state becomes infinite.

Another interesting result is presented in Fig. 7(a). which shows a gap at the small τ on moderate space distances, where the constituents are not overlapping. It means that for the bound non-overlapping particles, the 4-distances with $\tau < \chi$ are forbidden. In the case of small space intervals, where constituents overlap, the gap is blurring (see Fig. 7(b)).

EMC-effect One of the essential questions is - can the finite time-size of the bound state lead to any observable effects. Here we will consider one - the EMC-effect.

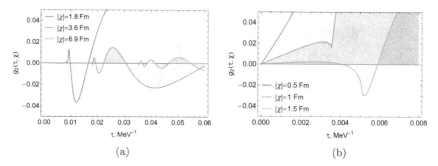

Fig. 7. (a) The small τ gap for the three values of space interval - $\chi = 1.8\,Fm$ (blue line), $\chi = 3.6\,Fm$ (orange line), $\chi = 6.9\,Fm$ (green line). The size of the gap corresponds the light cone condition $\tau^2 - \chi^2 = 0$. (b) The small τ area with overlaping constituents for the three values of space interval - $\chi = 0.5\,Fm$ (blue line), $\chi = 1\,Fm$ (orange line), $\chi = 1.5\,Fm$ (green line).

The EMC-effect discovered in the experiments of the EMC collaboration [21] shows that short-range structure of bound nucleon has observable differences with the free nucleon. This discovery clearly contradicts the traditional picture of bound states. The scales of nuclear and intranuclear forces differ so much that the former cannot affect the latter.

This contradiction can be solved if one considers the 4D structure of nucleon bound states.

Indeed, since the time distribution of a bound nucleon is not uniform and limited, the time translation invariance for an individual constituent is broken.[a] While for a free nucleon, it is conserved. Since the time-invariance is conserved for the whole bound state, the corresponding observables have physical meaning after integration over the bound state space-time.

The covariant treatment of the deep inelastic amplitude and integration with respect to zero-component of the relative momentum k_0 give the following relation between the bound and free nucleon structure functions:

$$F_2^{\tilde{N}}(x) = \int \frac{d^3\mathbf{k}}{(2\pi)^3} \left[\frac{E - k_3}{E} F_2^N(x_N) + \frac{2E - M}{E} x_N \frac{dF_2^N(x_N)}{dx_N} \right] \Phi^2(\mathbf{k}), \qquad (20)$$

where $x = Q^2/(2Mq_0)$ and $x_N = Q^2/(Eq_0 - k_3q_3)$ are Bjorken scaling variable for the deuteron and nucleon respectively. The $Q^2 = -q^2 = q_3 - q_0$ virtual photom space-like momentum. The first term in Eq. (20) is the contribution of the Fermi-motion. The second term gives finite time-size effects.

The previous numerical calculations presented in Fig. (8) show good agreement with data in the whole region of x values.

---

[a]This circumstance, by the way, explains why quasi-potential approaches were failed to explain the EMC-effect.

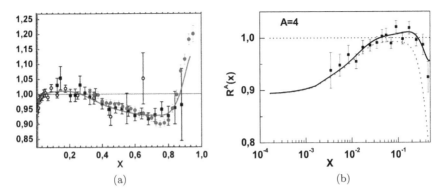

Fig. 8. (a) Ratio of the 4He and D structure functions.The experimental values are shown by the full squares [22], the light circles [23] and full circles [24]. (b) The ratio of the nuclear to deuteron structure functions at small Bjorken x for ^{4}He. The experimental values are shown by the solid squares [23].

4. Conclusion

The main question discussed in the presented article is the nature of the vacuum and matter restructuring in spaces with boundaries. We have considered several examples of systems with boundaries to clarify this question. The examples we have considered illustrate three different causes of vacuum and matter restructuring.

The strongly interacting fermion system demonstrates that the boundary effects restore the chiral symmetry in a chirally broken phase. Due to the boundary effects, the critical temperature decreases, and the phase transition becomes first order. The chiral symmetry is restored at a sufficiently small distance between the plates, even at zero temperature.

The compact QED with Casimir boundaries demonstrates a purely geometrical effect. The small interplate distance constrains the monopole density, leading to the early deconfinement phase transition.

The non-Abelian SU(2) gauge field theory demonstrates vacuum restructuring due to the non-perturbative dynamics of gluon fields. It leads to the shift of the scale defined by glueball mass to significantly smaller values. It also exhibits early deconfinement phase transition, which might have similar nature as in compact QED.

The two-fermion bound state, where the boundaries have a dynamical nature, exhibit mass scale shift due to the space-like time boundaries. It is a purely geometrical effect. The limited time-size of the bound state lead to the time translation invariance breaking for an individual constituent. What leads to the mass-scale shift, or in other words, off-mass-shell effects. One of the observable effects of the finite time-size of nucleon bound states is the EMC-effect.

Acknowledgments

I thank M. Chernodub, and V. Goy, for fruitful discussions and collaboration on related topics. This work has been supported by the Ministry of Science and Higher Education of Russia (Project No. 0657-2020-0015)

References

1. H. B. G. Casimir, Indag. Math. **10**, 261 (1948) [*Kon. Ned. Akad. Wetensch. Proc.* **51**, 793 (1948)].
2. M. N. Chernodub, V. A. Goy and A. V. Molochkov, *Phys. Rev. D* **95**, no. 7, 074511 (2017).
3. M. N. Chernodub, V. A. Goy and A. V. Molochkov, *Phys. Rev. D* **96**, no. 9, 094507 (2017).
4. M. N. Chernodub, V. A. Goy, A. V. Molochkov and H. H. Nguyen, *Phys. Rev. Lett.* **121**, no. 19, 191601 (2018).
5. A. Flachi, *Phys. Rev. Lett.* **110**, no. 6, 060401 (2013).
6. A. Flachi, *Phys. Rev. D* **86**, 104047 (2012).
7. B. C. Tiburzi, *Phys. Rev. D* **88**, 034027 (2013).
8. M. N. Chernodub and S. Gongyo, *Phys. Rev. D* **95**, no. 9, 096006 (2017).
9. M. Bordag, G. L. Klimchitskaya, U. Mohideen, and V. M. Mostepanenko, Advances in the Casimir Effect (Oxford University Press, New York, 2009).
10. K. A. Milton, The Casimir Effect: Physical Manifestations of Zero-Point Energy (World Scientific Publishing, Singapore, 2001).
11. S. K. Lamoreaux, *Phys. Rev. Lett.* **78**, 5 (1997); Erratum: [*Phys. Rev. Lett.* **81**, 5475 (1998)].
12. U. Mohideen and A. Roy, *Phys. Rev. Lett.* **81**, 4549 (1998); G. Bressi, G. Carugno, R. Onofrio, and G. Ruoso, *Phys. Rev. Lett.* **88**, 041804.
13. T. H. Boyer, *Phys. Rev.* 174, 1764 (1968).
14. K. Scharnhorst, *Phys. Lett. B* **236**, 354 (1990).
15. G. Barton and K. Scharnhorst *J. Phys. A: Math. Gen.* **26**, 2037 (1993).
16. V. L. Berezinskii, *Sov. Phys. JETP* **32**, 493 (1970); ... *II. Quantum Systems, Sov. Phys. JETP* **34**, 610 (1971); J. M. Kosterlitz and D. J. Thouless, *J. Phys. C* **6**, 1181 (1973).
17. M. N. Chernodub, E. M. Ilgenfritz and A. Schiller, *Phys. Rev. D* **64**, 054507 (2001).
18. M. J. Teper, *Phys. Rev. D* **59**, 014512 (1998); A. Athenodorou and M. Teper, *JHEP* **1702**, 015 (2017)
19. D. Karabali and V. P. Nair, *Phys. Rev. D* **98**, no. 10, 105009 (2018).
20. E. E. Salpeter and H. A. Bethe, *Phys. Rev.* **84**, 1232 (1951).
21. J. J. Aubert *et al.*, *Phys. Lett. B* **123**, 275 (1983).
22. J. Gomez *et al.*, *Phys. Rev. D* **49**, 4348 (1994).
23. NMC, P. Amaudruz, *et al.*, *Nucl. Phys. B* **441**, 3 (1995).
24. J. Seely *et al.*, *Phys. Rev. Lett.* **103**, 202301 (2009).

Emergent Electromagnetic Phenomena in Spin Chiral Matter

Yoshinori Tokura[1,2†]

[1] *RIKEN Center of Emergent Matter Science (CEMS),*
[2] *Tokyo College, The University of Tokyo*
[†] *E-mail: tokura@riken.jp*

Intriguing electromagnetic phenomena can show up in a solid when the electrons spins take a helical form. To name a few, the cycloidal order of the spin moments can produce the electric polarization and generate the multiferroics where the ferroelectric and magnetic orders coexist and enable the cross control of the magnetism with electric field. When spin helices hybridize to form with plural propagation directions, the topological spin textures such as magnetic skyrmion lattice emerge, hosting the large emergent magnetic field or Berry curvature acting on the conduction electrons. Furthermore, the dynamics of these helical spin forms, as excited by electric current flow, can generate the emergent electric field acting on the electrons and hence cause a sort of electromagnetic induction. Such emergent electromagnetic induction based on the spin chiral matter is applied to the design of nanometric inductor element.

Keywords: spin chiral matter, multiferroics, skyrmions, emergent electromagnetism.

1. Spin screw and cycloid

Helical spin orders in magnetic materials can always cause the inversion symmetry breaking, even starting from the centrosymmetric crystal form, due to the relativistic coupling between spin and electron cloud, termed spin-orbit coupling. The outcomes of the helical spin orders can generate a variety of emergent electromagnetic phenomena, since the breaking of the space-inversion symmetry as well as of the time-reversal symmetry occurs in spin chiral matter and therein the state of matter control is possible by external electric field and/or magnetic field.

Let us begin with representative helical forms of the spin order, (a) spin proper screw and (b) spin cycloid, as illustrated in Fig. 1. These helical forms appear ubiquitously also on the magnetic domain walls, respectively referred to as (a) Bloch wall and (b) Néel wall. In analogy to the case of the chiral molecule or chiral crystal, we can define the *helicity* of the spin chiral matter; for example, along the fixed propagation direction, the *left-handed* or *right-handed* screw rotation for (a) the spin screw while the *clockwise* (CW) or *counter-clockwise*(CCW) rotation for (b) the spin cycloid. The magnetic (spin) moment is an axial vector viewed as generated by the circulating current and hence show a different transformation from the case of polar vector upon space symmetry operations. The spin screw (a) gives rise to the electric/optical chirality of the host magnetic material, similarly to chemically-chiral molecules and lattice-chiral crystals, *e.g.* endowed with the natural optical activity. For the spin cycloid (b), it can generate the electric polarity; each pair of the cycloidally arranged (in-plane mutually canted) spins, say S_i and S_j, transform like a polarity vector p lying within the spiral plane, following the relation

$$p \propto e_{ij} \times (S_i \times S_j). \tag{1}$$

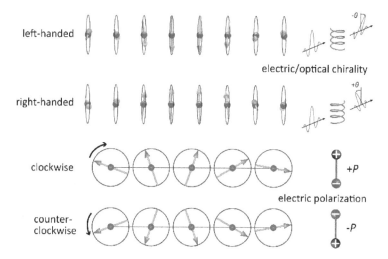

Fig. 1. Two representative helical spin structures: (a) spin screw and (b) spin cycloid. Those are characterized by spin helicity, right- or left-handedness for spin screw and clockwise (CW) or counter-clockwise (CCW) for spin cycloid. The spin screw order causes electric/optical chirality, while the spin cycloid order generates electric polarization, under the presence of relativistic spin-orbit coupling.

Here, e_{ij} is a unit vector connecting spins S_i to S_j and the quantity $(S_i \times S_j)$ is termed vector spin chirality. Every spin pair on the spin cycloid gives the same p, when the cycloidal pitch is constant, then giving a net polarization P whose direction (sign) depends on the spin helicity, namely CW or CCW. In fact, the microscopic theory [1] explicitly considering the spin exchange interaction under the relativistic spin-orbit coupling shows the presence of spin current flowing along e_{ij} with the spin moment along $S_i \times S_j$, which causes the electric polarization.

2. Multiferroics of spin origin

Multiferroics means the materials with simultaneous orders of magnetism and ferroelectricity, where the magnetoelectric (ME) effect, *i.e.* the magnetic (electric) control of the electric polarization (magnetization), can show up in an enhanced manner. The direct consequence of the spin cycloid order is the emergence of macroscopic electric polarization, ensuring the multiferroic state as well as the intrinsic gigantic ME effect. In 2003, one of the perovskite manganite family $TbMnO_3$ - was found to show the multiferroic phase transition, *i.e.* the cycloidal spin order and the simultaneous occurrence of the ferroelectric polarization [2]. $TbMnO_3$ and related perovskite manganite compounds are the Mott insulators showing the Mn $3d$-orbital order as shown in Fig. 2(a), which causes the frustration effect on the spin exchange interactions and eventually the modulated spin structure along the orthorhombic-lattice b axis. As indicated in Fig. 2(c), the compound at first undergoes the phase change to the collinear spin modulated structure at T_N ($\sim$42K) and

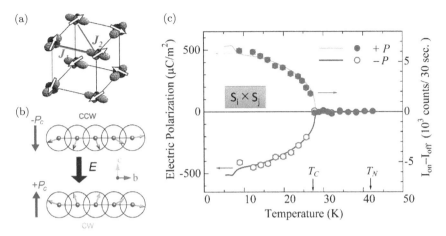

Fig. 2. Emergence of electric polarization by spin cycloidal structure in multiferroic perovskite TbMnO$_3$. (a) cycloidal spin habit produced by magnetic frustration in exchange interactions J_1 and J_2. (b) spin cycloid order and helicity (CCW or CW) dependent electric polarization P. (c) Temperature evolution of vector spin chirality ($S_i \times S_j$) below T_C, as determined by polarized neutron scattering [3] to be compared with the electric polarization.

then to the cycloidal structure on the bc plane (Fig. 2(b)) at T_C ($\sim$27K). The latter phase transition is also the ferroelectric transition with the electric polarization P along the c axis, the sign of which is controlled by the electric-field (E) cooling procedure; in some other manganite (DyMnO$_3$), the sign of P can be controlled also in terms of the E scan like conventional ferroelectrics.

According to the spin current model Eq. (1), the sign of P should show the one-to-one correspondence to the sign of vector spin chirality ($S_i \times S_j$) of the spin cycloid. Namely, the reversal of the ferroelectric polarization, $e.g.$ by E-scanning or E-cooling procedure, corresponds to the switching of the spin helicity between CCW and CW in the spin cycloid. This was clearly demonstrated by spin-polarized neutron scattering study on the E-controlled multiferroic state in TbMnO$_3$ [3]. The spin-helicity or vector-spin-chirality ($S_i \times S_j$) dependent intensity of the spin-cycloid superlattice peak is shown in Fig. 2(c). The temperature dependence of their sign and intensity variations coincides with those of the ferroelectric polarization, verifying the prediction by the spin current model. In fact, there have appeared an increasing number of the spin-cycloid multiferroics which have been explored by this powerful materials principle [4].

One important characteristic for the spin cycloid is that it can be transformed to the transverse conical form by the presence of the transverse magnetization (M) or in terms of the magnetic field applied perpendicular to the cycloid propagation direction q [5], as shown in Fig. 3. In this transverse conical order, the electric polarization arising from the relation eq.(1) shows the staggered pattern (p_i shown in Fig. 3(b)), but as a whole after summed up, the averaged electric polarization P turns parallel to the direction normal to q and M. In particular, the orthogonal

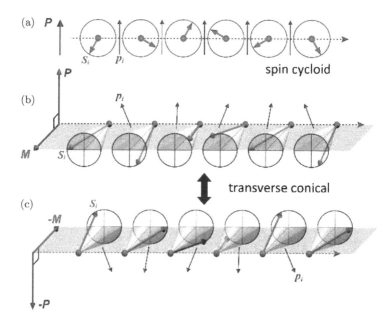

Fig. 3. (a) Spin cycloid order generates the macroscopic electric polarization P. (b) By adding the magnetization M or applying magnetic field, the transverse conical order shows up with the orthogonal M and P. (c) When M is reversed, most occasionally P is also reversed via the strong M-P coupling.

relation and strongly-tied coupling between P and M give an ideal material platform to test the P-M cross control in the multiferroics, namely E-control of M or H-control of P. For example, while the M is reversed by application of the transvers H, as shown in Fig. 3(c), P is possibly reversed simultaneously. Conversely, while the P is reversed by E, the M is possibly reversed simultaneously.

The transverse-conical type multiferroics have been extensively explored both for the intrinsically formed compounds, e.g. $CoCr_2O_4$ [6], and for the magnetic-field transformation from the screw or cycloid forms observed in many helimagnetic materials [4]. In most of those multiferroics, the H-induced reversal of P is ubiquitously observed, while the E-induced reversal of M is observed in a rather limited family in which the cycloidal spin induced P is large enough to respond to the external electric field magnitude. Among them, the E-control of the ferrimagnetic/ferromagnetic moment M at and above room temperature is quite important for the application to the multiferroic devices [7, 8].

3. Magnetic skyrmions

Skyrmion was originally proposed as a model of nucleon by Tony Skyrme [9]. The idea is to ensure the stability of a particle state by the topological protection; the particle is characterized by a topological integer that cannot be changed by a continuous deformation of the field configuration. The concept was later applied

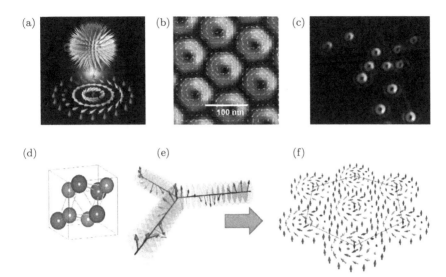

Fig. 4. Magnetic skyrmion. (a) Spin swirling object whose spin moments form the sphere when gathered, defining the topological number of unity. (b) Lorentz TEM image of the skyrmion lattice in a B20 type chiral magnet (Fe,Co)Si [12]. (c) Skyrmions can exist in form of independent particles. (d) B20 type crystal structure. (e) spin screw helix with possible three q-directions. (f) Triple-q state leads to the formation of skyrmion lattice.

to the topological spin object, termed magnetic skyrmion, in condensed matter by Bogdanov *et al.* [10]. A schematic example of a magnetic skyrmion configuration is shown in Fig. 1(a); with arrows depicting the direction of normalized spin moment $n(r)$ at spatial position $r = (x, y)$, the moments around the core point down, while those at the perimeter point up. The integer topological number N (skyrmion number, usually $N = -1$) is defined as a measure of the wrapping of $n(r)$ around a unit sphere (the upper part of Fig. 1(a)), such as

$$N = \frac{1}{4\pi} \int n \cdot \left(\frac{\partial n}{\partial x} \times \frac{\partial n}{\partial y} \right) dx dy. \tag{2}$$

The magnetic skyrmion was at first experimentally identified in a chiral-lattice magnet MnSi or related B20 compounds. In the noncentrosymmetric crystal structure, like a chiral lattice of B20 type compounds (Fig. 4 (d)), where the conventional ferromagnetic exchange interaction J competes with the antisymmetric exchange interaction D, Dzyaloshinskii-Moriya interaction (DMI), which tends to twist the neighboring spin configuration. Such competition in spin exchange interactions causes the helical spin structure, usually screw type but cycloid type in the polar lattice, with the modulation period of $\sim (J/D)a = 2\pi/q$, here a being the spin-site spacing. When the lattice structure is of high symmetry like cubic or hexagonal and some specific helix mode-mode coupling occurs like the two-dimensional (2D) triple-q state shown in Fig. 4(e), the 2D skyrmion lattice (SkL) can show up as schematically shown in Fig. 4(f); SkL was first identified for MnSi by small angle neutron scattering by Pfleiderer and collaborators [11]. The topological nature of the skyrmions constituting SkL was confirmed clearly by Lorentz transmission electron microscopy [12] which enables the imaging of the in-plane swirling spin-moment patterns (Fig. 1(b)) and thus confirms the skyrmion topology.

The magnetic skyrmion possesses a topologically protected particle nature beyond the triple-q density wave sate; this is clearly shown in the Lorentz transmission electron microscopy image (Fig. 4(c)) captured outside the thermodynamic equilibrium region of the SkL on the B20 compounds [20]. Such independent particle-like skyrmions are often seen in the bilayer system composed of the ferromagnetic metal (e.g. Co) layer and the heavy metal (e.g. Pt) layer with large spin-orbit coupling, where the DMI working at the interface can generate the cycloidal (Nèel-type) skyrmions. These systems are in particular important for the possible application to spintronics devices [13].

Inherent nature of skyrmions or SkL is listed as follows: (1) topological protection and metastability, (2) generation of emergent magnetic field, and (3) high drivability by spin current [14]. At first, (1) The topological protection, which prohibits the continuous deformation from the skyrmion ($N = -1$) to the non-topological ($N = 0$) state, gives unconventional metastability of the skyrmion or SkL out of the thermal equilibrium phase region. In the chiral-lattice magnet, in general, the triple-q state representing the SkL shows a thermal equilibrium phase only near below the magnetic transition temperature under a small/moderate magnetic field perpendicular the skrymionic 2D sheet. However, once the SkL is formed and then quickly quenched to lower temperature, the SkL survives as the metastable state even at the lowest temperature, as demonstrated for the MnSi SkL via rapid cooling (at a rate $\sim$700K/s) procedure [15]. Such a quenching-procedure condition is largely relaxed in the case of "dirty" crystals with abundant skyrmion trapping centers, where even a moderate or slow cooling rate is enough to maintain 100% of the once thermodynamically-generated SkL down to the lowest temperature as the metastabilized state. One such example is the room-temperature and zero-field SkL as realized in high-temperature chiral magnet of Co-Zn-Mn alloy (β-Mn type) through the cooling from the thermal equilibrium SkL phase around $\sim$400K [16].

(2) The emergent magnetic field (gauge field) acting on the conduction electrons is one of the most important characteristics of the skyrmions. When the conduction electron's spin couples strongly, like the case of Hund's-rule coupling, with the localized spins constituting the skyrmion, one skyrmion with the topological charge $N = -1$ can effectively produce twice of flux quantum (h/e); for example, when there is the SkL with skyrmion density of one per 10nm^2, the emergent magnetic field b which just works on the conduction electron like a conventional magnetic field, amounts to as large as 4000 tesla in the case of the strong coupling. This emergent magnetic field causes the Hall effect in a solid, which appears similar to

the normal Hall effect but obviously distinct from the conventional anomalous Hall effect: The anomalous Hall effect is the spontaneous Hall effect in a magnet via the spin-orbit coupling, *i.e.* the Berry curvature (emergent magnetic field) generated in the momentum-space band structure [17].

For the last decade, a number of magnetic skyrmion materials [18] have been developed not only in the noncentrosymmetric DMI system but also in the centrosymmetric magnetic-frustration systems. In particular, the magnetic system with the conduction-electron mediated (Rudderman-Kittel-Kasuya-Yosida or RKKY type) magnetic couplings gives the short-period (a few nanometers) helical spin orders including the SkL on the high-symmetry (hexagonal, tetragonal or cubic) lattices, which is favorable for the enhancement of the emergent magnetic field response due to much higher density of skyrmions than the chiral-lattice magnetic system with the skyrmion size $\sim (J/D)a$ exceeding 10nm. We show one such example of short-period SkL in Fig. 5. The compound is Gd_2PdSi_3 crystal which is composed of the stacking of Gd (magnetic ion) triangular-lattice and Pd/Si honeycomb lattice, as shown in Fig. 5(a). Gd moments are mutually coupled in a quasi-two-dimensional manner via the RKKY interaction mediated by the Pd/Si conduction electrons. The magnetic frustration on the Gd triangular lattice produces various in-plane helimagnetic structurers, such as labeled by IC-1, SkL, and IC2 in Fig. 5(b), with the nearly the same in-plane helix period of $\lambda \sim$3nm, as shown in the phase diagram in the temperature and magnetic field normal to the ab plane [19]. Among some helical spin orders, the intermediate-field phase SkL is assigned to the skyrmion lattice state as schematically shown in Fig. 5(d). Such a short-period SkL shows the large emergent magnetic field effect, which typically shows up as the gigantic *topological* Hall resistivity ρ_{yx} (Fig. 5(e)); the top-flat peaks are observed accompanying the field hysteresis (indicating the first-order transition) when the magnetic field B is within the range of the SkL region at respective temperatures. The large emergent magnetic field effect is also discerned in the Nernst effect as described by transverse thermoelectric conductivity α_{xy} (Fig. 5(f) [20]; here the Nernst effect represents the generation of the Hall like transverse electric field (E_y) with the thermal current or temperature-gradient along x (Fig. 5(c)). The observed *topological* Nernst effect is comparable with the largest anomalous Nernst response observed for some ferromagnetic metals. Here it is worth to note that the such a large topological Hall/Nernst effect from the short-period SkL may not be precisely described by the picture of the real-space effective magnetic field proportional to the skyrmion number densityN/λ^2 but rather by the momentum-space picture of the Berry curvature as in the case of the intrinsic, i.e. Karplus-Luttinger type, anomalous Hall effect [17].

Finally, the inherent nature (3) of the magnetic skyrmion is the high drivability of skyrmion by spin current. According to the generalized Thiele equation for the skyrmion's center of motion [14], the topology of skyrmion as described by the skrymion number N most essentially reacts to the spin current, enabling the efficient drive by spin-polarized electric current or the spin-momentum carrying thermal

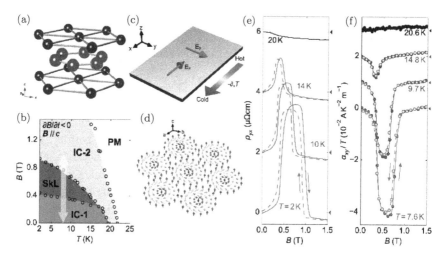

Fig. 5. Nanometric-sized skyrmion lattice. (a) Crystal structure of Gd_2PdSi_3. (b) Magnetic phase diagram of Gd_2PdSi_3. Arrow represents the field scan. (c) Configuration of Nernst effect. (d) Skyrmion lattice (schematic) observed in SkL state. (e), (f) Topological Hall and Nernst effects. Cited from Hirschberger et al. [20].

current. In particular, it was demonstrated that the current density necessary to drive the SkL [21, 22] in MnSi or FeGe is found to be as low as $10^6 A/m^2$, which is five or six orders of magnitude smaller than that needed for the magnetic domain wall motion. As schematically shown in Fig. 6(a), the skyrmion shows the combined longitudinal and transverse (Hall) motion under the steady flow of current. The skyrmion Hall motion can be viewed as the counteraction of the topological electron's Hall motion. The current-driven dynamics of the skyrmions play a crucial role in the spintonic application of skyrmion, i.e. *skyrmionics* [13].

4. Emergent electromagnetic induction

The generalized Faraday's law for the relation between the emergent magnetic field b and electric field e [23] tells that $\nabla \times e = -\partial b/\partial t$ or $e = -\partial a/\partial t$, where a is the Berry connection satisfying that $b = \nabla \times a$. Thus, the electric field e is induced by the current-driven motion of the skyrmion that carries b: Explicitly, $e = v_d \times b$; here v_d is the drift velocity of the skyrmion. This induced emergent electric field gives a Hall effect of skyrmion in the opposite direction of topological Hall effect (Fig. 6 (a)). In reality, the e accompanied by the skyrmion motion was successfully detected as the correction of the topological Hall effect when the current density exceeds the threshold value for the skyrmion drive [24]. This is one example of emergent electromagnetic induction phenomena.

On the other hand, to generate the emergent electric field on the spin helix via the relation that $e = -\partial a/\partial t$, the presence of the static b is not necessarily required but the time-dependent deformation of the spin helix by ac electric current is sufficient. Considering the spin transfer torque on the spin helix in the continuous

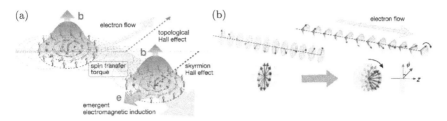

Fig. 6. (a) Current-driven motion of skyrmion with the emergent magnetic field b, generating the emergent electric field e. (b) Current-induced deformation of spin helix, the dynamics of which produced emergent electromagnetic induction.

limit, the emergent electric field is described as [23]

$$e_i = \frac{h}{2\pi e}\mathbf{n} \cdot (\partial_i \mathbf{n} \times \partial_{t_t} \mathbf{n}), \tag{3}$$

where $\mathbf{n}$, h and e are a unit vector parallel to the direction of spins, Planck's constant and bare electron charge, respectively. As opposed to the case of $\boldsymbol{b}$, $\boldsymbol{e}$ is related to the dynamics of spin structures and proportional to the solid angle dynamically swept by $\mathbf{n}(t)$ in the noncollinear spins, typically spin helix, as schematically shown in Fig. 6 (b). Nagaosa [25] proposed that in the linear spin helices like spin screw and cycloid (Fig. 1) the ac current induced deformation of spin spiral can generate the phase-π delayed component of the voltage along the current direction parallel to the helix $\boldsymbol{q}$, that is nothing but the electromagnetic inductance. In the proper screw helix case the emergent electric field is expressed as

$$e_x = \frac{Ph}{e\lambda}\partial_t \phi, \tag{4}$$

where P is a spin polarization factor and ϕ is the tilting angle of spin from spiral plane (Fig. 6(b)). Importantly, the magnitude of e is inversely proportional to the spin helix period λ.

The emergent electromagnetic inductance based on the spin chiral matter has recently been experimentally verified for $Gd_3Ru_4Al_{12}$ with Kagomé sublattice of Gd moments, which hosts short-λ($\sim$3nm) screw spin helices as well as the SkL state [26]. There the imaginary part (phase-π delayed component) of resistance, that is equal to $2\pi f L$ (f and L being ac current frequency and inductance, respectively), shows an enhanced value in the screw state even as compared with the SkL state; this is reasonable in terms of the magnitude of the induced noncollinear spin deformation. This emergent electromagnetic inductance phenomena have been further confirmed for the room-temperature screw spin magnet [27]. In those materials, in the current density of order of $10^8 A/m^2$, the large inductance value comparable with (or even larger than) classic commercial inductor elements were observed. Furthermore, the current density is inversely increased with the cross section area of the sample, which is contrary to the case of classical inductor whose inductance L is in proportion to the coil cross section. Thus the volume size of the emergent inductor is one-millionth reduced, implying the possibility of dramatically miniaturizing the classical inductor coil elements on the basis of emergent electromagnetic inductance, while a lot of issues about weird behaviors of the current induced dynamics of helix [27] should be solved toward the application.

5. Summary and outlook

Here we have shown some representative chiral spin textures and their emergent electromagnetic properties under the relativistic spin-orbit interaction in condensed matter. The key features of spin chiral matters originate from the simultaneous breaking of space-inversion and time-reversal symmetries. The multiferroics composed of spin helices (cycloid or screw) are typical examples, where the cross control of magnetization and electric polarization in terms of electric and magnetic fields, respectively, *i.e.* the magnetoelectric (ME) effect, is possible. Such an ME effect can be naturally extended to a high-frequency electromagnetic-field region, *i.e.* optical region, where the Maxell's wave equation should be modified to produce novel electro-optic and/or magneto-optical responses. Among them, the electrically-active magnetic excitations, so-called electromagnons, inherent to the multiferroics show remarkable ME optical effects in a terahertz region [4], such as the unidirectional light transmittance/absorption effect, called cloaking effect, whose proper direction is uniquely determined by the relative directions of P and M such as $P \times M$. The device concept based on multiferroics provides a nonvolatile information storage/carrier which can be directionally switched by the procedure with minimal energy consumption, *e.g.* by electric field causing merely the polarization current.

Spin chiral matter provides also a ubiquitous stage of noncollinear or noncoplanar spin textures which may host emergent electromagnetic field (real-space Berry curvature and Berry connection). The magnetic skyrmion is prototypical chiral matter which can generate an emergent magnetic field. The emergent magnetic field from the real-space topological state behaves just like an ordinary magnetic field as far as its action on the conduction electrons is concerned. Therefore, this can be utilized as a gigantic *synthetic* magnetic field. We have exemplified the high-density skyrmion lattice as formed by the hybridization of triple-q spin helices as the stage of the gigantic Hall and Nernst effects. A similar skyrmion-related lattice structure can be formed not only as the two-dimensional triple-q but also in the three-dimensional triple- and tetrahedral quadruple-q structures; in these three-dimensional topological magnetic lattices the magnetic hedgehog and anti-hedgehog like defects are regularly arranged and bridged by the skyrmion strings in analogy to the magnetic monopoles bridged by Dirac strings. Such emergent magnetic monopole-antimonopole lattices have been demonstrated to exist in the chiral cubic magnets and cause large anomalous responses in magnetotransport [18].

One important future direction of the research on the spin chiral matter is to explore its dynamics and discover the valuable electromagnetic response/function based on the emergent electromagnetism; as one such example was shown here the emergent electromagnetic induction enabled by ac current excitation on the spin helices. The aforementioned ME optical effect enhanced by the electromagnon excitation is also along this line. Some pulse electric, magnetic, and optical excitations may bring about large emergent electric field from the steep time-derivative of emergent magnetic field or Berry curvature, which may lead to new intriguing electromagnetic phenomena in the spin chiral matter. We have just encountered an abundant gold mine for the new nanoelectronics functionality.

Acknowledgements

The author would like to thank Naoto Nagaosa for enlightening discussions.

References

1. H. Katsura, N. Nagaosa, and A. V. Valatsky, Spin current and magnetoelectric effect in noncolinear magnets, *Phys. Rev. Lett.* **95**, 057205 (2005).
2. T. Kimura, T. Goto, H. Shintani, K. Ishizuka, T. Arima, and Y. Tokura, Mangetic control of ferroelectric polarization, *Nature*, **426**, 55 (2003).
3. Y. Yamasaki, H. Sagayama, T. Goto, M. Matusura, K. Hirota, T. Arima, and Y.Tokura, Electric control of spin helicity in a magnetic ferroelectric, *Phys. Rev. Lett.* **98**, 147204 (2007).
4. Y. Tokura, S. Seki, and N. Nagaosa, Multiferroics of spin origin, *Rep. Prog. Phys.* **77**, 076501 (2014).
5. Y. Tokura, Multiferroics as quantum electromagnets, *Science* **312**, 1481 (2006).
6. Y. Yamasaki, S. Miyasaka, Y. Kaneko, J. P. He, T. Arima, and Y. Tokura, Magnetic reversal of the ferroelectric polarization in a multiferroic spinel oxide, *Phys. Rev. Lett.* **96**, 207204 (2006).
7. T. Kimura, Y. Sekio, H. Nakamura, T. Siegrist, and A. P. Ramirez, Cupric oxide as a induced-multiferroic with high-T_C, *Nat. Mater.* **7**, 291 (2008).
8. V. Kocsis, T. Nakajima, M. Matsuda, A. Kikkawa, Y. Kaneko, J. Takashima, K. Kakurai, T. Arima, F. Kagawa, Y. Tokunaga , Y. Tokura, and Y. Taguchi, Magnetization-polarization cross-control near room temperature in hexaferrite single crystals, *Nat. Commun.* **10**, 1247 (2019).
9. T. H. R. Skyrme, A unified field theory of mesons and baryons. *Nucl. Phys.* **31**, 556 (1962).
10. A. N. Bogdanov and D. A. Yablonskii, Thermodynamically stable "vortices" in magnetically ordered crystals — the mixed state of magnets. *Sov. Phys. JETP* **68**, 101 (1989).
11. S. Mühlbauer, B. Binz, F. Jonietz, C. Pfleiderer, A. Rosch, A. Neubauer, R. Georgii, and P. Boni, Skyrmion lattice in a chiral magnet, *Science* **323**, 915 (2009).
12. X. Z. Yu, Y. Onose, N. Kanazawa, J. H. Park, J. H. Han, Y. Matsui, N. Nagaosa, and Y. Tokura, Real-space observation of a two-dimensional skyrmion crystal, *Nature* **465**, 901 (2010).
13. A. Fert, V. Cross, and J. Sampaio, Skyrmions on the track, *Nat. Nanotechol.* **8**, 152 (2013).
14. N. Nagaosa and Y. Tokura, Topological properties and dynamics of magnetic skyrmions, *Nat. Nanotechol.* **8**, 899 (2013).

15. H. Oike, A. Kikkawa, N. Kanazawa, Y. Tauchi, M. Kawasaki, Y. Tokura, and F. Kagawa, Interplay between topological and thermodynamic stability in a metastable magnetic skyrmion lattice, *Nat. Phys.* **12**, 62 (2016).

16. K. Karube, J. S. White, D. Morikawa, M. Bartkowiak, A. Kikkawa, Y. Tokunaga, T. Arima, H. M. Ronnow, Y. Tokura, and Y. Taguchi, Skyrmion formation in a bulk chiral magnet at zero magnetic field and above room temperature, *Phys. Rev. Mater.* **1**, 74405 (2017).

17. N. Nagaosa, J. Sinova, S. Onoda, A. H. MacDonald, and N. P. Ong, Anomalous Hall effect, *Rev. Mod. Phys.* **82**, 1539 (2010).

18. Y. Tokura and N. Kanazawa, Magnetic skyrmion materials, *Chem. Rev.* **121**, 2857 (2021).

19. T. Kurumaji, T. Nakajima, M. Hirschberger, A. Kikkawa, Y. Yamasaki, H. Sagayama, H. Nakano, Y. Taguchi, T. Arima, and Y. Tokura, Skyrmion lattice with a giant topological Hall effect in a frustrated triangular-lattice magnet, *Science* **365**, 914 (2019).

20. M. Hirschberger *et al.*, Topological Nernst effect of two-dimensional skyrmion lattice, *Phys. Rev. Lett.* **125**, 076602 (2020).

21. F. Jonietz, S. Muhlbauer, C. Pfleiderer, A. Neubauer, W. Munzer, A. Bauer, T. Adams, R. Georgii, P. Boni, R. A. Duine, K. Everschor, M. Garst, and A. Rosch, Spin Transfer Torques in MnSi at Ultralow Current Densities, *Science* **330**, 1468 (2010).

22. X. Z. Yu, N. Kanazawa, W. Z. Zhang, T. Nagai, T. Hara, K. Kimoto, Y. Matsui, Y. Onose, and Y. Tokura, Skyrmion flow near room temperature in an ultralow current density, *Nature Communications* **3**, 988 (2012).

23. G. E. Volovic, Linear momentum in ferromagnets, *J. Phys. C: Solid State Phys.* **20**, L83 (1987).

24. T. Schulz, R. Ritz, A. Bauer, M. Halder, M. Wagner, C. ranz, C. Pfleiderer, K. Everscjpr, M. Garst, and A. Rosch, Emergent electrodynamics of skyrmions in a chiral magnet. *Nat. Phys.* **8**, 301 (2012).

25. N. Nagaosa, Emergent inductor by spiral magnets, *Jpn. J. Appl. Phys.* **58**, 120909 (2019).

26. T. Yokouchi, F. Kagawa, M. Hirschberger, Y. Otani, N. Nagaosa, and Y. Tokura, Emergent electromagnetic induction in a helical-spin magnet, *Nature* **586**, 232 (2020).

27. A. Kitaori, N. Kanazawa, T. Yokouchi, F. Kagawa, N. Nagaosa, and Y. Tokura, Emergent elecytomagnetic induction beyond room temperatuer, *PNAS* **118**, e2105422118 (2021).

Dynamics of Chiral Fermions in Condensed Matter Systems*

Qiang Li

*Department of Physics and Astronomy, Stony Brook University,
Stony Brook, NY 11794-3800, USA
Condensed Matter Physics & Materials Science Division, Brookhaven National Laboratory,
Upton, New York 11973-5000, USA
E-mail: qiang.li@stonybrook.edu, or qiangli@bnl.gov
www.stonybrook.edu, www.bnl.gov*

In condensed matters, such as 3D Dirac and Weyl semimetals, fermions with linear energy-momentum dispersion gain chirality (handedness). The chiral anomaly produces an imbalance between the densities of right- and left-handed fermions, leading to generation of electric current in parallel electric and magnetic fields. This is called the chiral magnetic effect. Coupling of circular polarized light to chiral fermions breaks the chiral symmetry, and can generate chirality-dependent photocurrent. In this article, we review dynamics of chiral fermions in condensed matter systems to explain the theory of chiral magnetic effect and describe experimental signatures of the chiral anomaly. We then summarize recent theoretical and experimental studies of topological phase transition involving Dirac and Weyl semimetals, and detections of chirality and chiral photocurrent in static and dynamically-driven Weyl states. We conclude with potential uses of chiral fermions in quantum information systems

Keywords: Chiral Fermions; Chiral Anomaly; Chiral Magnetic Effect; Dirac and Weyl Semimetals

1. Introduction

The last decade has seen a dramatic shift in the focus of condensed matter physics. Topology has emerged as an organizing principle of states of matter that transcends the conventional classification in terms of phase transitions and symmetry breaking, as expressed in Landau's theory. A unique feature that makes many topological materials' electronic properties different from all other condensed matter systems is the linear dispersion of their low-energy excitations. This has led to a fascinating convergence between condensed matter physics and high energy nuclear physics.

Massless Dirac particles having linear energy-momentum dispersion are described by the Dirac equation $\hat{H} = c\vec{\sigma} \cdot \hat{p}$, where c is the speed of light, $\vec{\sigma}$ is the Pauli matrix, and $\hat{p}$ is the momentum operator. In condensed matters, linear dispersion makes electrons near the Fermi energy behave like massless Dirac fermions, which is different from all other condensed matter systems described by the non-relativistic Schrödinger equation and quadratic dispersion of electrons – $\hat{H} = \hat{p}^2/2m^*$, where m^* is the effective mass [1]. To describe linear dispersion in condensed matter systems, an analog to the Dirac equation is used, where the speed of the light is replaced by the Fermi-velocity of the quasi-particles [1].

*This work is supported by the U.S. Department of Energy, Office of Basic Energy Sciences, Division of Materials Sciences and Engineering, under Contract No. DE-SC0012704.

Materials having states with linear dispersion include graphene [2, 3], topological insulators [4, 5], and Dirac and Weyl semimetals [6–9].

Topological insulators (TI, e.g., HgTe and Bi_2Se_3) display the intriguing property of being insulating in the bulk but conducting on the surface [4, 5]. This feature is enabled by Dirac-like surface states that arise from strong spin-orbit coupling. The spin in these topological states is locked at a right angle to the charge carriers' momentum. Closely related to TIs are topological semimetals, such as *Dirac semimetals* (DSM e.g., Na_3Bi,Cd_3As_2) and *Weyl semimetals* (WSM, e.g., TaAs) [6–11], which are three dimensional (3D) analogs of graphene. The bulk states – gapless fermions – in 3D topological semimetals are described by eigenstates of the Weyl or Dirac equation with a definite projection of spin onto momentum. This fermion is right-handed (parallel) or left-handed (anti-parallel), dictated by relativistic quantum field theory for a system with even spacetime dimension. This "handedness" is called chirality and is depicted in Fig. 1b, along with a description of chiral materials such as DSMs and WSMs (Fig. 1a), a schematic of the chiral anomaly (Fig. 1d), and examples of chiral applications (Fig. 1c, 1e, 1f).

In this article, we review dynamics of chiral fermions in condensed matters, a rapidly developing field. We begin in Sec. 2 with an introduction to the chiral

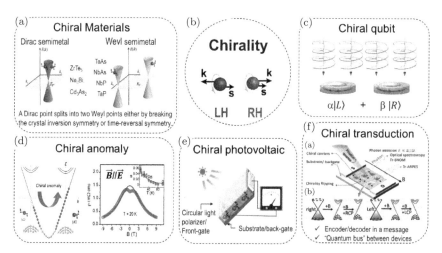

Fig. 1. Chiral fermions in condensed matter systems: (a) A Dirac semimetal (purple) hosts both left-handed (L)-and right-handed (R)-fermions on the same band. A Weyl semimetal (blue and orange) hosts L-and R-fermions on separate bands. (b) The definition of chirality in relativistic quantum field theory. (c) Chiral qubits have two base states describing chiral fermions circulating clockwise and counter-clockwise [13]. (d) The chiral anomaly is manifested, e.g., as the chiral magnetic effect, generation of an electric current by external gauge fields with non-trivial topology (e.g. by parallel electric and magnetic fields). The signature of the chiral magnetic effect is the negative longitudinal magnetoresistance discovered in the Dirac semimetal $ZrTe_5$ [12]. (e) The chiral photovoltaic cell enables a production of electric current in a Weyl semimetal via circular photogalvanic effect. (f) Chiral transduction uses chirality as an encoder/decoder in transmitting a message and serves as a "quantum bus" between quantum devices operating at different frequencies.

magnetic effect and chiral anomaly. Sec. 3 describes the topological phase transitions between TIs and DSMs/WSMs and discusses the static and dynamic controls of the band topology transition. Sec. 4 describes chiral photocurrent generation and discusses potential uses of chiral fermions for quantum information science and technology.

2. Chiral magnetic effect and chiral anomaly

The conservation of chiral symmetry implies that populations of right- and left-handed massless fermions are separately conserved. However, coupling of the fermions to a vector gauge potential breaks the chiral symmetry, leading to the appearance of a chiral anomaly. The chiral anomaly was discovered in 1969 based on the observation of the rapid decay of the neutral pion π^0 into photons [14]. In 1983, Nielsen and Ninomiya pointed out a similarity between the fermion system of lattice gauge theories and the electron system of gapless semiconductors, in which energy bands have point-like degeneracy [16], known now as Weyl semimetals (WSM). Nielsen and Ninomiya predicted that longitudinal magnetoconductivity becomes large for Weyl fermions in parallel electric and magnetic fields.

In relativistic heavy-ion collisions, the chiral anomaly manifests as the chiral magnetic effect that transforms the chiral asymmetry, generated by topological transitions in hot QCD matter, into an electric current, where magnetic field is produced by the colliding ions. The chiral magnetic effect in hot QCD matter was first predicted by Kharzeev and colleagues [17] and since has been intensely studied in heavy ion collision experiments at Relativistic Heavy Ion Collider (RHIC) at BNL and the Large Hadron Collider (LHC) at CERN [18]. Of note, when chiral asymmetry is produced by parallel electric and magnetic fields, either in condensed matters (e.g., WSMs) or in quark gluon plasmas, the chiral anomaly and the chiral magnetic effect describe the same phenomenon. Here, we will use the terms interchangeably.

The recent experimental discovery of DSM/WSM has opened unprecedented opportunities to study the quantum dynamics of relativistic field theory in condensed matter systems. The chiral magnetic effect was discovered experimentally in the DSM ZrTe$_5$ by the author and collaborators [12] in 2014. Below, we explain the chiral magnetic effect using a DSM ZrTe$_5$ as an example.

In the absence of external fields, each Dirac point contains left- and right-handed fermions with equal chemical potentials, $\mu_L = \mu_R = 0$. Applying a magnetic field breaks time reversal symmetry that can transform a DSM into a WSM. If the energy degeneracy between the left- and right-handed fermions is broken, quantum electrodynamics dictate that the parallel external electric and magnetic fields will generate the chiral charge ρ_5 with the rate given by

$$\frac{d\rho_5}{dt} = \frac{e^2}{4\pi^2\hbar^2 c}\vec{E}\cdot\vec{B}. \tag{1}$$

The ρ_5 is controlled by the chiral chemical potential $\mu_5 \equiv (\mu_R - \mu_L)/2$ and given by [17]

$$\rho_5 = \frac{\mu_5^3}{3\pi^2 v^2} + \frac{\mu_5}{3v^3}\left(T^2 + \frac{\mu^2}{\pi^2}\right), \tag{2}$$

where $\mu = (\mu_R + \mu_L)/2$, and v is the Fermi velocity. The left- and right-handed fermions in chiral matters can mix through chirality-changing scattering. Introducing the chirality-changing scattering time τ_V, we thus get the equation

$$\frac{d\rho_5}{dt} = \frac{e^2}{4\pi^2 \hbar^2 c}\vec{E}\cdot\vec{B} - \frac{\rho_5}{\tau_V}. \tag{3}$$

The solution of equation (3) at $t >> \tau_V$ is

$$\rho_5 = \frac{e^2}{4\pi^2 \hbar^2 c}\vec{E}\cdot\vec{B}\tau_V. \tag{4}$$

Assuming that $\mu_5 << \mu, T$, we obtain the non-zero chiral chemical potential μ_5 as

$$\mu_5 = \frac{3}{4}\frac{v^3}{\pi^2}\frac{e^2}{\hbar^2 c}\frac{\vec{E}\cdot\vec{B}}{T^2 + \frac{\mu^2}{\pi^2}}\tau_V. \tag{5}$$

The corresponding chiral magnetic current can be computed [17] and is given by

$$\vec{J}_{CME} = \frac{e^2}{2\pi^2}\mu_5 \vec{B} = \sigma_{CME}\vec{E}. \tag{6}$$

When the electric and magnetic fields are parallel, the CME conductivity is given by

$$\sigma_{CME} = \frac{e^2}{\pi\hbar}\frac{3}{8}\frac{e^2}{\hbar c}\frac{v^3}{\pi^3}\frac{\tau_V}{T^2 + \frac{\mu^2}{\pi^2}}B^2. \tag{7}$$

Magnetoconductance σ_{CME} induced by the chiral magnetic effect has a characteristic quadratic dependence on magnetic field in weak field limits. It is precisely this dependence of longitudinal magnetoconductance that was observed in ZrTe$_5$ (Fig. 2d) [12]. Fig. 2a, 2b, and 2c show the linearly dispersed electronic bands near Fermi level, the measurement configuration, and the orientation dependence of magnetoresistance, respectively. In this experiment, the current and voltage contacts on the single crystal ZrTe$_5$ (Fig. 2b) provided a uniform flow of electric current and uniform voltage detection. Thus, the data is free of measurement artifacts such as the current-jetting found in other measurements [19]. The chiral anomaly was also detected in another DSM Na$_3$Bi in 2015, then subsequently in a dozen more materials, including WSM TaP [20] and the half-Heusler GdPtBi [21]. Recently, using magneto-terahertz spectroscopy (a non-contact method), Cheng and coworkers investigated epitaxial DSM Cd$_3$As$_2$ films, in which the conductivities $\sigma(\omega)$ as a function of $\vec{E}\cdot\vec{B}$ showed a sharp Drude response over the background. This behavior was taken as a definitive signature of the chiral anomaly [22].

The chiral anomaly is a dynamic behavior of chiral fermions. In the absence of external fields, left-handed and right-handed fermions are present in the ground

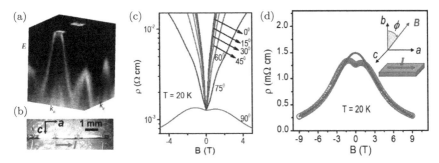

Fig. 2. (a) Angular resolved photoemission spectroscopy (ARPES) of a DSM ZrTe5 showing linearly dispersed electronic bands near Fermi level. (b) Current-voltage configurations and a single crystal ZrTe5 used for detecting the chiral magnetic effect, free of measurement artifacts, e.g., current-jetting. (c) Angular dependent longitudinal magnetoresistance (LMR) of ZrTe5 at 20 K showing the negative LMR in parallel electric and magnetic fields (i.e. $\phi = 90°$). (d) A quadratic magnetic field dependence of negative LMR in parallel $\vec{E}$ and $\vec{B}$ fields – a characteristic property of the chiral anomaly in Weyl systems. (b-d from Ref. [12]).

state in equal numbers. There is no chiral current as expected for a system in thermodynamic equilibrium. Because of the anomaly, the chiral charge is not conserved, and can decay in the presence of parallel electric and magnetic fields. The chiral anomaly, sourced by non-zero $\vec{E} \cdot \vec{B}$, generates a non-zero chiral chemical potential difference that produces so-called charge pumping, shown in the left panel of Fig. 1d. The generation of this chiral current is the chiral magnetic effect. The chiral magnetic effect is not a thermodynamic phenomenon or ground state property. Its direct consequence is the large negative longitudinal magnetoresistance predicted by Nelson and Ninomiya. Notably, the chiral magnetic effect was previously misunderstood by some researchers as a thermodynamic phenomenon, where the associated chiral current was suggested to be proportional to the energy difference between the left-handed and right-handed Weyl nodes, rather than proportional to the chiral chemical potential difference (as it should be).

3. Topological phase transitions

The topological states in TIs and DSMs/WSMs are largely due to strong spin-orbit interactions. Lead chalcogenides Pb(Te, Se) are classic semiconductors that were extensively investigated for thermoelectric power generation in the 1950's. In 1966, a band inversion for $Pb_xSn_{1-x}Te$ alloys was reported [24]. The valence and conduction bands of SnTe are inverted from that of PbTe, with the band gap initially decreasing with rising Sn content, going to zero at an intermediate Sn content, then increasing again with rising Sn content [24]. The electronic band topology transforms from a trivial band insulator Pb(Te, Se) at one end to a non-trivial topological crystal insulator Sn(Te, Se) at the other end. A gapless DSM is at an intermediate Sn content [23]. Another example of composition controlled topological phase transition exists in $Bi_{1-x}Sb_x$. Elemental Bi is a classic semimetal,

88

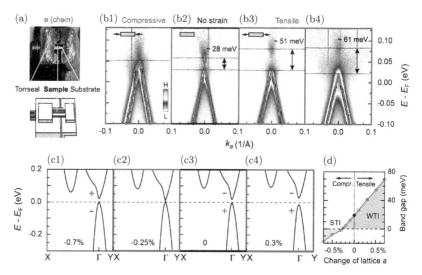

Fig. 3. Control of the topological phase in ZrTe₅ by external strain. (a) The strain device. (b) ARPES results: bulk band gap changes with compressive (Dirac state) and tensile strain (weak topological insulator). (c) The band structure calculations with different lattice constant a. + and − signs indicate the parity of the two bands. (b), (c) The blue (red) frames correspond to compressive (tensile) strain. (d) Calculated phase diagram with different lattice constant (strain). Solid markers roughly indicate the experimental values (from Ref. [25]).

and adding Sb (x in at. %) in $Bi_{1-x}Sb_x$ closes the band gap near x ≈ 3.5 − 6%. Alloys with x > 8.6% produce a direct-gap TI. A DSM emerges at a doping level in the range of 3.5 – 8.6 %.

The gapless Dirac points in DSMs are protected by time-reversal symmetry and crystal inversion symmetry. A small distortion in the crystal lattice can open a gap leading to DSM to TI transition [25, 26]. Taking ZrTe₅ as an example, Fig. 3 shows that a small amount of uniaxial strain can drive the material from a strong topological insulator (STI) to weak topological insulator (WTI) with a DSM in between [25]. Using laser-based angular resolved photoemission spectroscopy (ARPES), the topological states in ZrT₅ can be visualized under external strain. In this experiment, the uniaxial strain along the high conducting Te-chain direction was applied by using a screw to compress or stretch the substrate on which the single crystal sample was glued (Fig. 3a). With compressive strain, the gap was closed, reaching a DSM state. With tensile strain (Fig. 3b3–3b4), the band gap became larger, stabilizing the WTI state [25]. + and − signs in Fig. 3c1-3c4 indicate that the parity of the two bands are reversed in the STI and WTI phases. The example in Fig. 3 demonstrates the control of topological states via static strain. This control can also be achieved using a dynamic approach (e.g. Raman active phonons), which is described next.

We used first-principles methods to demonstrate that in ZrTe₅, atomic displacements corresponding to five of the six zone-center A_g (symmetry-preserving) phonon

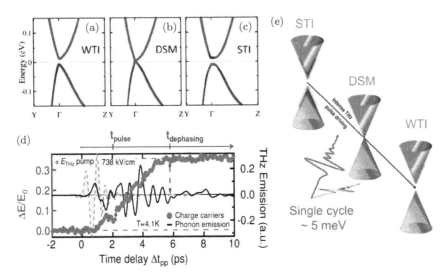

Fig. 4. (a)–(c) Evolution of the band structure and of the orbital content of the bands in ZrTe$_5$ forming the Dirac cone around the Γ point for different values of the normal coordinate Q corresponding to the A$_g$-27 Raman-active phonon mode. The Q values are (a) -0.6, (b) -0.25, and (c) 0.3. (d) THz differential transmission ($\Delta E/E_0$, red circles) as a function of the pump-probe time delay (Δt_{PP}) for a peak pump E field of 736 kV/cm. The black trace shows the simultaneously measured phonon emission. The THz-pump trace is shown in gray. The THz differential transmission starts to build up during the coherent phonon emission at much longer times than the pump THz pulse. (e) Schematic of the topological switching driven by the THz coherent excitation in Dirac semimetals. (a–c from Ref. [27], d–e from Ref. [28]).

modes can drive a topological phase transition from STI to WTI, with a DSM state emerging at the transition [27]. Shown in Fig. 4a–4c are the evolution of the band structure and orbital content of the bands in ZrTe$_5$ forming the Dirac cone around the Γ point for different values of the normal coordinate Q, corresponding to the A$_g$-27 Raman-active phonon mode. These topological phases in ZrTe$_5$ can be realized with many different settings of external stimuli that can penetrate through the phonon-space Dirac surface without breaking crystallographic symmetry.

In Fig. 4d, 4e, a few-cycle THz-pulse-induced topological phase transition in ZrTe$_5$ is depicted, driven by the lowest Raman active mode A$_{1g}$ [28]. Above a critical THz-pump field threshold, a long-lived metastable phase of approximately 100 ps emerged. These results, taken together with first-principles modeling, identified a mode-selective Raman coupling that drives the system from STI to WTI, with a DSM phase established at a critical atomic displacement, controlled by the phonon coherent pumping. Harnessing vibrational coherence to steer symmetry-breaking transitions (i.e., Dirac to Weyl transitions) has implications for THz topological quantum gate and error correction applications in quantum information science and technology.

4. Chiral photocurrent and chiral qubits

While chiral current can be generated by a static magnetic field via the chiral magnetic effect, it can also be generated by electromagnetic waves. Circular polarized light (CPL) carries the left or right helicity of electromagnetic fields, which can couple to the chirality of charge carriers in Weyl semimetals. The current generated in the chiral fermionic system by the CPL, via circular photogalvanic effect, carries the chirality information of the system. This was the approach used by Q. Ma and collaborators to directly detect the chirality of the Weyl fermions in WSM TaAs by measuring chirality dependent photocurrent induced by circularly polarized mid-infrared light [29].

Chirality dependent photocurrent can also be generated in DSMs dynamically if its crystal inversion symmetry is broken, thus transforming DSMs to metastable WSMs. We used first-principles and effective Hamiltonian methods to show that in central-symmetric DSM ZrTe$_5$, lattice distortions corresponding to all three types of zone-center infrared (IR) optical phonon modes can drive the system from an STI or WTI to a WSM by breaking the global inversion symmetry [30]. Fig. 5a shows the "twisting" IR phonon mode B$_{1u}$-4 of ZrTe$_5$ projected onto the b − c plane, with the vectors indicating the normal atomic displacement. Such dynamic crystal

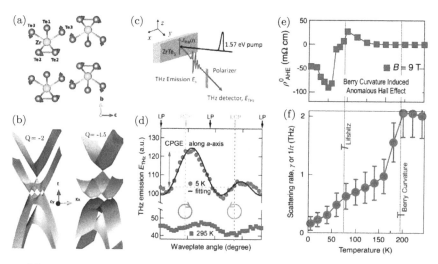

Fig. 5.   (a) IR phonon mode B$_{1u}$-4 of ZrTe$_5$ projected onto the b–c plane with the vectors showing the normal atomic displacement. (b) Bands forming the Weyl points on the $k_x - k_y$ ($k_a - k_b$) plane for different Q values corresponding to the B$_{1u}$-4 phonon mode. (c) Experimental schematics for polarized THz emission spectroscopy to detect chiral photocurrent from THz fields E_s and E_{THz}. (d) Helicity dependent THz emission peak-to-peak amplitude along the a- axis, orthogonal to the light-induced symmetry-breaking axis, as a function of the quarter waveplate angle at 5 K and 295 K; 0, 90 and 180 degrees represent linear polarization (LP) of the pump; 45 and 135 degrees represent right (RCP) and left (LCP) circular polarization of the pump, respectively, and are marked on the top. (e) Temperature-dependent anomalous Hall saturation resistivity measured at 9 T. (f) Temperature-dependent chiral photocurrent scattering rate. (a–b from Ref. [30], c–f from Ref. [31])

"twisting" gives rise to electronic bands forming the Weyl points, shown in Fig. 5b. The achieved Weyl phases are robust, highly tunable and one of the cleanest Weyl systems due to the proximity of the Weyl points to the Fermi level [30].

Giant anisotropic THz photocurrents with vanishing scattering were discovered in $ZrTe_5$ by using femto-second laser induced coherent phonons to break inversion symmetry [31]. The experimental schematics are shown in Fig. 5c. This light-induced phononic symmetry switching leads to formation of Weyl points, whose chirality manifests in a transverse and helicity-dependent photocurrent. This chiral photocurrent is orthogonal to the broken inversion symmetry axis, generated via the circular photogalvanic effect, shown in Fig. 5d. The chiral photocurrent has two distinct temperature–dependent features: i) Berry curvature creation by the separated Weyl nodes and strongly–suppressed scattering (Fig. 5f), with the same onset temperature $T_{Berry\ curvature}$, at which anomalous Hall resistivity was detected in the crystals (Fig. 5e) and ii) particle-hole reversal near the temperature-induced Lifshitz transition [32, 33]. Together with first-principles modeling, these results identify two pairs of Weyl points dynamically created by broken inversion-symmetry phonons of the B_{1u} mode, which are driven by the photoexcited and non-equilibrium population of spatially separated charges between the inverted bands of the Te pair. The chiral current appears dissipationless towards low temperature as the scattering rate of photocurrent (Fig. 5f) approaches zero. Berry curvature dominance below $T_{Berry\ curvature}$ marks the sharp suppression of the scattering rate. In addition to revealing dissipationless photocurrents with remarkable ballistic chiral fermion transport lengths of $\sim 10\,\mu$m, the phononic symmetry switching principle lays the groundwork for the coherent manipulation of Weyl states without application of any static electric or magnetic fields.

In the absence of chirality flipping, the chiral anomaly produces a flow of dissipation-free charge, which is useful for electronics and optoelectronics requiring ultralow energy consumption. Chiral photocurrents offer a new type of light-electricity conversion device, which we termed a chiral photovoltaic cell, depicted in Fig. 1e. This device converts light into electricity using a set of optical lenses and does not suffer from the electron-hole recombination process that limits conversion efficiency in a conventional solar cell with semiconductor p-n junctions.

Chiral materials make good candidates for topological qubits and quantum transducers, because of the chiral current and its ability to couple the helicity (CPL) of an electromagnetic field to the chirality of charge. This is depicted in Fig. 1c and Fig. 1f, respectively. The chiral qubit [33] we proposed is enabled by the chiral anomaly and constructed from a micron-meter-scale ring made of a WSM. The $|0\rangle$ and $|1\rangle$ states correspond to the symmetric and antisymmetric superpositions of quantum states describing chiral fermions circulating along the ring clockwise and counter-clockwise. The entanglement of qubits can be implemented through the application of a circularly-polarized THz frequency electromagnetic wave to the system. Chiral qubits are advantageous because they can be potentially operated at room temperature, at THz frequencies, and with a large ratio of the coherence to gate time on the order of 10^4.

In quantum communication applications, chirality can function as an encoder and a decoder for a transmitted message, as well as a "quantum bus" between devices operating at different frequencies. Examples include GHz superconducting qubits and optical communication channels. A chiral photocurrent-based transducer would be revolutionary as it could have the capabilities of generation, transduction (high fidelity down-conversion and up-conversion), and coherent control of the photon polarization at GHz to UV frequency ranges. This can be achieved using chirality, which is intrinsic to the chiral fermions in 3D DSMs/WSMs, without invoking traditional methods such as waveplates, polarizers, or low-efficiency nonlinear crystals.

In conclusion, the powerful notion of chirality, originally discovered in high energy and nuclear physics, underpins a wide range of new phenomena and useful applications. Recent discoveries of the chiral anomaly in a variety of materials have made it possible to study the quantum dynamics of relativistic field theory in condensed matters. The observation of chiral anomaly-enabled photocurrent generation opens a new avenue for novel electronics requiring ultra-low-energy dissipation that we call "chiraltronics". The potential of chiraltronics may be further enhanced by our abilities to statically and dynamically control the topological phase transitions between TIs and DSMs/WSMs. The proposed exploration of chiral qubits and chiral transduction may help to answer a grand challenge underlying the field of topology-enabled quantum information science and technology: how to establish principles of topological control, driven by quantum coherence, and understand its time-dependent effects. The advances described in this article make the "dynamics of chiral fermions in condensed matter systems" an exciting and potentially transformative field.

Acknowledgments

The author would like to express his sincere gratitude for Prof. Dmitri Kharzeev for generously giving of his knowledge in chiral magnetic effect, and many fruitful collaborations. The author would also like to acknowledge collaborations with Drs. Cheng Zhang, Genda Gu, Tonica Valla, Jigang Wang, and Mengkun Liu.

References

1. A. K. Geim, "Graphene: Status and Prospects" *Science* **324**, 1530 (2009).
2. A. K. Geim and K. S. Novoselov, "The rise of graphene" *Nat. Mater.* **6**, 183 (2007).
3. A. H. Castro Neto, F. Guinea, N. M. R. Peres, K. S. Novoselov, and A. K. Geim "The electronic properties of graphene" *Rev. Mod. Phys.* **81**, 109 (2009).
4. M. Z. Hasan and C. L. Kane " Colloquium: Topological insulators" *Rev. Mod. Phys.* **82**, 3045 (2010).
5. X. L. Qi and S. C. Zhang, "Topological insulators and superconductors" *Rev. Mod. Phys.* **83**, 1057 (2011).

6. Z. Wang, *et al.* "Dirac semimetal and topological phase transitions in A_3Bi (A = Na, K, Rb)". *Phys. Rev. B*, **85**, 195320 (2012).

7. Z. Wang, *et al.* "Three-dimensional Dirac semimetal and quantum transport in Ca_3As_2" *Phys. Rev. B*, **88**, 125427 (2013).

8. S. M. Young, *et al.* "Dirac semimetal in three Dimensions" *Phys. Rev. Lett.* **108**, 140405 (2012).

9. B. Yang and N. Nagaosa "Classification of stable three-dimensional Dirac semimetals with nontrivial topology" *Nature Communications* **5**, 4898 (2014).

10. S. Xu, *et al.* "Discovery of a Weyl fermion semimetal and topological Fermi arcs" *Science* 349, 613–617 (2015).

11. B. Q. Lv. *et al.* "Experimental discovery of Weyl semimetal TaAs" *Phys. Rev. X* **5**, 031013 (2015).

12. Q. Li, *et al.* "Chiral magnetic effect in $ZrTe_5$" arXive:1412.6542 (2014), *Nature Physics* **12**, 550 (2016).

13. D. Kharzeev and Q. Li "Quantum computing using chiral qubits" US Patent #10,657,456 (2020); D. Kharzeev and Q. Li "The Chiral Qubit: quantum computing with chiral anomaly" arXiv:1903.07133.

14. S. L. Adler, "Axial-vector vertex in spinor electrodynamics" *Phys. Rev.*, **177**, 2426 (1969).

15. J. S. Bell and R. Jackiw "A PCAC puzzle: $\pi^o \rightarrow \gamma\gamma$ in the σ-model" *Il Nuovo Cimento A*, **60**, 47–61 (1969).

16. H. B. Nielsen and M. Ninomiya. "The Adler-Bell-Jackiw anomaly and Weyl fermions in a crystal" *Physics Letters* B, **130**, 389–396 (1983).

17. K. Fukushima, D. Kharzeev, and H. Warringa. "Chiral magnetic effect" *Phys. Rev. D*, **78**, 074033 (2008); D. E. Kharzeev "The chiral magnetic effect and anomaly-induced transport" *Progress in Particle and Nuclear Physics*, **75**, 133–51 (2014); D. Kharzeev "Parity violation in hot QCD: Why it can happen, and how to look for it". *Physics Letters* B, **633**, 260–264 (2006).

18. D. E. Kharzeev and J. Liao, "Chiral magnetic effect reveals the topology of gauge fields in heavy-ion collisions" *Nature Reviews Physics* **3**, 55 (2021).

19. N. P. Ong and S. Liang "Experimental signatures of the chiral anomaly in Dirac–Weyl semimetals" *Nature Reviews Physics* **3**, 394 (2021).

20. F. Arnold, *et al.* "Negative magnetoresistance without well-defined chirality in the Weyl semimetal TaP" *Nat. Communications* **7**, 11615 (2016).

21. M. Hirschberger, *et al.* "The chiral anomaly and thermopower of Weyl fermions in the half-Heusler GdPtBi" *Nature Materials* **15**, 1161 (2016).

22. B. Cheng, *et al.* "Probing charge pumping and relaxation of the chiral anomaly in a Dirac semimetal" Science Advances, abg0914 (2021).

23. Q. Li "Thermoelectrics with a twist" *Nature Materials* **18**, 1267 (2019).

24. J. O. Dimmock, I. Melngailis, and A. J. Strauss "Band structure and laser action in $Pb_xSn_{1-x}Te$". *Phys. Rev. Lett.* **16**, 1193–1196 (1966).

25. P. Zhang, *et al.* "Observation and control of the weak topological insulator state in $ZrTe_5$" Nature Communications **12**, 406 (2021).

26. J. Mutch, *et al.* "Evidence for a strain-tuned topological phase transition in $ZrTe_5$" Science Advances eaav9771 (2019).

27. N. Aryal, *et al.* "Topological phase transition and phonon-space Dirac topology surfaces in $ZrTe_5$" *Phys. Rev. Lett.* **126**, 016401 (2021).

28. C Vaswani, *et al.* "Light-driven Raman coherence as a nonthermal route to ultrafast topology switching in a Dirac semimetal" *Phys. Rev. X* **10**, 021013 (2020).

29. Q. Ma, *et al.* "Direct optical detection of Weyl fermion chirality in a topological semimetal" *Nature Physics* **13**, 842 (2017).

30. N. Aryal, *et al.* "Robust and tunable Weyl phases by coherent infrared phonons in ZrTe$_5$" *npj Comput Mater* **8**, 113 (2022).

31. L. Luo, *et al.* "A light-induced phononic symmetry switch and giant dissipationless topological photocurrent in ZrTe$_5$. *Nature Materials* **20**, 329 (2021).

32. H. Chi, *et al.* "Lifshitz transition mediated electronic transport anomaly in bulk ZrTe$_5$" *New J. Phys.* **19**, 015005 (2017).

33. P. Lozano, *et al.* "Anomalous Hall effect at the Lifshitz transition in ZrTe$_5$" *Phys. Rev. B*, **106**, L081124 (2022).

The Chiral Anomaly in Dirac and Weyl Semimetals

N. Phuan Ong

*Department of Physics, Princeton University,
Princeton, NJ 08544, U. S. A.*
* E-mail: npo@princeton.edu

The prediction that the chiral anomaly may be observable as a negative longitudinal magnetoresistance (LMR) in Weyl semimetals has attracted considerable interest in the condensed-matter physics community researching topological materials. I review transport experiments that address this prediction, focussing on the two semimetals, Na$_3$Bi and GdPtBi, and emphasizing symmetry-protection of the Dirac nodes in these systems. I will describe tests that can distinguish the chiral anomaly LMR from artefacts caused by field-induced inhomogeneous current flow.

Keywords: Dirac and Weyl semimetals, chiral anomaly, longitudnial magnetoresistance, symmetry protection, current jetting

1. Introduction

The two major fields of physics, high-energy physics and condensed-matter physics, have greatly benefitted from the exchange of fundamental notions and paradigms. Examples are spontaneously broken symmetry, Goldstone modes, the renormalization group, Nambu's formulation of the BCS ground state and the rich phenomenology associated with Dirac excitations in graphene and topological matter. Here I describe recent experimental progress in my lab towards the confirmation of a famous prediction made many decades ago [1], namely the anticipation that the chiral anomaly [2, 3], which dictates the decay of neutral pions [4–6], is observable in Weyl semimetals.

In addition to summarizing the results, I will discuss the experimental challenges, and stress the importance of starting with materials in which the Dirac and Weyl nodes are rigorously symmetry protected, as well as implementing tests that guard against artefacts arising from inhomogeneous current flow. A recent extensive review that includes discussion of complementary experiments in other labs is given in Ref. [7]. A review of Dirac semimetals is in Ref. [8]. See also the talk by Qiang Li in these proceedings.

2. The chiral anomaly

An anomaly in quantum field theory (QFT) is the breaking of a classical symmetry of the starting Lagrangian $\mathcal{L}$ when quantum fluctuation effects are switched on [4–6]. The best known example, the chiral anomaly, may be illustrated by considering a system that harbors two populations, which we call left- and right-moving fermions. We will see below (see Fig. 1) that this situation is realized when Weyl fermions in a semimetal are confined to the lowest Landau level in a magnetic field. In the chiral representation, the two populations are described by the two-spinor field $\hat{\Psi} = (u_L, u_R)^T$, whose components u_L and u_R destroy left- and right-moving fermions, respectively [$(\cdots)^T$ stands for transpose]. If we rotate the phase of both components by the same angle (called a "vector" rotation), $\mathcal{L}$ is left invariant. Noether's theorem then yields the familiar conservation of matter current. Now, when the fermions are massless, a new symmetry emerges. If we rotate the phases of u_L and u_R by equal but opposite angles ("axial" rotation), $\mathcal{L}$ is also left invariant. The new symmetry is called axial or chiral symmetry. The corresponding conserved current, called J^5, is the difference of the left and right-moving currents. The chiral symmetry is broken by quantum effects when we allow the massless fermions to couple to the electromagnetic field. In parallel electric and magnetic fields ($\mathbf{E} \parallel \mathbf{B}$), the conservation of J^5 is ruined by the appearance of a source term called the anomaly $\mathcal{A}$. In $3 + 1$D spacetime, $\mathcal{A}$ is proportional to $\mathbf{E} \cdot \mathbf{B}$. This constitutes the chiral anomaly [4–6].

The chiral anomaly was discovered by Adler [2] and Bell and Jackiw [3] (ABJ) in their quest to calculate the anomalous rate of decay of neutral pions π^0 into two photons. As π mesons are the lightest hadrons, their decay channels are purely leptonic (to $\mu^\pm$ and ν) [4]. This results in relatively long lifetimes for $\pi^\pm$ ($\sim 10^{-18}$ s). Because the leptonic channels are unavailable to the neutral pion π^0, its lifetime is even longer (by a factor of 300 million). Its decay channel is purely electromagnetic: $\pi^0 \to 2\gamma$. Throughout the 1960's, this low decay rate posed a formidable challenge to theory (standard current-algebra calculations consistently yielded zero [6]). In 1969, ABJ identified the correct diagram as an anomaly process whereby $\mathcal{A}$ appears as a triangular fermion loop in which an axial current (π^0) enters at one vertex and two vector currents (photons) exit at the other two. In addition to solving the π^0 decay problem, the ABJ theory provided the earliest quantitative confirmation of color, as well as the earliest evidence for topological effects in quantum field theory. Subsequently, anomalies have been identified in diverse phenomena that span a vast range of energy scales.

In 1983, Nielsen and Ninomiya [1] (NN) predicted that the chiral anomaly should be observable in Weyl semimetals as a negative longitudinal magnetoresistance (LMR) measured with $\mathbf{E} \parallel \mathbf{B}$. The chiral anomaly is specific to systems in even spacetime dimensions ($1 + 1$D or $3 + 1$D). Thus, graphene ($2 + 1$D) is not a candidate material. The discovery in $\sim$2011 of Dirac-Weyl semimetals [9, 14, 15] that feature protected Dirac nodes led to a surge of experimental interest in testing NN's prediction in 3+1D materials. (I first learned of NN's prediction in the late 80's because of speculations that sliding charge-density-wave conductivity could be anomaly related.)

3. Dirac and half-Heusler semimetals

The chiral anomaly experiments were enabled by the discovery of materials featuring bulk, three-dimensional (3D) Dirac or Weyl nodes that are rigorously protected by internal and lattice symmetries against gap formation.

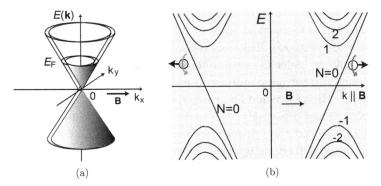

Fig. 1.   Sketch of 3D Dirac cone in zero B, and Landau levels in a Weyl semimetal in intense B. Panel (a): The Dirac node is the superposition of two symmetry-protected Weyl nodes that have opposite chiralities, $\chi = \pm 1$. In applied $\mathbf{B}$, the Weyl nodes separate in $\mathbf{k}$ space. Panel (b) shows the Landau levels (LLs) of the two Weyl nodes. The $N = 0$ levels (0LL) are chiral; they disperse with velocity $\mathbf{v}$ parallel (antiparallel) to $\mathbf{B}$ if $\chi = +1$ (-1). Electrons occupying the two chiral branches segregate into right and left-moving fermions.

As an example of symmetry protection, we describe the role of time-reversal invariance (TRI) in protecting 2D Dirac nodes on the surface of a topological insulator (TI) [10]. In a molecule, the time-reversal operator Θ transforms a state ψ to its time-reversed pardner ϕ, i.e. $\Theta\psi = \phi$ and $\Theta\phi = -\psi$. By TRI, the pardners are degenerate. The anti-unitarity of Θ immediately implies that $(\psi, \phi) = -(\Theta\phi, \Theta\psi) = -(\phi^*, \psi^*) = -(\psi, \phi) = 0$. Hence the two states are orthogonal and distinct (a Kramers doublet).

To generalize this to Bloch states on the TI surface, we must restrict the Dirac node to momenta $\mathbf{K}$ that are invariant under Θ (up to an Umklapp process). The set $\{\mathbf{K}\}$ includes the zone center and high-symmetry points on the Brillouin zone (BZ) boundary. Then the time-reversed pardners $\psi_{\mathbf{K}}$ and $\phi_{\mathbf{K}}$ form a Kramers doublet [10]. As a consequence, all matrix elements of TRI potentials formed between them must vanish. TRI alone suffices to protect a Dirac node at $\mathbf{K}$ against gap formation. As already noted, the 2D Dirac nodes on the surface of a TI are not candidates for the chiral anomaly.

To protect 3D nodes in the bulk of a crystal, however, more symmetries are needed. Initial searches [11] were based on protection by the combination of TRI and inversion symmetry, which led to candidate materials with nodes pinned at the boundaries of the BZ. This constraint was too restrictive. Subsequently, the insight of including the discrete point-group rotations C_n allowed protected nodes to lie anywhere on a symmetry axis [12, 13]. This soon led to the prediction by Wang *et al.* of protected Dirac nodes in the semimetals Na_3Bi [14] and Cd_3As_2 [15].

We focus on the workhorse Na_3Bi, whose crystal lattice is invariant under the 3-fold rotation C_3 about the axis Γ-A (parallel to k_z). The states near E_F are primarily derived from Na-3s orbitals (for the conduction band) and Bi-6$p_{x,y}$ orbitals (valence band) [14]. Band inversion caused by strong spin-orbit coupling forces the

Na-$3s$ band to lie below the Bi-$6p_{x,y}$ band at Γ. The resulting crossings lead to the formation of two Dirac nodes at $\mathbf{k}_D^{\pm} = (0, 0, \pm 0.26\pi/c)$ on the the k_z axis (c is the lattice parameter along k_z). The nodes are symmetry protected because the $3s$ and $6p$ bands transform under C_3 with different irreducible representations (hence all matrix elements formed between them vanish).

At Princeton, Bob Cava and I had forged a team dedicated to exploring the topological insulators and other novel materials. Following the publication of Refs. [9], [14] and [15], we immediately switched gears towards Na$_3$Bi and Cd$_3$As$_2$. Thanks to brilliant crystal-growth efforts by Satya Kushwaha and Quinn Gibson, we learned to grow crystals of ever higher quality. Na$_3$Bi was quite problematical because crystals oxidized in moist air within a minute. Jun Xiong solved the problem by cutting and mounting them in a glove box and sealing the assembly inside an epoxy cell filled with argon. Initially, our experiments were hampered by unintended doping (holes added due to Na vacancies in the case of Na$_3$Bi) which situated the Fermi energy E_F well below the node energy. In the early crystals, the profile of the resistivity ρ vs. T in zero magnetic field was metallic with a modest RRR (residual resistivity ratio). As the stoichiometry improved, E_F edged closer to the nodes.

Finally, in optimal crystals with greatly reduced vacancy population (and carrier mobility μ_e of 2,400 to 3,000 cm^2/Vs), Xiong observed that the profile of ρ vs. T switched to a highly unusual non-metallic form [16]. As T was lowered from 300 K, ρ increased 10-fold to saturate to a constant below 4 K (solid curve in Fig. 2a). The Hall coefficient R_H changed sign near 60 K and saturates to a positive value at 4 K (red circles). In a magnetic field $\mathbf{B}$ applied parallel to $\mathbf{E}$, the longitudinal

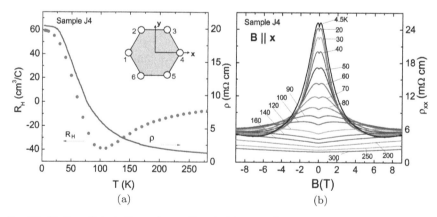

Fig. 2. The Hall coefficient R_H and zero-B resistivity ρ and negative LMR in a crystal of Na$_3$Bi (J4) with E_F very close to the Dirac node energy. Panel A: From the curve of R_H vs. T (red circles), the carriers are predominantly p-type at $T = 0$, but the n-type population dominates above $\sim$60 K. The unusual non-metallic profile of ρ (purple curve) suggests dominance of low-mobility holes below 50 K and high-mobility electrons above. Inset shows the 6 contracts attached to the hexagonal crystal. Panel B: LMR profiles measured with $\mathbf{B} \parallel \mathbf{E} \parallel \hat{\mathbf{x}}$ at selected T showing the emergence of negative LMR below 100 K. At 4.5 K, longitudinal resistivity ρ_{xx} decreases by a factor of 4 when B exceeds 6 T. Adapted from Xiong et al. [16].

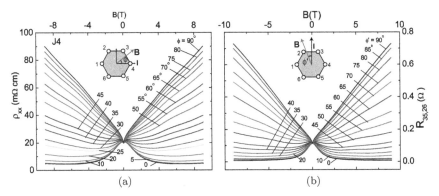

Fig. 3.   Magnetoresistance (MR) curves measured in Na$_3$Bi with **B** tilted to current **I** ∥ **E** in the a-b plane. (a) Resistivity ρ_{xx} versus B at selected field-tilt angles ϕ to the x-axis (inferred from resistance $R_{14,23}$; see inset). For $\phi = 90°$, ρ_{xx} displays a B-linear, positive MR. However, as $\phi \to 0°$ (**B** ∥ $\hat{x}$), ρ_{xx} displays negative LMR (decreases as B increases). (b) $R_{35,26}$ measured with **E** rotated by 90° relative to (A) (**B** makes an angle ϕ' relative to $\hat{y}$; see inset). The resistance $R_{35,26}$ changes from a positive MR to negative as $\phi' \to 0°$. In both configurations, the negative MR appears only when **B** approaches alignment with **E**. Adapted from Xiong et al. [16].

resistivity ρ_{xx} immediately revealed a striking negative LMR, first apparent near 100 K, and becoming increasingly prominent at lower T (Fig. 2b). As shown in Fig. 3, when **B** was rotated over nearly 4π in solid angle (keeping **E** ∥ **x**), Xiong observed that the axial current induced by the chiral anomaly was confined to a narrow plume directed parallel to **E** [16].

4. Weyl landau levels and chiral anomaly

From these experiments, we derived the following picture. In zero B, the Dirac node at k_D^+, say, is the superposition of two Weyl nodes of opposite chirality $\chi = \pm1$, both situated at k_D^+ (Fig. 1a). As B increases, the Zeeman field causes the Weyl nodes to move apart in **k**-space while their orbitals become Landau quantized. A distinguishing feature of the Weyl-Landau spectrum is that the $n = 0$ Landau level (hereafter, 0LL) is strictly chiral, dispersing with a group velocity **v** ∥ **B** if $\chi = +1$, but $-$**v** ∥ **B** if $\chi = -1$ (see Fig. 1b). Crucially, because the daughter Weyl nodes inherit the symmetry protection conferred on the original Dirac node, the Weyl nodes are also protected; they cannot be removed except by mutual annihilation (coalescing of nodes with opposite χ).

We now have a platform for observing the chiral anomaly. With E_F lying in the 0LL (in quantizing B, E_F represents the chemical potential), the two chiral 0LL branches represent the left- and right-moving fermions discussed in the introduction. In the presence of applied **E** ∥ **B**, the current conducted by, say, the right-moving fermions, is not conserved. Instead it receives an additional contribution from carriers "pumped" by **E** from the left-moving 0LL. This source term constitutes the anomaly term $\mathcal{A}$.

We may obtain $\mathcal{A}$ by a simple argument. With $\mathbf{B} \parallel \hat{\mathbf{x}}$, the density of states in a single LL is $\mathcal{D} = L_y L_z eB/(2\pi\hbar)$ where L_i is the sample dimension along axis i, e the elemental charge, $\hbar = h/2\pi$ and h is Planck's constant. In applied $\mathbf{E}$, the rate of state accumulation $\dot{n}_x$ equals $L_x eE_x/(2\pi\hbar)$. Hence the rate $\dot{Q}$ at which charge is pumped from the left-moving 0LL to the right-moving 0LL is $e\dot{n}_x\mathcal{D} = (L_x L_y L_z/4\pi^2)(e^3\mathbf{E}\cdot\mathbf{B}/\hbar^2)$. This gives for the anomaly term

$$\mathcal{A} = \frac{\dot{Q}}{eL_x L_y L_z} = \frac{e^2\mathbf{E}\cdot\mathbf{B}}{4\pi^2\hbar^2}, \tag{1}$$

a result identical to the Feynman diagram resullt of ABJ [2, 3]. The chiral-anomaly induced LMR, initially feeble in weak B, becomes very large once E_F enters the chiral 0LL in the quantum limit (this occurs at the field B_Q). For $B > B_\mathrm{Q}$, ρ_{xx} saturates to a constant value. These predictions agree with the observed ρ_{xx} vs. B in Na$_3$Bi.

When $\mathbf{B}$ is tilted by an angle θ to $\mathbf{E}$, the LMR expressed as ρ_{xx} is expected to vary as $\cos^2\theta$. In the experiment of Xiong $et\ al.$, however, the angular width of the plume was observed to be much narrower. The narrowing was an early hint that current jetting effects, although subdominant in Na$_3$Bi, can distort the true LMR signal (see below).

A year later, our group found a second candidate that displays the chiral anomaly [17]. This was a fortuitous finding by Max Hirschberger and Carina Belvin (an undergrad intern) who were investigating crystals of the half-Heusler GdPtBi grown by Cava's group. In this material (with typical mobility $\mu_e \sim 1{,}600$ cm^2/Vs), Dirac nodes are absent in zero B. Instead, the valence and conduction bands touch quadratically at the center of the BZ, Γ (as sketched in Fig. 4a). A strong B lifts the degeneracy at Γ, causing the bands to overlap and cross at a finite $\mathbf{k}$ to define Weyl nodes. Thus, unlike Na$_3$Bi in which two Dirac nodes already exist in zero B, the nodes in GdPtBi have to be created by a large Zeeman field [18]. They then emerge as protected Weyl nodes (Fig. 4a).

Again, the profile of ρ vs. T in zero B was observed to be non-metallic as in Na$_3$Bi (Fig. 4b). In a field $\mathbf{B} \parallel \mathbf{E}$, ρ_{xx} displayed a large negative LMR similar in pattern to that observed in Na$_3$Bi (Fig. 4c). The prominent difference is the absence of Weyl nodes for $B < 2$ T. When the nodes appear, the LMR is similar to that in Na$_3$Bi including the appearance of a narrow plume when $\mathbf{B}$ is rotated over the 4π solid angle (Fig. 4d). Unlike Na$_3$Bi, GdPtBi can be doped intentionally (with Au) to tune E_F. As shown by Hirschberger $et\ al.$ [17], the non-metallic profile of ρ was most conspicuous in samples with E_F closest to the node (Fig. 4b). The negative LMR amplitude was also largest in these samples. Both features were strongly suppressed when E_F was moved away from the node energy [17]. This provided complementary evidence that the chiral anomaly is associated with states closest to the node energy.

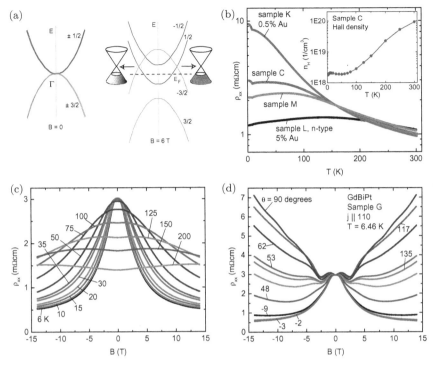

Fig. 4. Field-induced emergence of Weyl nodes and the chiral anomaly in the half-Heusler GdPtBi. Panel (a) is a sketch of the quadratic touching of conduction and valence bands at Γ in zero B. At 6 T, the Zeeman energy splits the two bands (the split in the valence band is 3× larger). The overlap creates two protected Weyl nodes. Panel (b) plots the non-metallic resistivity ρ vs. T (in zero B) for Sample K (with E_F nearest to the Weyl node energy) and Samples C, M and L (E_F progressively shifted farther away). The Hall density n_H in Sample C falls by a factor of 50 between 300 K and 2 K (inset). Panel (c) shows the negative LMR curves representing the appearance of the chiral anomlay (Sample G). Panel (d) shows the effect on ρ_{xx} when $\mathbf{B}$ is tilted in the polar plane x-z at angle θ to current $\mathbf{J}$ (Sample G). As θ increases from $0°$, the bell-shaped profile broadens. For $\theta \sim 90°$, the MR is positive. Adapted from Hirschberger et al. [17].

5. Current jetting

An important experimental pitfall confronting all LMR experiments is the problem of current jetting. In any conventional semimetal, with $\mathbf{B} \parallel \mathbf{E} \parallel \hat{\mathbf{x}}$, the conductivities in directions transverse to $\mathbf{B}$, $\sigma_{yy}(B)$ and $\sigma_{zz}(B)$, decrease steeply with B as $\sigma_0/[1+(\mu_e B)^2] \sim B^{-2}$ when $\mu_e B \gg 1$, where σ_0 is the isotropic conductivity in zero B. By contrast, the longitudinal conductivity σ_{xx} remains at the B-independent value σ_0 (assuming the chiral anomaly is absent). As B increases, the divergent anisotropy $\sigma_{xx} \gg \sigma_{yy}, \sigma_{zz}$ causes the current density $\mathbf{J}(x, y, z)$ to be strongly focussed into a narrow beam parallel to $\mathbf{E}$ (jetting). As a result, $|\mathbf{J}(x, y, z)|$ sharply decreases all along the edge of the sample. In a LMR measurement, voltage probes placed along the edge will record a sharply falling effective resistance that mimics rather convincingly the negative LMR predicted for the chiral anomaly (even though $\sigma_{xx} = \sigma_0$ at all B). The apparent negative LMR is purely artefactual, a result of strongly inhomogeneous current flow.

After the publication of Ref. [16], numerous experiments on the chiral anomaly in a slew of semimetals (SMs) were reported. These may be grouped into two classes. The first included conventional high-mobility SMs, notably elemental bismuth, Sb and InAs, which do not even feature protected Weyl nodes. Their negative LMR are artefactual (see below). The second class is comprised of LMR experiments on the Weyl semimetals TaAs, NbAs and NbP. These have 24 Weyl nodes, but their mobilities are so high ($\mu_e > 150{,}000\,\mathrm{cm}^2/\mathrm{Vs}$) that current jetting effects onset at the "cyclotron" field-scale $B_{\mathrm{cyc}} \ll B_{\mathrm{Q}}$, i.e. the current density becomes strongly inhomogeneous long before E_{F} enters the 0LL. The experiments on negative LMR in Weyl semimetals are actually detecting dominant current jetting effects. Consistent with this conclusion, the overall decrease in ρ_{xx} is typically less than 1% (compared with 10-fold total decrease in Na$_3$Bi). Moreover the apparent negative LMR in ρ_{xx} often vanished when the voltage probes were reconfigured. The fragility is a hallmark of current jetting.

The current-jetting concerns [20, 21] led to much confusion in the community. To address them, Sihang Liang and I performed extensive numerical simulations of current jetting for many sample geometries under various boundary conditions in a strong B. Guided by the simulations, we devised a test (the "squeeze test") [19], in which we measured the minimum and maximum of the distribution $J(x_0, y, z)$ within the plane $x = x_0$ transverse to $\hat{\mathbf{x}}$ while B was swept. This was achieved using micro, local voltage probes placed along the spine (the mid-line joining current contacts) and along one edge of a plate-like crystal (the plate geometry is chosen to accentuate the current jetting effects). When current jetting effects dominate, E measured on the spine (expressed as a resistance R_{spine}) increases steeply with B whereas E on the edge (R_{edge}) decreases sharply. The jetting, which onsets at B_{cyc}, reflects pronounced concentration of $J(x_0, y, z)$ along the spine occuring concurrently with a steep withdrawal from the edges. In any semimetal, the emergence of divergent profiles raises a red flag that current jetting effects are dominant and the observed LMR is artefactual. In Liang et al. [19], we applied the test to pure Bi (see Figs. 5a and b) and the Weyl SM, TaAs (Fig. 6). In both cases, R_{spine} and R_{edge} diverge from each other at all B, confirming that the observed LMR was artefactual. As shown in Fig. 6, the early onset of current jetting at $B_{\mathrm{cyc}} \sim 0.3\,\mathrm{T}$ effectively precludes experimental observation in TaAs of chiral anomaly contributions which are prominent only above 6 T.

6. Intrinsic LMR curve

Applying the squeeze test to the semimetals Na$_3$Bi and GdPtBi (in which $B_{\mathrm{Q}} \ll B_{\mathrm{cyc}}$), we observed that both R_{spine} and R_{edge} decrease with increasing B (Figs.

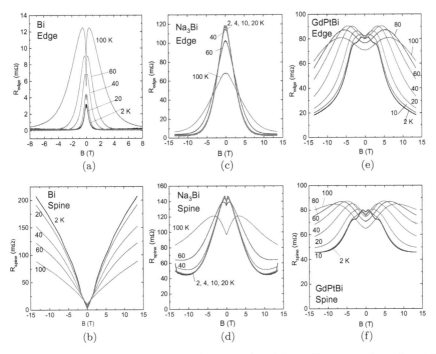

Fig. 5. The measured field profiles of R_{edge} (upper row) and R_{spine} (bottom row) at selected T in elemental Bi (Panels a and b), Na$_3$Bi (c and d) and GdPtBi (e and f). In Bi, the divergence between the field profiles in Panels a and b provides direct evidence for dominance of current jetting. In both Na$_3$Bi and GdPtBi, R_{edge} and R_{spine} decrease steeply with increasing B, implying that the intrinsic LMR dominates a sub-dominant current jetting distortion. The latter causes R_{spine} to lie higher than R_{edge}, and also leads to a weak upturn for $B > 8\,\text{T}$ (curves below 40 K in Panel d). From Liang *et al.* [19].

5c–5f). This confirmed that the negative LMR is intrinsic. However, even when current jetting effects are sub-dominant, they still lead to several observable effects. First, the curve of R_{spine} lies above R_{edge} at all B (compare 5c and 5d). Secondly, the competition between current jetting and the intrinsic LMR leaves an imprint on $R_{\text{spine}}(B)$. As B increases, R_{spine} first decreases steeply to a broad minimum, then it edges upwards at large B because the intrinsic contribution has saturated but current focussing continues to grow (see curves at and below 40 K in Fig. 5d). This non-monotonic profile is apparent in both Na$_3$Bi and GdPtBi. A third effect — the distortion of the angular spread of the axial current — is discussed below.

In Fig. 7a, we display the field profiles of R_{spine} (black curve) and R_{edge} (red curve) measured in Na$_3$Bi. Using a numerical deconvolution scheme, Liang *et al.* [19] extracted the intrinsic $R_{\text{int}}(B)$ (i.e. the curve with all current-jetting distortions removed) starting from the measured profiles of $R_{\text{spine}}(B)$ and $R_{\text{edge}}(B)$. The curve of $R_{\text{int}}(B)$ (blue curve) lies between them (but it is not just the average of the two outer curves). Now, with the distortions removed, we may regard $R_{\text{int}}(B)$ as a clean depiction of how the chiral anomaly alters the longitudinal resistivity in

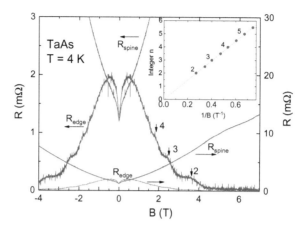

Fig. 6. Application of squeeze test to the Weyl semimetal TaAs at 4 K. For clarity, the curves of R_{edge} (blue) and R_{spine} (red) are plotted twice with different vertical scales (indicated by horizontal arrows). When B exceeds $B_{cyc} \sim 0.3$ T, R_{spine} increases steeply (upper red curve), maintaining its rising trend to 7 T (lower red curve). In contrast, R_{edge} falls monotonically, reaching values close to zero above 5 T. Shubnikov de Haas oscillations are observable in the blue curve. The corresponding Landau indices (2–5) are determined in the index plot (inset). The lowest LL is reached above 6 T. The divergent profiles are signatures of strong current jetting artefacts. Adapted from Liang et al. [19].

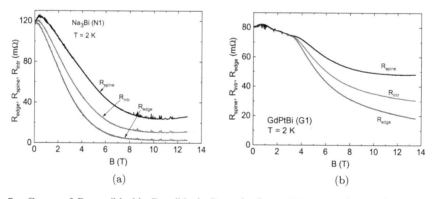

Fig. 7. Curves of R_{spine} (black), R_{int} (blue), R_{edge} (red) vs. B in Na$_3$Bi (Panel a) and GdPtBi (b). Using the measured R_{spine} and R_{edge}, the intrinsic profile $R_{int}(B)$ is obtained by numerical integration of the Laplace equation [19]. The curves R_{spine} and R_{edge} bracket R_{int} at all B. In Panel a, once B exceeds B_Q, R_{int} becomes independent of B. Adapted from Liang et al. [19].

Na$_3$Bi. When the 0LL is accessed ($B_Q = 5$–6 T) $R_{int}(B)$ saturates to a uniform "floor" value that is 10% of the initial value at $B = 0$. The axial current is roughly 10× larger than the Drude current in zero B. The corresponding curves for GdPtBi are shown in Fig. 7b.

Finally, we comment on the angular width of the LMR. Physical reasoning suggests that current jetting effects, even when subdominant, cause the angular width of the plume to be narrower than the expected form $\cos^2 \theta$. As an extreme example, we have found from LMR experiments on needle-like samples [22] that current jetting causes the observed angular width of the LMR plume to be less than 1°. In needle samples, the artefactual LMR becomes apparent only when **B** is exquisitely aligned with the needle axis; the current jet is able to travel the length of the needle without striking a side wall. A misalignment of ±1° suffices to squelch the artefactual LMR. Numerical simulations in a tilted **B** are more demanding. Nonetheless, to us, the observation of ultra-narrow angular widths of the LMR in long-needle samples raises serious concerns regarding artefactual LMR.

7. Perspective

In QFT, anomalies occur in many quantum phenomena that span a vast energy scale, extending from superfluid helium to energy scales in QED, QCD and gravitation physics [4–6]. In the recent LMR experiments on Dirac/Weyl semimetals, the thread of anomaly observations has been extended to topological materials. The prediction of NN [1] has now received confirmation in semimetals that feature symmetry-protected Dirac nodes. However, experiments involving the longitudinal MR require considerable care to preclude artefacts caused by strong current density inhomogeneities. When the carrier mobility is high, current jetting becomes dominant long before the lowest LL is accessed [19]. In these samples, the early onset of current jetting renders the LMR technique inapplicable. Even when the mobility is quite modest (as in Na$_3$Bi and GdPtBi), current jetting effects can still distort the intrinsic LMR signal. In these cases, the distortions can be simulated and removed to reveal the intrinsic negative LMR attributed to the chiral anomaly. The curve of the intrinsic LMR shows that the anomaly induced conductance can be 10× larger than the Drude conductance in zero B. When the chemical potential enters the lowest LL, the anomaly term saturates to a B-independent value [19]. Optical reflectivity and pump-probe experiments, which are unencumbered by current jetting effects, provide valuable alternate approaches that complement LMR experiments [7].

Acknowledgments

It is a pleasure to acknowledge the crucial contributions of my former students Jun Xiong, Max Hirschberger, Tian Liang, Jingjing Lin and Sihang Liang in the entire effort described here. I am also indebted to Bob Cava, Andrei Bernevig, Zhijun Wang, Jen Cano, Xi Dai, Liang Fu, Boris Spivak, Stephen L. Adler and Satya Kushwaha who patiently shared their expertise in diverse, often orthogonal, disciplines. The research was supported by the U.S. Army Research Office (Grant No.W911NF-16-1-0116) and the U.S. National Science Foundation (under MRSEC Grant No. DMR 1420541). I gratefully acknowledge generous support from the EPiQS program of the Gordon and Betty Moore Foundation (Grant GBMF4539).

References

1. H. B. Nielsen and M. Ninomiya, "The Adler-Bell-Jackiw Anomaly and Weyl Fermions in a Crystal," *Phys. Lett.* B **130**, 389-396 (1983).
2. Stephen L. Adler, "Axial-Vector Vertex in Spinor Electrodynamics," *Phys. Rev.* **177**, 2426 (1969).
3. J. S. Bell and R. Jackiw, A PCAC Puzzle: $\pi^0 \rightarrow \gamma\gamma$ in the σ-Model, *Nuovo Cimento* **60A**, 4 (1969).
4. *Introduction to Quantum Field Theory*, Michael E. Peskin and Dan V. Schroeder, (Westview Press, 1995), Ch. 19.
5. *Anomalies in Quantum Field Theory*, Reinhold A. Bertlmann (Clarendon Press, 2011).
6. *The Quantum Theory of Fields, Vol. II*, Steven Weinberg (Cambridge Univ. Press, 2005).
7. N. P. Ong and Sihang Liang, Experimental signatures of the chiral anomaly in Dirac-Weyl semimetals, *Nature Rev. Phys.* **3**, 394-404 (2021). DOI: 10.1038/s42254-021-00310-9.
8. N. P. Armitage, E. J. Mele, Ashvin Vishwanath, Weyl and Dirac semimetals in three-dimensional solids, *Rev. Mod. Phys.* **90**, 015001 (2018). DOI: 10.1103/RevModPhys.90.015001
9. Xiangang Wan, Ari M. Turner, Ashvin Vishwanath, and Sergey Y. Savrasov, Topological semimetal and Fermi-arc surface states in the electronic structure of pyrochlore iridates, *Phys. Rev.* B **83**, 205101 (2011). DOI: 10.1103/PhysRevB.83.205101
10. Liang Fu and C. L. Kane, Topological insulators with inversion symmetry, *Phys. Rev.* B **76**, 045302 (2007). DOI: 10.1103/PhysRevB.76.045302
11. S. M. Young, S. Zaheer, J. C. Y. Teo, C. L. Kane, E. J. Mele, and A. M. Rappe, Dirac Semimetal in Three Dimensions, *Phys. Rev. Lett.* **108**, 140405 (2012). DOI: 10.1103/PhysRevLett.108.140405
12. Chen Fang, Matthew J. Gilbert, Xi Dai, and B. Andrei Bernevig, Multi-Weyl Topological Semimetals Stabilized by Point Group Symmetry, *Phys. Rev. Lett.*, **108**, 266802 (2012). DOI: 10.1103/PhysRevLett.108.266802
13. Bohm-Jung Yang and Naoto Nagaosa, Classification of stable three-dimensional Dirac semimetals with nontrivial topology, *Nat. Commun.* 5:4898. doi: 10.1038/ncomms5898 (2014).
14. Zhijun Wang, Yan Sun, Xing-Qiu Chen, Cesare Franchini, Gang Xu, Hongming Weng, Xi Dai, and Zhong Fang, Dirac semimetal and topological phase transitions in A_3Bi (A = Na, K, Rb), *Phys. Rev.* B **85**, 195320 (2012). DOI: 10.1103/PhysRevB.85.195320
15. Zhijun Wang, Hongming Weng, Quansheng Wu, Xi Dai, and Zhong Fang, Three-dimensional Dirac semimetal and quantum transport in Cd3As2, *Phys. Rev.* B **88**, 125427 (2013). DOI: 10.1103/PhysRevB.88.125427
16. Jun Xiong, Satya K. Kushwaha, Tian Liang, Jason W. Krizan, Max Hirschberger, Wudi Wang, R. J. Cava, N. P. Ong, Evidence for the chiral anomaly in the Dirac semimetal Na_3Bi, *Science* **350**, 413 (2015). 10.1126/science.aac6089
17. Max Hirschberger, Satya Kushwaha, Zhijun Wang, Quinn Gibson, Sihang Liang, Carina A. Belvin, B. A. Bernevig, R. J. Cava and N. P. Ong, The chiral anomaly and thermopower of Weyl fermions in the half-Heusler GdPtBi, *Nat. Mater.* **15**, 1161 (2016). DOI: 10.1038/NMAT4684
18. J. Cano *et al.*, Chiral anomaly factory: Creating Weyl fermions with a magnetic field, *Phys. Rev.* B **95**, 161306 (2017). DOI10.1103/PhysRevB.95.161306
19. Sihang Liang, Jingjing Lin, Satya Kushwaha, Jie Xing, Ni Ni, R. J. Cava, and N. P. Ong, Experimental Tests of the Chiral Anomaly Magnetoresistance in the Dirac-Weyl

Semimetals Na_3Bi and GdPtBi, *Phys. Rev. X* **8**, 031002 (2018). DOI: 10.1103/Phys-RevX.8.031002

20. R. D. dos Reis, M.O. Ajeesh, N. Kumar, F. Arnold, C. Shekhar, M. Naumann, M. Schmidt, M. Nicklas and E. Hassinger, "On the search for the chiral anomaly in Weyl semimetals: the negative longitudinal magnetoresistance," *New J. Phys.* **18** (2016) 085006, doi:10.1088/1367-2630/18/8/085006

21. F. Arnold *et al.*, Negative magnetoresistance without well-defined chirality in the Weyl semimetal TaP, *Nat. Commun.* 7:11615 doi: 10.1038/ncomms11615 (2016).

22. Tian Liang, Jingjing Lin, Quinn Gibson, Satya Kushwaha, Minhao Liu, Wudi Wang, Hongyu Xiong, Jonathan A. Sobota, Makoto Hashimoto, Patrick S. Kirchmann, Zhi-Xun Shen, R. J. Cava and N. P. Ong, Anomalous Hall effect in $ZrTe_5$, *Nat. Phys.* **14**, 451 (2018). 10.1038/s41567-018-0078-z

Chiral "Graviton" and Fractional Quantum Hall Effect

Dung Xuan Nguyen

Brown Theoretical Physics Center and Department of Physics,
Brown University
Providence, Rhode Island 02912, USA

Dam Thanh Son*

Kadanoff Center for Theoretical Physics,
University of Chicago
Chicago, Illinois 60637, USA
* E-mail: dtson@uchicago.edu

We explain how the conservation laws satisfied by electrons on a single Landau level suggest that the lowest neutral excitation in a fractional quantum Hall liquid carries spin 2 or −2. We show how the spin of the magnetoroton can be determined by polarized Raman scattering. We also argue that there must be more than one magnetoroton in the Jeans sequences near $\nu = 1/4$.

Keywords: Fractional quantum Hall effect; magnetoroton; Raman scattering.

1. Introduction

The fractional quantum Hall effect [1, 2] is a remarkable physical phenomenon that occurs with two-dimensional electrons in a high magnetic field. At the microscopic level, the Hamiltonian describing this electron system is

$$H = \sum_a \frac{(\mathbf{p}_a + e\mathbf{A}_a)^2}{2m} + \sum_{\langle a,b \rangle} \frac{e^2}{|\mathbf{x}_a - \mathbf{x}_b|^2} \tag{1}$$

In real systems there are also terms corresponding to the interaction of the electrons with impurities. Despite the simplicity of eq. (1), a multitude of interesting ground states are realized with this Hamiltonian, corresponding to a large number of quantum Hall plateaus.

The fractional quantum Hall effect occurs when the filling factor ν, defined as the ratio of the number of electrons and the degeneracy of a Landau level, is fractional. In particular, when $\nu < 1$, the lowest Landau level is only partially filled, and without interaction between the electrons the ground state is exponentially degenerate. The interaction between the electron, thus, is decisive for determining the properties of the ground state. From experiment we know that, e.g., at $\nu = 1/3$ the ground state is gapped and at $\nu = 1/2$ is is ungapped.

In a gapped state the lowest charge neutral excitation is called the "magnetoroton." The magnetoroton was first studied variationally by Girvin, MacDonald, and Platzman [3] in 1986. The method was similar to what Feynman employed in his study of superfluid helium, with some modifications specific for the FQHE. Namely, they took the ground state $|0\rangle$ of the $\nu = 1/3$ state and acted on it the operator of electron density, projected onto the lowest Landau level, and use it as the variational ground state. The resulting spectrum is always gapped, but exhibits a minimum near a momentum of order of the inverse magnetic length. The magnetoroton excitation was observed experimentally by Pinczuk *et al.* using Raman scattering [4].

2. Conservation laws and the spin of the magnetoroton

There is a curious, but important fact pertaining to states on the lowest Landau level. Namely, the density-density correlation function (the imaginary part of which is usually called the dynamic structure factor $S(q, \omega)$) behaves like q^4 at small q and finite ω:

$$S(q, \omega) \to q^4 S(\omega), \quad q \to 0. \tag{2}$$

In particular, for gapped state $S(q, \omega)$ is nonzero only for ω larger than a finite gap, so the static structure factor $S(q) = \int_0^\infty d\omega \, S(q, \omega)$ is also $O(q^4)$. This fact was important in the GMP variational calculation for the energy of the magnetoroton to remain finite at $q \to 0$.

We recall that charge conservation requires only that $S(q, \omega)$ behaves like q^2 in the small q, fixed ω limit, so there must some additional reason for the q^4 behavior. Let us recall the connection between the behavior of the density-density correlator at small q and charge conservation. The conservation law for charge reads

$$\frac{\partial \rho}{\partial t} + \boldsymbol{\nabla} \cdot \mathbf{j} = 0 \tag{3}$$

In momentum space it means that $n = (\mathbf{q} \cdot \mathbf{j})/\omega$, so $\langle \rho\rho \rangle \sim q^2 \langle jj \rangle / \omega^2$. So in the limit $q \to 0$, ω =fixed, $\langle \rho\rho \rangle \sim q^2$.

For electrons in a magnetic field, one can write a second "conservation" law for momentum:

$$\partial_t(m j_i) + \partial_k T_{ik} = (\mathbf{j} \times \mathbf{B})_i \tag{4}$$

We have used Galilean invariance when writing the momentum density as $m\mathbf{j}$. The lowest Landau level limit can be taken as the $m \to 0$ limit. In this limit the above equation becomes one of balance of forces: $\partial_k T_{ik} = (\mathbf{j} \times \mathbf{B})_i$, which can be solved to express $\mathbf{j}$ in terms of the stress tensor T_{ij}. Schematically, $j \sim \frac{1}{B}\partial \cdot T$, and substituting this equation into the equation for charge conservation we find $\partial_t \rho + \frac{1}{B}\partial^2 T = 0$. That means $\rho \sim q^2 T/\omega$, hence $\langle \rho\rho \rangle \sim q^4 \langle TT \rangle / \omega^2$.

Now we note that the two-point function of ρ can be expanded as a sum over intermediate states:

$$\langle \rho\rho \rangle \sim \sum_n \langle 0|\rho|n \rangle \langle n|\rho|0 \rangle \tag{5}$$

Excitations with zero momentum ($q = 0$) can be classified by the spin, so one can assign to each branch of excitations the spin of that excitation at $q = 0$. Due to rotational invariance,

$$\langle q, s | \rho | 0 \rangle \sim (q_x + iq_y)^s \sim q^s \tag{6}$$

Hence $\langle \rho\rho \rangle \sim q^4$ can be satisfied if the intermediate states contributing to the two-point function has spin 2 or -2. This observation, together with some other arguments, lead Haldane to suggest [5] that the long-wavelength limit of the magnetoroton is a quantum Hall "graviton." In fact, one can write down a "bimetric" model in which the magnetoroton looks exactly like a massive graviton [6].

One can try to visualize the magnetoroton as bound states of charge quasiparticles (Laughlin's quasiholes and quasiparticles). According to the picture often drawn, a magnetoroton near the minimum of the dispersion curve is a pair of a quasiparticle and a quasihole, while the magnetoroton at zero momentum is a quartet of two quasiparticles and two quasiholes [7].

3. Sum rules

The argument above does not fix the *sign* of the spin of the magnetoroton: does it carry spin 2 or -2? To settle this question, one can appeal to an exact sum rule [8]. We introduce $T = \int d\mathbf{x}\, T_{zz}(\mathbf{x})$ and $\bar{T} = \int d\mathbf{x}\, T_{\bar{z}\bar{z}}(\mathbf{x})$, where T_{zz} and $T_{\bar{z}\bar{z}}$ are the spin-2 and spin-(-2) components of the stress tensor and their spectral densities,

$$\rho(\omega) = \frac{1}{N_e} \sum_n |\langle n | T | 0 \rangle|^2 \delta(\omega_n - \omega) \tag{7}$$

$$\bar{\rho}(\omega) = \frac{1}{N_e} \sum_n |\langle n | \bar{T} | 0 \rangle|^2 \delta(\omega_n - \omega) \tag{8}$$

$$\tag{9}$$

By definition the spectral sum rules are non-negative functions. The sum rule reads

$$\int\limits_0^\infty \frac{d\omega}{\omega^2} [\rho(\omega) - \bar{\rho}(\omega)] = \frac{\mathcal{S} - 1}{8} \tag{10}$$

where $\mathcal{S}$ is the shift, which is a topological property of a quantum Hall state. For example, the shift of the Laughlin $\nu = 1/3$ state is $\mathcal{S} = 3$. In this case the sum rule implies that spin-2 excitations (ρ) is preferred over spin-(-2). If one assume, following GMP [3], that there is only one single magnetoroton (the "single-mode approximation"), then the spin of this mode has to be $+2$. On the other hand, the shift of the state $\nu = 2/3$, which is the particle-hole conjugate of the $\nu = 1/3$

Laughlin state, is zero. In Ref. [9] it was showed that for the "model" Hamiltonians one of the spectral densities is identically zero.

4. Determining the spin of the magnetoroton experimentally

We do not have a spin-2 probe like a graviton to create a magnetoroton from the ground state. However, one can still probe the spin-2 magnetoroton experimentally by using polarized Raman scattering. In Raman scattering, an incident photon with energy ω falls perpendicularly onto the electron layer, then from the system a photon with a slightly smaller energy ω' comes out. The difference between the energies of the incoming and outgoing photons $\Delta = \omega - \omega'$ is the energy of excitation leaved behind. But Raman scattering can determine not only the energy of the magnetoroton, but also its spin. Recall that a photon with circular polarization carries spin pointing either along or opposite to the direction of its momentum; in a Raman scattering experiment, a circularly polarized photon can flip the direction of its spin and transfer an angular momentum ± 2 to the system under study. If at a given energy we have a magnetoroton with spin of a certain sign, then only one direction of photon spin flip can occur at that energy.

One immediate issue with the above proposal is that it relies on rotational invariance. In the case of real GaAs, the lattice respects only the discrete C_4 symmetry, which means angular momentum is preserved only mod 4 and spin $+2$ and -2 cannot be distinguished. The issue was carefully analyzed in Ref. [10]. Experiments are usually conducted in the regime of resonant Raman scattering where the frequency of light is tuned to the band gap between the conduction and valence zones. In Ref. [10] it was shown that in resonant Raman scattering, if the frequency detuning is larger than the energy scales of the Hall effect, one can "integrate" out the trivial physics of the band structure by using the Luttinger Hamiltonian and arrive at a formula relating the intensity of the Raman scattering with the matrix element of a certain lowest-Landau-level operator. The result can be summarized by an effective coupling of the Raman photons with the FQH degrees of freedom:

$$E^z E^z (T'_{zz} + 0.16 T'_{\bar{z}\bar{z}}) \tag{11}$$

where T' is a tensor that is similar to the stress tensor, though not identical to it. In terms of the density operator projected to the lowest Landau level ρ,

$$T_{zz} = \sum_{\mathbf{q}} \frac{q_z^2}{q} \frac{\partial}{\partial q} e^{-q^2 \ell_B^2 / 2} V(q) \rho(\mathbf{q}) \rho(-\mathbf{q}) \tag{12}$$

The coefficient 0.16 in Eq. (11) comes from the numerical values of the Luttinger parameters in the Luttinger Hamiltonian. Equation (11) tells us that Raman scattering is predominantly spin-conserving; the probability of spin change by 4 is suppressed by $0.16^2 \approx 1/40$. Thus, Raman scattering can be used to determine the spin of the magnetoroton.

5. Possible applications of Raman scattering: the $\nu = 5/2$ quantum Hall plateau

One area where Raman scattering may have an important role to play is in the determination of the nature of the $\nu = 5/2$ state. There have been many theoretical proposals, including the original proposal of the Pfaffian (or Moore-Read) state, the anti-Pfaffian state (which is the particle-hole conjugate of the Pfaffian state), and the recently proposed PH-Pfaffian state. Numerical calculations generally favor the anti-Pfaffian state while some experimental results seem to favor the PH-Pfaffian. We can use the sum rule to determine the likely spin of the lowest-energy magnetoroton in each of these candidate states. The Pfaffian state has $\mathcal{S} = 3$, so according to the sum rule (10) spin-(+2) modes dominate over spin-(−2) modes. The situation is reversed for the anti-Pfaffian state ($\mathcal{S} = -1$). For the PH-Pfaffian state $\mathcal{S} = 1$, so one expects both $s = 2$ and $s = -2$ magnetorotons.

6. Multiple magnetorotons near $\nu = 1/4$

The story of the magnetoroton near $\nu = 1/4$ turns out to be more complicated that the situation near $\nu = 1/2$. From the original proposal by Girvin, MacDonald, and Platzman, the implicit assumption has been that the response functions of a FQH states are dominated by one excitation—the so-called "single-mode approximation." This is a reasonable approximation for Jain's states near half filling ($\nu = N/(2N+1)$ and $\nu = (N + 1)/(2N + 1)$), supported by both composite fermion theory and numerical simulations. However, it turns out that this is not a good approximation for Jain's states near $\nu = 1/4$ ($\nu = N/(4N+1)$ and $\nu = N/(4N-1)$). One indication of the problem has arisen from the composite fermion theory near $\nu = 1/4$: the projected structure factor fails to satisfy the Haldane bound. The Haldane bound states that for gap FQH states, the coefficient s_4 of the leading small-q asymptotics of the projected structure factor $s_4 q^4$ is bounded from below

$$s_4 \geq \frac{\mathcal{S} - 1}{8} \tag{13}$$

In the composite fermion theory, one can compute s_4 for the $\nu = N/(4N + 1)$ state. One obtains $s_4 = (N + 1)/8$ [11]. On the other hand, the right-hand side of the inequality (13) is however $(N + 3)/8$, thus the Haldane bound is violated. In order to solve this problem, Ref. [11] postulates the existence of an additional magnetoroton with energy comparable to the energy scale of Coulomb interaction. This mode contributes to the projected structure factor and brings s_4 to a value consistent with the Haldane bound. This extra high-energy magnetoroton mode has recently been seen in numerical simulations [12, 13].

7. Conclusion

To summarize, the lowest neutral excitation of a fractional quantum Hall liquid is the magnetoroton. At zero momentum, the magnetoroton can have spin +2 or −2, depending on the FQH state in question. The $q = 0$ magnetoroton thus can be considered a chiral massive "graviton" of the FQHE. Polarized Raman scattering provides an experimentally accessible "gravitational" probe of the FQH states and, in principle, can help distinguishing different topological orders, for examples, different candidates for the $\nu = \frac{5}{2}$ plateau. The Jain's states near $\nu = 1/4$ contain two, instead of one, magnetorotons.

References

1. D. C. Tsui, H. L. Stormer and A. C. Gossard, Two-Dimensional Magnetotransport in the Extreme Quantum Limit, *Phys. Rev. Lett.* **48**, 1559 (1982).
2. R. Laughlin, Anomalous Quantum Hall Effect: An Incompressible Quantum Fluid with Fractionally Charged Excitations, *Phys. Rev. Lett.* **50**, 1395 (1983).
3. S. M. Girvin, A. H. MacDonald and P. M. Platzman, Magneto-roton theory of collective excitations in the fraction al quantum Hall effect, *Phys. Rev. B* **33**, 2481 (1986).
4. A. Pinczuk, B. Dennis, L. Pfeiffer and K. West, Light scattering by collective excitations in the fractional quantum Hall regime, *Physica B: Condens. Matter* **249-251**, 40 (1998).
5. F. D. M. Haldane, Geometrical Description of the Fractional Quantum Hall Effect, *Phys. Rev. Lett.* **107**, p. 116801 (2011).
6. A. Gromov and D. T. Son, Bimetric Theory of Fractional Quantum Hall States, *Phys. Rev. X* **7**, p. 041032 (2017).
7. S.-C. Zhang, The Chern-Simons-Landau-Ginzburg theory of the fractional quantum Hall effect, *Int. J. Mod. Phys. B* **6**, 25 (1992).
8. S. Golkar, D. X. Nguyen and D. T. Son, Spectral sum rules and magneto-roton as emergent graviton in fractional quantum Hall effect, *J. High Energy Phys.* **1601**, p. 021 (2016).
9. D. X. Nguyen, D. T. Son and C. Wu, Lowest Landau Level Stress Tensor and Structure Factor of Trial Quantum Hall Wave Functions (11 2014).
10. D. X. Nguyen and D. T. Son, Probing the spin structure of the fractional quantum Hall magnetoroton with polarized Raman scattering, *Phys. Rev. Res.* **3**, p. 023040 (2021).
11. D. X. Nguyen and D. T. Son, Dirac composite fermion theory of general Jain sequences, *Phys. Rev. Res.* **3**, p. 033217 (2021).
12. D. X. Nguyen, F. D. M. Haldane, E. H. Rezayi, D. T. Son and K. Yang, Multiple Magnetorotons and Spectral Sum Rules in Fractional Quantum Hall Systems (11 2021).
13. A. C. Balram, Z. Liu, A. Gromov and Z. Papić, Very high-energy collective states of partons in fractional quantum Hall liquids (11 2021).

Topology and Chirality

C. Felser

Max Planck Institute for Chemical Physics of Solids,
01187 Dresden, Germany
E-mail: Claudia.Felser@cpfs.mpg.de
www.cpfs.mpg.de

J. Gooth

Max Planck Institute for Chemical Physics of Solids,
01187 Dresden, Germany
E-mail: Johannes.Gooth@cpf.mpg.de
www.cpfs.mpg.de

Topology, a well-established concept in mathematics, has nowadays become essential to describe condensed matter. At its core are chiral electron states on the bulk, surfaces and edges of the condensed matter systems, in which spin and momentum of the electrons are locked parallel or anti-parallel to each other. Magnetic and non-magnetic Weyl semimetals, for example, exhibit chiral bulk states that have enabled the realization of predictions from high energy and astrophysics involving the chiral quantum number, such as the chiral anomaly, the mixed axial-gravitational anomaly and axions. The potential for connecting chirality as a quantum number to other chiral phenomena across different areas of science, including the asymmetry of matter and antimatter and the homochirality of life, brings topological materials to the fore.

Keywords: Topology, Chirality.

1. Introduction

Electronic properties of solids play a central role in our everyday life. One recent research area, topology, became a major direction in condensed matter physics, solid state chemistry and materials science. It has led to a fundamental new understanding of solids mainly due to relativistic effects in compounds made of heavier elements [1]. Before topology in 2005 turned to the center stage, it was assumed that the electronic properties of superconductors, metals, insulators and semiconductors are completely described by the energy-momentum relations of the electrons in them, the so-called "band structures", which in turn are defined by the symmetries of the underlying crystal lattice of the host solid. However, this description turned out to be incomplete with the prediction of the quantum spin Hall effect in 2005 [2, 3], which may be viewed as the quantum Hall effect without an external magnetic field, but with strong spin-orbit coupling instead. In the Quantum Hall regime, that is when electrons are restricted to two dimensions and are exposed to a strong magnetic field, condensed matter systems can exhibit completely different electrical properties without any additional symmetries being broken [4]. Due to the quantization of electronic states in the magnetic field, such a system has an energy gap at the Fermi energy. Despite this gap, this state is not a conventional insulator, but has metallic chiral edges and a quantized Hall conductivity. Observing

116

Fig. 1. (a) Electrons are chiral, if spin and momentum are locked. (b) Topology is a simple concept, dealing with the surfaces of objects. The topology of a mathematical structure is identical if it is preserved under continuous deformation. A pancake has the same topology as a cube, a donut as a coffee cup and a pretzel as a board with three holes.

such a system for the first time, the German physicist Klaus von Klitzing received the Nobel Prize in 1985. This was the beginning of the topological classification of solids. However, it took more than 25 years and new predictions [2, 3] before the concept of topology became a main stream direction in solid state research and was associated to an intrinsic material property: the Berry curvature — a quantum mechanical property that is related to the phase of electrons wavefunction and is particularly relevant in condensed matter systems exposed to strong magnetic fields or strong spin-orbit coupling. Meanwhile, all non-magnetic inorganic compounds are characterized by the topology of their band structures based on group and graph theory based on a single particle picture [5, 6] as well as some of the most import antiferromagnetic materials classes [7]. Surprisingly, it was found that more than 20% of all materials that we know today have electronic properties that are governed by the topology of their electron wave functions, i.e. Berry curvature effects.

One of the most intriguing consequences of the topological description of materials are chiral states on the bulk, surfaces and edge of topological condensed matter systems. This intimate connection goes even far beyond the characterization of electron states and comprises today (quasi-)particle states in condensed matter systems, such as chiral electron states, chiral photons, chiral magnons, chiral plasmons, chiral spins, to name a few of them. In the context of electrons in solids, chirality χ is defined in reciprocal (momentum) space as the handedness of the spin of the electrons relative to their direction of motion. Right-handed electrons have a chirality of $\chi = -1$ and left-handed electrons have a chirality of $\chi = +1$, see Fig. 1a.

The concept of chirality is an overarching theme in physics, chemistry and biology, permeating much of modern science. It links the properties of the universe and its constituent elementary particles, through organic stereochemistry, to the structure and behavior of the molecules of life, with much else besides (chemical crystallography, chiroptical spectroscopy, nonlinear optics, nanoscience, materials, electrical engineering, planar plasmonic metamaterials, spintronics, molecular motors, pharmaceuticals, astrobiology, origin of life, etc.). However, the connections between the different concepts of chirality (an overview over a few definitions of chirality in various contexts is given in box 1) all over the sciences remains elusive.

This article gives the authors' personal perspective on how chirality in topological materials is potentially related to chirality as a general concept.

Keypoints and definitions in the context of topology and chirality:

Chiral object: has a geometric property of being non-superposable on its mirror image; such an object has no symmetry elements of the second kind (a mirror plane, a center of inversion, a rotation-reflection axis) [8].

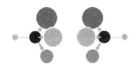

Chiral molecules: are chiral, if the point group contains only symmetry operations, of the first kind (rotation and translation). Both enantiomers consist of identical chemical compositions but are distin-guished in their light–matter interaction and catalytic reactivity. Chiral molecules can crystallize in chiral and non-chiral crystal structures [9].

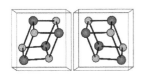

Chiral space groups: contain symmetry operations of the first kind. There are 11 pairs of enantiomorphic space groups (e.g. $P6_1$ and $P6_5$) which are chiral. 43 achiral space groups can host a chiral crystal structure. A crystal structures in space group $P2_13$ for example, both enantiomeric crystals crystallize in the same space group.

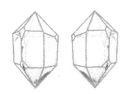

Chiral crystals: is an inorganic or organic material crystallizing in one of the 65 Söhncke symmetry space group. If the space group contains only proper operations, the crystal is chiral and obey a well-defined handedness. Of all inorganic compounds in the inorganic database ICSD approximately 20% are chiral [16]. Recently, it was observed that spin-polarized currents in chiral crystals can propagate over tens of micrometers [10].

Chiral crystal surface: a crystal with a surface is periodic in only in two dimensions, one consequence is that the inversion symmetry is lost. Surfaces of chiral crystals are intrinsically chiral, achiral crystal have chiral surfaces, if its surface normal does not lie in any of the mirror planes of the bulk crystal lattice; and are typically high-Miller-index surfaces [11].

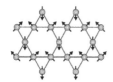

Chiral magnetic structure: Magnetism breaks time reversal symmetry. In helical magnets inversion symmetry is broken, these magnets are abundant, in metals alloys, semiconductors and multiferroics. Mn_3Z (Ge, Sn, Ir, Pt) were identified to be non-collinear antiferromagnets with a chiral spin arrangement and a large anomalous Hall effect (AHE) arising from the topologically non-trivial spin texture [12–14].

Topology: is a mathematical description of the properties that are preserved through deformations. The concept in condensed matter includes symmetry, and relativistic effects such as spin orbit coupling (SOC) and band inversion [1]. All chiral crystals with large SOC are Kramers – Weyl Fermions, with monopols and antimonopols. B20 compounds have a maximal Chern equal 4 [15–17].

Berry phase: is a geometric phase in a quantum mechanical system, when the system does not return to its initial. It gives rise to observable effects such as the anomalous Hall effect and the orbital magnetization [18]. A new concept proposes a connection of the sign of the Berry curvature with the orbital angular momentum around Weyl points in semimetals and in chiral new Fermions [19].

Spin momentum locking: can be observed in topological insulator surface states, the spin is locked at a right angle to their momentum [20]. In chiral systems (chiral molecules) the spin is coupled to the electron linear momentum, the origin of the CISS effect [21].

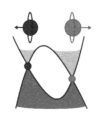

Chiral anomaly: is subject of research in high-energy, condensed matter, and nonequilibrium physics via parity-breaking of chiral currents along a magnetic field [22]. It can be broken in a quantum world, in a quark-gluon plasma created in heavy-ion collisions, Floquet systems, and non-Hermitian systems. In solids the chiral anomaly can be observed in Weyl systems, when an electric field is applied parallel to a magnetic field, and charges are pumped between two Weyl points of opposite chirality [23–25].

 **Chiral quasiparticle:** The quasiparticle concept was developed by Landau to describe Fermions (electrons or holes) interacting with other particles. Bosons (phonons or plasmons) are named collective excitations. Quasiparticles or collective excitations, which behave more like non-interacting particle are easier to describe. Graphene can be described by chiral quasiparticles [26].

2. Topology

Topology is the branch of mathematics that deals with the deformation of objects. Objects that can be continuously transformed into one another belong to a certain category — they are said to have a certain topological order, which is characterized by a so-called topological invariant. In contrast to this, objects that can only be reshaped by cutting or breaking into one another have different topological orders. A classic example of topological classification is by the number of holes in an object. A cup and a donut have the same topological order, but a Brezel has a different one, see Fig.1 b. Such objects can also be quantum mechanical. Electron states can be classified topological on the basis of a property that is the sum of the so-called "Berry curvature" along a closed path in momentum space. Admittedly, it is a very abstract quantity that sounds quite constructed at first. The "Berry curvature" is a quantum mechanical property that is closely linked to the phase of the electron wave function, which an electron picks up after having traveled a closed path when it has returned to the origin. This phase is called the "Berry Phase" [18]. It may sound a little intuitive that the property of an object changes when it returns to its starting point after having traveled a closed path, but systems with such properties also exist in the classical world. A well-known example is the Foucault pendulum, whose direction of oscillation deviates after one day from the value of the previous day. The same applies to the "Berry Phase", which defines the topology of the electronic states in a solid. Solid-state crystals have the same topological order if the sum of the "Berry curvature" of their electrons along a closed path in momentum space is the same and solid-state crystals, for which this sum is different, have a different topological order. It turns out that different topological orders in solid-state crystals, despite their abstract nature, are associated with different optical and electrical properties that can actually be observed in experiments. Historically, it was believed that only a change in the system's symmetry (that is the lattice symmetry or time-reversal symmetry) can lead to different optical and electrical properties. The significance of these different properties for the interpretation of the quantum Hall effect described above for example is that electronic states of different topological orders can merge into one another by "breaking" the band structure by means of the magnetic field without breaking an additional symmetry in the solid-state system. In other words, the quantized Hall conductivity in a two-dimensional electron system that is exposed to a strong magnetic field represents the topological invariant of the system and thus marks the topological order. A two-dimensional quantum Hall system is therefore also called a topological insulator. A characteristic property of topological insulators is the presence of metallic, i.e. dissipationless, chiral states at the sample boundary, in which the spin and momentum of the electrons is locked. Such states always occur at the spatial interface between regions that are in different topological orders. The easiest way to understand this is to note that different topological orders in insulators are associated with the parity of band gap, that defines which particular band is the conduction and which is the valance band. Imagine now a smooth boundary between two systems of opposite band parity (in one of the systems band A is the conduction band and band B is the valance band, and in the second system band B is the conduction band and band A is the valance band). At this boundary, the band structure slowly interpolates as a function of position between the two systems. Somewhere along the way, the energy gap has to disappear; otherwise both sides would be in the same class. Dissipationless chiral states are therefore bound to the interface. The surface of a quantum Hall system or a solid-state crystal can be viewed as the interface to the vacuum, which, like a conventional insulator, belongs to the trivial topological class. This guarantees the existence of gapless states on the surface (or edge) of a topological insulator. One of the most important discoveries in recent years is that topological order also occurs in some three-dimensional (3D) materials. In these materials the role of the magnetic field of a 2D-qantum Hall system is taken over by the mechanism of spin-orbit coupling, like in the 2D quantum spin Hall systems. These materials have been called 3D topological insulators, because they are insulators inside them, but because of the topological order they have exotic chiral 2D metallic surface states. Topological states of matter are so far identified via a simple single electron picture, which explains also the fast success of the field in condensed matter physics.

Recently the 3D-Quantum Hall effect was observed in a single crystal of $ZrTe_5$ [27]. Bertrand Halperin had proposed long ago in 1987 that it should be possible to realize the QHE in a three-dimensional semimetal or doped semiconductor with a particular instability in the Fermi surface [28]. The considerable challenge was to realize a single crystal, which fulfills all necessary conditions for such a 3D QHE: namely, high mobility, an adjusted charge carrier concentration and impurity level with the Fermi level tuned to be in an energy gap to the applied magnetic field. In two-dimensional electron gas (2DEG) systems, the charge carrier concentration, the mobilities and the Fermi energy can be varied by gating. Tantalizing signatures of the 3D QHE were seen in step-like anomalies in the Hall conductivity in the extreme quantum limit in Bismuth and graphite, and in some semimetallic compounds with the extreme quantum limit reached at low magnetic fields. However, no convincing plateaus as in the classical 2DEG nor finite conductivity perpendicular to the QHE

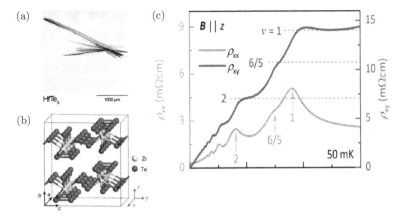

Fig. 2. Figure according to [29], (a) HfTe$_5$ crystals (b) ZrTe$_5$ and HfTe$_5$ crystal structure and (c) Longitudinal electrical resistance ρ_{xx} (left axis) and Hall resistance ρ_{xy} (right axis) as a function of B at 2 K with B applied along the b-axis of single crystal HfTe$_5$.

were realized. Last year, however, strong evidence for a 3D QHE was found in single crystals of ZrTe$_5$ in small magnetic fields of just 2 Tesla. The electron density modulation in the direction of the magnetic field of the ZrTe$_5$ single crystals was accounted for by a charge density wave instability. The results were soon reproduced by other groups and, moreover, evidence for a 3D version of the fractional QHE was seen in single crystals of HfTe$_5$, see Fig. 2 [29].

As already noted, although related, trivial and topological materials cannot be distinguished directly from the band structure itself. This requires the information from which atomic orbitals the individual bands come from. From a chemical point, in trivial insulating and semiconducting materials, the s-electrons usually form the conduction band, while the p-electrons form the valence band (Fig. 3 a). If the materials consist of heavy elements (atomic number $Z > 32$), the bands of the outer s-electrons are lower in energy compared to light elements, so that they can appear below or just at the Fermi energy. This can result in the minimum of the 5 s- or 6 s-bands of the electrons being below the maximum of the p-bands of the electrons, and s- and p-bands to intersect at points of intersection or nodal lines in three dimensions. One then speaks of band inversion. However, due to strong spin-orbit coupling, such intersection points and node lines can be forbidden and split (Fig. 3b). An energy gap is created again, with parts of the s-electrons now forming the conduction band, while parts of the p-electrons also form the valence band. With regard to the band gaps of trivial isolators, one speaks in this case of a negative or inverted band gap. Materials that have such inverted band gaps are topological insulators (TI). The inversion of the 6s-states of Hg/Bi and 5p-states of Te in HgTe/Bi$_2$Te$_3$ are prominent examples of such inverted bands with a negative band gap. The electrons are close to the nucleus and do not like to be ionized. Examples of topological elements are bismuth from group 5, as well as phosphorus.

Phosphorus can have a formal valence of +5 in phosphates, for example, while bismuth only occurs as Bi^{3+}, the outer s-electrons are not available as valence electrons. Many of the heavy element compounds exhibit unusual electronic and optical properties due to the topology of their electronic structure, e.g. one finds extremely high values or anomalies in certain experiments that can only be explained with the help of the topology. In fact, one can find metallic surfaces with three-dimensional topological insulators (Fig. 3 b, right side). In this case it is the chiral two-dimensional surfaces of the solid-state crystal that arise for the same reasons as for quantum Hall systems.

3. Weyl semimetals

So far, we have limited our discussion to isolators and semiconductors. However, the electronic structure of a solid can also have crossings of inverted bands, which are stable if they are not prohibited for reasons of symmetry. Such material systems are topological semimetals, in which exotic properties are derived from the crossing points of the electron bands (Fig. 3 c). Ordinary metals and semimetals mostly always have a curved — mostly parabolic — dispersion, which describes non-relativistic electrons with finite rest mass. In contrast to this, topological semimetals in the vicinity of their crossing points often show a linear relationship between energy and momentum. Such a linear dispersion describes relativistic electrons

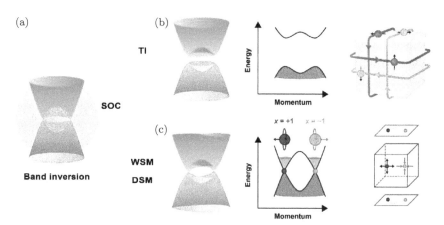

Fig. 3. Figure according to [31], displaying the topological classification of solids. The conduction band often s-electrons are colored yellow and those of the valence band often p-electrons are blue. (a) The band inversion, overlap between conduction and valence electrons, leads to a nodal line. Strong spin-orbit coupling can lead to the splitting of the bands (b). The resulting band gap is inverted and represents a topological isolator (TI), sketch of the surface states, connecting valence and conduction band (dotted lines) and spin–momentum locked surface electrons for the two spin directions on the surface of a crystal (c) sketch of a Dirac semimetal (DSM) and a Weyl semimetal (WSM). The band structure of a Weyl semimetal consists of pairs of linear crossing points with chirality $\chi = +1$ and $\chi = -1$. (b) These crossing points behave like magnetic monopole and anti-monopole in momentum space, from which special topological ones Form surface conditions, so-called Fermi arcs, which connect the Weyl crossing points.

without rest mass, which move at about a thousandth of the speed of light — just like the elementary particles in high-energy physics.

Topological semimetals with 4-dimensional crystal symmetry are called Dirac semimetals (DSM, Fig. 3 c). Graphene, a layer of carbon atoms in graphite, is the most prominent 2D example of such a material, but they also exist as three-dimensional materials. The electrons in such Dirac semimetals the energy-momentum relation is linear and the electrons behave like relativistic Dirac Fermions without rest-mass. This is in contrast to normal metals, which exhibit parabolic energy-momentum relations that describe massive electron (quasi-)particles. About 5 years ago, a new class of materials with new topological properties was identified — the so-called Weyl semimetals (WSM, Fig. 3 c), named after the physicist Hermann Weyl. At first glance, the electronic structure of a Weyl semimetal looks like that of a Dirac semimetal, with intersecting linear bands in the electronic structure of the solid. However, Dirac and Weyl semimetals differ in their symmetry and therefore also in their properties. A Dirac semimetal is centro-symmetrical, while a Weyl semimetal has a lower two-dimensional symmetry. It has no inversion center in the crystal structure — which breaks the inversion crystal symmetry, or a ferromagnetic order — which breaks the time-reversal symmetry. In contrast to Dirac semimetals, Weyl semimetals exhibits pairs of non-degenerate band crossings. The electrons at these crossing points behave like a special solution of the Dirac equation for massless particles — the so-called Weyl Fermions [22, 30, 31]. Weyl Fermions are characterized by the unique property that their intrinsic angular momentum or spin is inextricably linked with their linear momentum. What is special about Weyl materials is that in each pair of their crossing points, there is one crossing point with left and one crossing point with right-handed electrons. As such, the crossing points in a Weyl semimetal behave like magnetic monopole and anti-monopole in momentum space (Berry phase). The chiral volume properties result in special topological surface states, the Fermi arcs that connect the Weyl points (Fig. 3 c).

The relativistic equations that describe these linear dispersions have one mathematically inevitable feature: negative energy solutions (Fig. 4). The interpretation of these filled vacuum states of negative energy in high-energy physics has always been a bit controversial, as it contradicts our intuition of an "empty" vacuum. However, their physical reality is beyond question, as one of their direct consequences — the existence of antimatter — is confirmed by experiments. On the other hand, filled states of negative energy are a very natural concept from the point of view of solid-state physics: filled valence bands in the electronic structure of topological semimetals. The upper cone is the conduction band and represents the analogue to matter, the lower one is the valence band and represents the antimatter analogue.

As an example of a Weyl semimetal from our work, we discuss briefly NbP, a Weyl semimetal, which is characterized by breaking the inversion symmetry. We found all the typical characteristics of a Weyl semimetal in single crystals of NbP. For example, some electrons behave as if they were almost massless, which leads to extremely high mobility and a large magneto resistance. The key signature of a Weyl

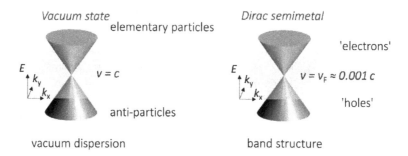

Fig. 4. In a quantum system the charge parity can be broken (charge parity violation). Comparison of the vacuum state with an imbalance between particles and anti-particles and a semimetal with an imbalance of electrons and holes, as a consequence of the chiral anomaly.

semimetal is evidence of a Fermi arc. As with a camera, the electronic structure and thus the sought-after Fermi arc can be photographed using angle-resolved photo emission (ARPES) [32, 33].

4. Chiral and axial-gravitational anomaly in Weyl semimetals

Classically, the chirality is a strictly conserved physical quantity — such as angular momentum, energy or electrical charge. Chirality must be preserved in the sum, *i.e.* there cannot simply be more particles of one chirality than of the other. In the context of an accelerator experiment in the 1970s, however, it was discovered that the conservation law of chirality is broken in parallel electric and magnetic fields at the quantum level. The observed decay of a neutral pion into two photons should actually be prohibited by the conservation law of chirality, but it turned out that this is not the case. Even if, in this case, it is not a matter of massless Fermions, but of bosons, the chirality is defined — however, as for massless Fermions, it can no longer be equated with the handedness of the particles. In 1969, this "chiral anomaly" was explained by the theorists Stephen Adler [34] as well as John Stewart Bell and Roman Jackiw [35] independently of one another by going from the classical to a quantum field theoretical description. Therefore, the chiral anomaly is now also known as the Adler-Bell-Jackiw anomaly. In the standard model of particle physics, certain proportions of the particle-antiparticle asymmetry are believed to be attributed to such quantum anomalies, the violation of classical conservation laws due to quantum fluctuations. The chiral anomaly, in particular, is said to play an important role in the standard model of particle physics, but Weyl Fermions have not yet been detected as elementary particles. In addition, an underlying curved spacetime is predicted to make a significant contribution to the chiral imbalance. This effect is known as the mixed axial-gravitational anomaly. In extreme gravitational fields, which correspond to a strongly curved spacetime, the axial-gravitational anomaly might thus also contribute to the particle-antiparticle asymmetry.

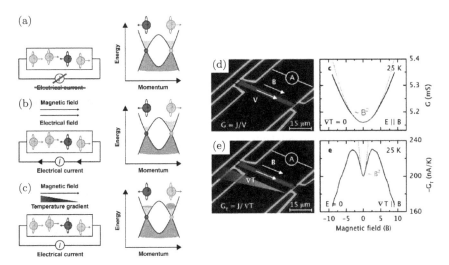

Fig. 5. Chiral and axial-gravitational anomaly in Weyl semimetals. (a) Circuit sketch (right) and band structure diagram of a Weyl semimetal (right) without a magnetic field. (b) Circuit sketch (right) and band structure diagram of a Weyl semimetal (right) in a parallel electric and magnetic field. (c) Circuit sketch (right) and band structure diagram of a Weyl semimetal (right) in parallel temperature gradient and magnetic field Figure (a-c) according to [36]. Figure (d) and (e) according to [38]. Experiments on the chiral and axial-gravitational anomaly in the Weyl semimetal NbP. Colored optical recordings of the measuring arrangement of the electric current I as a response to the electric (V) and magnetic field (B). (b) Magnetic field dependence of the electric current for different angles (color legend) between V and B. (d) Colored optical recordings of the measuring arrangement of the electric current I as a response to the temperature gradient (Δ T) and magnetic field (B). The red and green ends of the color gradient indicate the hot and cold side of the device. (e) Magnetic field dependence of the electric current for different angles (color legend) between Δ T and B.

In Weyl semimetals, the conservation law of chirality leads to a suppression of the electric current. This can be understood using a simple Weyl band structure: Figure 5 a shows a pair of Weyl crossing points on the right side, with the left crossing point representing electrons moving to the left and the right crossing point moving to the right — represents moving electrons. The spin of the electrons is the same for both crossing points. Every electron at the left crossing point has a chirality of $\chi = +1$ and every electron at the right crossing point has a chirality of $\chi = -1$. In thermal equilibrium, *i.e.* when no electrical energy flows through the Weyl semimetal, the Fermi energy (the filling level of the cones) is the same in both Weyl cones. There are just as many electrons moving to the right as there are electrons moving to the left — so there is no net charge transport and, therefore, just as many right-handed as left-handed Weyl Fermions. The chirality of the entire system is zero. If you apply an electric field to a normal metal, e.g. from left to right, an electric current flows from left to right, because there are more electrons moving to the right. But not so in a Weyl semimetal. A current flow from left to right would mean a higher filling level in the right Weyl cone than in the left and thus to the generation of more right-handed Weyl Fermions in the semimetal than left-handed. However, according to the conservation law, this is prohibited. Only the application of an additional magnetic field parallel to the electric field, which is so high that it quantizes the electron states (Figure 5b), leads, analogously to high-energy physics, to the breaking of the law of conservation of chirality and enables a constant redistribution of electrons between the left and right Weyl cones. This induced breaking of the chiral symmetry is a macroscopic manifestation of the chiral anomaly in the relativistic field theory and leads in Weyl semimetals to a positive longitudinal magnetic field-dependent electric current [36]. However, care has to be taken in experiments, because a positive magnetoconductance can also be cause by extrinsic effects, such as by a magnetic field-induced inhomogeneous current distribution inside the sample [37]. Samples with lower mobilities, well defined shape and additional cross-check experiments, like probing the mixed axial-gravitational anomaly in thermal and thermoelectric experiments performed under open electrical circuit conditions (no net charge current flow) are required to exclude such extrinsic effects. One might think that since the gravitational fields on our earth are relatively weak and experiments on Weyl semimetals always take place in relatively flat spacetime, that the mixed axial-gravitational anomaly does not play a role in solid-state physics. In fact, however, the mixed axial-gravitational anomaly leads to measurable effects in magnetic field-dependent thermal and thermoelectric experiments through a "back door". You have to know that temperature gradients in relativistic systems act like a gravitational field. This goes back to the American theorist Luttinger, who introduced statistical quantities such as entropy and temperature into quantum field theory in 1964. At a similar time, Tolman and Ehrenfest, who argued that temperature is not constant is space at thermal equilibrium, but varies with curved space time. Simply put, one can say that mass and energy are the same in relativistic systems. As is well known, masses are moved in gravitational fields and energy as heat in temperature gradients. Consequently, temperature gradients in relativistic systems represent analogues for gravitational fields. If a temperature gradient is applied to a normal metal, e.g. from left (warm) to right (cold), an electric current flows from left to right because the hot electrons diffuse from left to right. But not so in a Weyl semimetal. Corresponding to an electric field, a current flow from left to right would mean a higher filling level in the right Weyl cone than in the left and thus for the generation of more right-handed Weyl Fermions in the semimetal than left-handed. However, according to the conservation law, this is prohibited. Only the application of an additional magnetic field parallel to the temperature gradient, which is so high that it quantizes the electron states (Figure 5c), leads, analogous to high-energy physics, to the breaking of the chirality conservation law and enables a constant redistribution of electrons between the left and right Weyl cones. This induced breaking of the chiral symmetry is a macroscopic form of the mixed axial-gravitational anomaly in the relativistic field theory and leads to a positive longitudinal magnetic field-dependent thermoelectric current in Weyl semimetals. We were able to use these relationships to demonstrate the chiral (Figure 5d) and axial-gravitational anomaly (Figure 5e) in the Weyl semimetal NbP by measuring the electrical and thermoelectric currents in the magnetic field in 2017 [38]. The relative orientation of the electric field or temperature gradient to the magnetic field is decisive in these experiments. Both the chiral and axial-gravitational anomaly are experimentally reflected in an electrical current that increases with the magnetic field; this is only the case if the magnetic field and the electric field current are parallel to one another, otherwise the electric and thermoelectric current will decrease as the magnetic field increases. A critical component in these experiments is the ability to synthesize high quality samples. Therefore, groups capable of synthesizing high-quality single crystal solids are leaders in this field.

In case of chiral particles, not only the chiral symmetry itself, but also the energy-momentum tensors of the chiral quasiparticles are separately conserved. The energy-momentum tensor encodes the density and flux of energy and momentum, i.e. measures the contributions of quasiparticle currents and heat currents to the total energy currents in the system. However, in strong gravitational fields, in addition to the conservation law of chirality also this conservation law of the momentum tensors should be violated by quantum fluctuations. The separate conservation of the energy-momentum tensors of the chiral particles represents another consequence of the gravitational anomaly, but is elusive today. It is the emergent separate conservation at the Weyl cones at Weyl semimetals that makes us confident that it will be possible to probe the gravitational anomaly in the thermal transport of Weyl semimetals in the future exposed. In such systems, the gravitational anomaly is expected to cause a positive contribution to the longitudinal magneto-thermal conductivity in such systems. In thermal equilibrium, both left and right-movers have the same temperature, and are, hence, described by a similar energy-momentum tensor. If a temperature gradient is applied to a normal metal, e.g. from left (warm) to right (cold), a heat current flows from left to right because the hot electrons diffuse from left to right. But not so in a Weyl semimetal. A heat current flow from left to right would mean a higher temperature in the right Weyl cone than in the left and thus hotter right-handed Weyl Fermions in the semimetal than left-handed ones. However, according to the conservation law, this is prohibited. Only the application of an additional magnetic field parallel to the temperature gradient, which is so high that it quantizes the electron states, leads, analogous to the predictions high-energy physics, to the breaking of the separate conservation law for the energy momentum tensors and enables a constant redistribution of heat between the left and right Weyl cones. This induced breaking of the separate conservation of the energy momentum tensors in a chiral electron system is a macroscopic form of the gravitational anomaly. We have planned such experiments and performed first test measurements, which make us optimistic that we can probe this consequence of the gravitational anomaly in the near future.

5. Magnetic Weyl semimetals

The first predicted Weyl semimetals were magnetic, namely the pyrochlore iridate $Y_2Ir_2O_7$ and $HgCr_2Se_4$ [39, 40]. An overview with all refences about the research area is given in [41]. In Figure 6 a single crystals of important magnetic Weyl semimetal are displayed and the characteristic experimental methods are summarized. Angle resolved photoemission (ARPES) and Scanning Tunneling Microscopy (STM) allow for the direct monitoring of the bulk and surface electronic structure. For the ideal topological insulator, the surface state is a Dirac cone while the bulk electronic structure is gapped. In a magnetic TI the surface states are gapped, and properties such as axion insulator or quantum anomalous Hall can be observed. In magnetic Weyl and Dirac semimetals linear dispersion can be seen in bulk and Fermi arcs at the surfaces, if the material is magnetized. In other topological magnetic topological materials, the bulk spectra consist of nodal lines and more complex surface states are obtained as for example drum head states. Transport measurements with typical external stimuli such as electric and magnetic fields, light, temperature, pressure, strain are available for manipulating electronic properties of the magnetic topological materials. The classical anomalous Hall Effect (AHE) exists in nearly all ferromagnetic semimetals and metals. In magnetic topological materials the Berry curvature plays an important role. Magnetic Weyl semimetals are common: every crossing point in the band structure of a ferromagnetic centrosymmetric compound is related to nodal lines or Weyl points. An enhanced Berry curvature and a strong linear electromagnetic response leads to a large AHE and a large anomalous Nernst effect (ANE). The measurement set up of the Nernst effect is related to the AHE setup, instead of a current a thermal gradient is applied. The ANE is a transverse thermomagnetic effect, the anomalous transverse voltage is generated perpendicularly to the temperature gradient and the magnetization.

The chiral anomaly and the gravitational anomaly are also expected for magnetic Weyl semimetals. As an example, the chiral anomaly of the magnetic Weyl semimetal $Co_3Sn_2S_2$ is displayed in Figure 6 b [43]. However, similar to the chiral anomaly, circularly polarized light can induce an asymmetry between left- and right-handed chiral quasiparticles in nonmagnetic and magnetic Weyl semimetals. The light matter interaction with magnetic topological materials is an unexplored area, which we plan carefully investigate in near future.

In retrospect, this is a comprehensible development, since magnetic Weyl metals are close relatives of quantum Hall systems, which, as described above, represent the first topological solid-state systems. Quantum Hall systems always contain externally applied magnetic fields, i.e. in experiments, and therefore do not appear as a natural solid. One of the first ideas for a topological solid without an external magnetic field were two-dimensional crystals, which have a similar band structure as quantum Hall systems without a magnetic field, but are also ferromagnets. In such materials, the role of the external magnetic field is taken over by the intrinsic magnetization. Here, too, the topological invariant can be measured directly as a so-called quantum anomalous Hall effect — abnormal due to the lack of an external magnetic

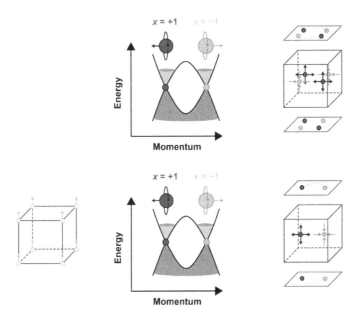

Fig. 6. Comparison of the simplified electronic structure of a non-magnetic Weyl semimetal and a ferromagnetic Weyl semimetal and the corresponding chiral Weyl points in the bulk and Fermi arcs on the surface of a crystal.

field. If such two-dimensional quantum anomalous Hall systems combine to form a three-dimensional solid crystal, a Weyl semimetal is created. Starting from a layered system of individual quantum anomalous Hall systems, as the coupling of the two-dimensional layers becomes stronger, the inverted band gap closes in the coupling direction and creates Weyl crossing points at the edge of the band structure, which move further and further into its center. In general, one can even say that every crossing point of electronic tapes in ferromagnetic materials is a Weyl point. In the simplest case, ferromagnetic crystals have at least four Weyl points Figure 7.

In antiferromagnetic compounds Weyl physics was recognized first [13, 14], since an anomalous Hall effect (AHE) which is typical for ferromagnets was theoretically found based on an unusual Berry curvature. A common understanding was that the anomalous Hall effect is proportional to the magnetization. All antiferromagnets that have zero magnetization should not have an AHE. However, the non-collinear triangular antiferromagnetic arrangement in Mn_3Z (Z = Ge, Sn) [12] lead to a non-vanishing Berry curvature in momentum space and Weyl points at the Fermi energy. Shortly after the prediction a large AHE at room temperature was observed in Mn_3Sn [13, 44] and Mn_3Ge [14]. However, it is known that numerous ferromagnetic Heusler compounds are still ferromagnetic even at room temperature and even up to over 900 °C. Therefore, it was not surprising that the first predicted ferromagnets were Co-Heusler compounds: Co_2YZ (X = V, Zr, Nb, Ti, Hf, Z=Si, Ge, Sn) [45, 46]. However, here the crossings are far above the Fermi energy and therefore not reachable in transport. With Co_2MnAl [47, 48] ferromagnets were predicted to

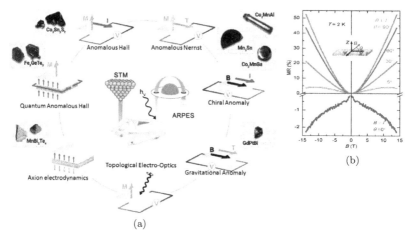

(a)

(b)

Fig. 7.   Figure (a) according to [41]. In the outer circle, single crystals of the ferromagnetic Weyl semimetals Co_2MnGa, Co_2MnAl and $Co_3Sn_2S_2$, the antiferromagnetic Weyl semimetals Mn_3Sn and GdPtBi and the topological insulator $MnBi_2Te_4$ and the topological ferromagnetic metal Fe_3GeTe_2 are displayed. Physical investigation methods and characteristic properties of a magnetic Weyl semimetals are shown in the inner circle: Angle resolved photoemission (ARPES) and Scanning Tunneling Microscopy (STM) allow for the direct monitoring of the bulk and surface electronic structure. In a magnetic topological insulator some of the surface states are gap, and properties such as axion insulator or quantum anomalous Hall can be observed. In magnetic Weyl and Dirac semimetals linear dispersion can be seen in bulk and Fermi arcs at the surfaces, if the materials is magnetized. In other topological magnetic topological materials such as Co_2MnAl and Co_2MnGa, the bulk spectra consist of nodal lines and more complex surface states are obtained as for example drum head states. Transport measurements with typical external stimuli such as electric and magnetic fields, light, temperature, pressure, strain are available for manipulating electronic properties of the magnetic topological materials. The classical anomalous Hall Effect (AHE) exists in nearly all ferromagnetic semimetals and metals. In magnetic topological materials the Berry curvature plays an important role. Magnetic Weyl semimetals are common: every crossing point in the band structure of a ferromagnetic centrosymmetric compound is related to nodal lines or Weyl points. An enhanced Berry curvature and a strong linear electromagnetic response leads to a large AHE and a large anomalous Nernst effect (ANE). The measurement set up of the Nernst effect is related to the AHE setup, instead of a current a thermal gradient is applied. The ANE is a transverse thermomagnetic effect, the anomalous transverse voltage is generated perpendicularly to the temperature gradient and the magnetization. The chiral anomaly is a smoking gun experiment for all Weyl semimetals. The gravitational anomaly is related to the chiral anomaly and should be observed in magnetic Weyl semimetals. A thermal gradient substitutes a current and a positive longitudinal magneto-thermoelectric conductance is measured for collinear temperature gradients and magnetic fields. The light matter interaction with magnetic topological materials is an unexplored area. Figure (b) according to [43]. The chiral anomaly for the ferromagnetic Weyl semimetal $Co_3Sn_2S_2$.

be Weyl semimetals with Weyl points near the Fermi energy, with, as a consequence, giant anomalous Hall effects (AHE) due to the large Berry phase. In thin films of Co_2MnAl a large AHE, in excellent agreement with theory, has already be found in [49]. Co_2MnGa and Co_2MnAl were found to be very large anomalous Hall values, in excellent agreement with the theory. The anomalous Hall angle of up to 12% in Co_2MnGa is also promising [50]. The proof that these materials are magnetic Weyl semimetals was provided by electronic structural studies of Co_2MnGa. In this case, too, the expected surface conditions could be observed using ARPES [51]. Realizing the QAHE at room temperature would be revolutionary as it would overcome the limitations of many of today's data-based technologies that are affected by large electron scattering-induced power losses. This would pave the way for a new generation of quantum electronics and spintronics devices with low energy consumption.

In 2018 we realized an intrinsic hard magnetic Weyl semimetal with Weyl crossing points near the Fermi energy [43]. The Shandite crystals contain transition metals on a quasi-two-dimensional Kagome lattice. One of the most interesting candidates is $Co_3Sn_2S_2$, which has the highest magnetic order temperature within this family and in which the magnetic moments on the Co atoms are oriented in a direction perpendicular to the Kagome plane. Magnetic field-dependent electrical transport measurements indicate the chiral anomaly [43] and ARPES measurements clearly demonstrate the existence of magnetic Weyl Fermions and very long Fermi arcs [52]. $Co_3Sn_2S_2$ shows a huge anomalous Hall effect up to temperatures of $150\,K$ and a huge Hall angle overall, which indicates a Weyl semimetal that is still very close to layered anomalous quantum Hall systems. For a large Hall angle, two conditions must be met: first, a large Hall conductivity and, second, a small number of electrons. These conditions are met in Weyl semimetals in which the Weyl crossing points are close to the Fermi energy. In other words, the coupling of the Kagome planes in $Co_3Sn_2S_2$ is weak and it has a two-dimensional magnetic and electronic structure. In the compound, three cobalt atoms share a free electron. In fact, we have found that if you divide the measured anomalous Hall effect by the thickness of the crystal, you get a value that is roughly expected to be the quantum anomalous Hall effect per layer. We were also able to observe quantization of the edge states on cobalt edges of the material with scanning tunneling microscopy [53]. Subsequent band structure calculations actually showed the presence of Weyl nodes near the Fermi energy, calculations of the transport properties indicate a direct connection between the Weyl nodes and the increased anomalous Hall effect. In addition, strongly increased thermoelectric could be predicted and shown experimentally. In addition to the interesting quantum effects, magnetic Weyl semimetals also have a high potential in thermoelectric applications. In principle, the Fermi energy can be shifted in every ferromagnet (by substituting elements) so that the Weyl points are located at the Fermi energy. The observation of the quantum anomalous Hall effect at room temperature would enable novel computer technologies including quantum computers. To realize this possibility, our strategy is to look for quasi-two-dimensional magnetic materials with topological band structures — that is, three-dimensional crystals that look almost like individually layered anomalous quantum Hall systems — and to apply these materials as monolayers or very thin films synthesize. So far, however, no magnetic materials are known that could lead to a quantum anomalous Hall effect with a higher temperature. With a magnetic transition temperature of $150\,K$ in $Co_3Sn_2S_2$ we are still a long way from potential room temperature effects.

6. Chiral electrons beyond Weyl semimetals

As already mentioned, electrons in conventional Weyl semimetals are described by relativistic equations and were originally proposed in the field of high energy physics. But there are also solid-state crystals, in which chiral electrons have no corresponding analogues in high-energy physics and are therefore described as 'New Fermions' [15–17, 54]. In contrast to the vacuum considered in high energy physics, electronic (quasi)particles in condensed-matter systems are not constrained by Poincare symmetry. Instead, they must only respect the crystal symmetry of one of the 230 space groups. Hence, there is the potential to find and classify free fermionic excitations in solid-state systems that have no high-energy counterparts. While Dirac Fermions are four times degenerate and Weyl Fermions are degenerate twice, the new Fermions even show six- and eight-times degeneracy. Particular examples, in which inversion symmetry is broken, are crystals with a chiral crystal lattice. Such chiral crystals are examples, in which the concept of chirality appears in real space. In this context, chirality refers to the property of these crystals that their atoms follow a spiral, step-like pattern as in the biological systems, such as DNA. While the staircase rotates clockwise in one system, it rotates counterclockwise in the opposite system (Figure 8), but both systems have identical composition. These systems, called "enantiomers", are mirror images of each other (Figure 8 a). Chiral crystals with heavy elements, topological chiral Fermions are of particular interest. In the family of the B20 structure type, the $P2_13$ (198) space group (Figure 8 a, shows both enantiomers), a new type of electron, the so-called Rarita-Schwinger Fermions, was confirmed. The two band crossing points in crystals with broken inversion symmetry, are at different energies (Figure 8 b) and show six- and four-times degeneracy. This leads to several notable properties, including: a huge quantized circular photogalvanic current [56–58], a chiral magnetic effect, and other novel transport and optical effects not observable in Weyl semimetals. Various candidate materials such as PdGa were selected for ARPES studies because the spin orbit coupling is very large and both enantiomers could be synthesized from this compound. The complex band topology of the two enantiomers leads to Fermi arcs (Figure 8 c and d), [16, 54, 55, 63–65] which, however, are also mirror images of each other (Figure 8 d). The high spin-orbit coupling is responsible for a strong splitting of the bands so that the experimental resolution was sufficient to confirm the four bands representative of the topological number of the Rarita-Schwinger Fermions. The experiments also show the expected band degeneracies at the highly symmetrical points of the Brillouin zone with the crossing points at different energies. Fermi arcs that run across the entire Brillouin zone and are also chiral, with different chirality in the lattice structure for the two enantiomers, are shown in Figure 8 c and d.

In general, chiral inorganic or organic materials show exceptional optical properties and crystallize in one of the 65 Sohncke symmetry space group [59–61]. If the space group contains only proper operations, the crystal is chiral and obey a well-defined handedness. There are more than 100 000 non-biological chiral crystals. Of all inorganic compounds in the inorganic database ICSD approximately 20% are

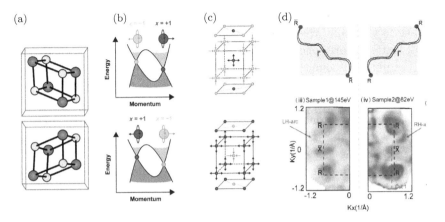

Fig. 8. Figure according to [16]. Experiments on chiral Weyl semimetals. (a) Sketch of the atomic structure of PdGa (red spheres represent Ga atoms and silver spheres represent Pd atoms), enantiomer A and enantiomer B. (b) Sketch of the electronic structure of enantiomer A and B, and (c) its distribution of the Weyl crossing points and the corresponding chiral surface states. (d) Images of the angle-resolved emission spectroscopy of the surface of PdGa enantiomer A and enantiomer B.

chiral [59]. Beside the structural chirality in materials itself, materials can host chiral quasiparticles and excitations such as chiral electrons, chiral photons, chiral magnons, chiral plasmons, chiral spins etc. In a few cases it is speculated whether there is a relationship between the handedness of structure and handedness of the chiral objects such as Skyrmions, even on different length scales (atomic distances of Å and Skyrmions nm–μm) [62]. The underlying correlation between a chiral lattice structure and the chiral quantum number of the electrons remains largely unexplored to date. However, it has been suggested that while the band structure in the bulk of the two enantiomers is identical, the chirality of the electrons and the Berry curvature behave like image and mirror image.

7. Weyl semimetals with strongly interacting electrons

While Weyl systems with free electrons are experimentally well investigated today, materials with strongly interacting electrons are largely unexplored. By switching on strong interactions between electrons with vibrations of the crystal lattice, so-called phonons, a charge density wave can be induced in Weyl semimetals, which connects the two Weyl crossing points and opens an energy gap at them. The resulting material is a so-called axion isolator (Figure 9 a). A charge density wave is the energetically preferred ground state of a strongly coupled electron-phonon system in certain quasi-one-dimensional metals and semi-metals at low temperatures. It is characterized by a gap in the dispersion of the free electrons and by a collective metallic mode which, like a superconductor, is formed by electron-hole pairs. The electron distribution and the position of the lattice atoms are periodically modulated in real space with a period that is greater than the original lattice constant

(Figure 9 b). The electric current-carrying collective mode, the so-called phason, is a solid-state version of the axion particle, which is traded as a possible candidate for dark matter in high-energy physics, but has not yet been observed. In most real solids, the phase is bound to impurities. Therefore, it can only "slide" freely over the grid and contribute to the flow of electrical current when a certain electrical threshold field is applied (above which the electrical force overcomes the binding forces). The resulting conduction behavior of the solid is strongly non-linear — as in the case of a diode, for example — and the electric current increases as the electric field increases. The signature of an "axionic" charge density wave is then, corresponding to the chiral anomaly in a Weyl semimetal with free electrons, a positive longitudinal magnetic field-dependent electrical conductivity. However, like its elementary particle analogue, this axionic charge density wave was not experimentally demonstrated until recently.

Only in 2019, we succeeded in measuring a large positive contribution to the magnetic field-dependent electrical conductivity of the phason [66, 67] of the charge carrier density wave in the Weyl semimetal $(TaSe_4)_2I$ for collinear electric and magnetic fields (Figure 9 d). We were able to show that this positive contribution results from the anomalous axionic contribution of the chiral anomaly to the electrical current flow and that this is linked to the parallel alignment of the electrical

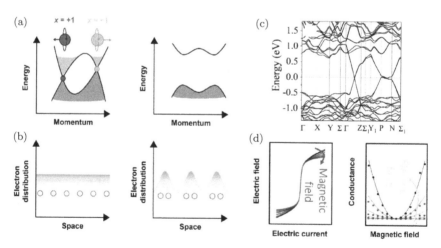

Fig. 9. Axion charge carrier density wave in a Weyl semimetal. (a) Band structure of the free electrons (top) and electron distribution in real space for a Weyl semimetal. (b) Periodic modulation of the electron distribution (charge carrier density wave) through strong electron-phonon interaction in real space creates a band gap at the Weyl crossing points. The crystal lattice is also modulated. The collective mode of the charge carrier density wave, the phason, is an axion. Figure (c) according to [68], the band structure of the Weyl semimetal $(TaSe_4)_2I$. (d) Transport experiment on the axion charge carrier density wave. In the case of high electric fields, where that of the phason dominates the electric current flow, the current is increased by a magnetic field applied parallel to the electric field. The magnetic field dependence of the longitudinal conductivity agrees with the anomalous transport of an "axionic" charge density wave.

and magnetic fields (Figure 9d). By rotating the magnetic field, we showed that the angular dependence of the phason conductivity is consistent with the anomalous transport of an axionic charge density wave. Our results show that it is possible to find experimental evidence for axions in highly correlated topological systems of condensed matter that were previously not found in any other context.

8. Outlook

It is intriguing that nature provides us with Chirality and symmetry breaking at several levels, ranging from the most fundamental standard model in particle physics to homochirality in biomolecules. While many hypotheses have been provided, and intriguing experiments have been performed, most fundamental questions on chiral symmetry breaking remain largely unanswered. Already Pasteur conjectured that spatial crystal chirality (which he called 'dissymmetry') in living systems [68–70] like as in DNA molecular may be induced by some universal chiral force or influence in nature. The chiral anomaly, in principle provides such an "influence" on a fundamental quantum level, but whether there is such a connection between the chiral quantum number and spatial structures is not known to date in the high energy context. Key insights may arise from other areas of physics, e.g. an analogy has been found in solid-state systems hosting Weyl Fermions [22, 30–32] but also can exhibit chiral crystal structures. In fact, we believe, the topological materials discussed above might allow to test such a connection in future experiments.

On the one hand, the chiral and gravitational anomalies have been observed on the electron quasi-particle level in Weyl semimetals and on the other hand, we have observed first signatures of an imprint of the chiral crystal structure in non-centrosymmetric topological semimetals on the distribution of the chiral quantum number of their electron states in momentum space. Specifically, one possible root to explore this connection further in Dirac, Weyl and new Fermion semimetals might be to externally twist or apply torsional torque to materials. In particular, the chiral anomaly in the quasiparticle framework could be studied in materials with structural chirality and even in twisted 2D materials [71–73]. Controlling and promoting chirality in real and momentum space could be a future direction in topological material science. Light matter interaction is another important and unexplored direction which may couple topological band structures and spatial chirality. And as Frank Wilczek, who named the axion [74], said very recently: "If we know that there are some materials that host axions, well, maybe the material we call space also houses axions [75]."

On the other hand, molecular catalysis and the selective adsorption of helical molecules can happen on surface states, where high mobilities, spin momentum locking and chirality might be important criteria for efficiency and selectivity of molecular handedness. In this regard, topological surface states may favor heterogeneous catalysis processes such as the hydrogen evolution reaction (HER). Recently, we started to perform experiments that can potentially build a bridge between chiral

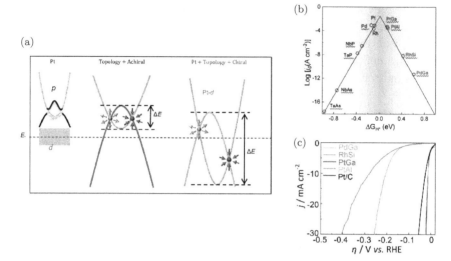

Fig. 10. Figure according to [76]. Hydrogen evolution reaction with Pt-group metal based chiral crystals. (a) Illustration of the band inversion mechanism and topologically nontrivial energy window in pure Pt, non-chiral topological semimetals, and chiral B20 compounds. (b) Volcano plot of PtAl, PtGa, RhSi, PdGa, the related metal catalysts and Weyl semimetals NbP, TaP, NbAs, and TaAs are also presented for comparison. (c) Polarization curves of the chiral crystal catalysts: PdGa, RhSi, PtAl, PtGa, and 20% Pt/C catalysts.

surface states and chiral molecules. We observed for example that the efficiency for HER on PdGa, PtAl and PtGa metals (space group of $P2_13$ (#198)) on surfaces with a long topological Fermi arc (Figure 10) is unexpectedly high, with turnover frequencies as high as 5.6 and 17.1 s^{-1} and an overpotential as low as 14 and 13.3 mV at a current density of 10 mA cm^{-2}. These chiral crystals outperform commercial Pt and nanostructured catalysts [76, 77]. First experiments show evidence that these materials allow enantiospecific control and the asymmetric on-surface synthesis of homochiral molecules [78].

In parallel, the so called CISS effect, that is chirality-induced spin selectivity (CISS) [79–81] has recently attracted attention, but is still not fully understood. We believe at the core of all these open problems is the question "how the information of a chiral quantum number transfers to a structural chirality". Possible options under discussion are the interplay of spin-orbit coupling, spin momentum locking and a geometric phase similar to the Berry phase in real space and brings chiral molecules close to topology. As explained above, we have made the first steps to understand this connection. Although our results are in part preliminary and cannot yet be generalized, they give already such strong indications for the connection of chirality in momentum and real space that we are excited about working on topology and chirality in the future!

References

1. N. Kumar, S. N. Guin, K. Manna, C. Shekhar, and C. Felser, Topological Quantum Materials from the Viewpoint of Chemistry, *Chem. Rev.* **121**, 2780–2815 (2021).
2. C. L. Kane and E. J. Mele, Quantum Spin Hall Effect in Graphene, *Phys. Rev. Lett.* **95**, 226801 (2005).
3. B. A. Bernevig, T. L. Hughes, and S.-C. Zhang, Quantum Spin Hall Effect and Topological Phase Transition in HgTe quantum wells, *Science* **314**, 1757 (2006).
4. K. von Klitzing, New Method for High-Accuracy Determination of the Fine-Structure Constant Based on Quantized Hall Resistance, *PRL* **45**, 494 (1980).
5. B. Bradlyn, L. Elcoro, J. Cano, M. G. Vergniory, Z. Wang, C. Felser, M. I. Aroyo, and B. A. Bernevig, Topological Quantum Chemistry, *Nature* **547**, 298 (2017).
6. M. G. Vergniory, L. Elcoro, C. Felser, N. Regnault, B. A. Bernevig, and Z. Wang, A Complete Catalogue of High-Quality Topological Materials, *Nature* **566**, 480 (2019).
7. Y. Xu, L. Elcoro, Z.-D. Song, B. J. Wieder, M. G. Vergniory, N. Regnault, Y. Chen, C. Felser, and B. A. Bernevig, High-throughput calculations of magnetic topological materials, *Nature* **586**, 702 (2020).
8. IUPAC. Compendium of Chemical Terminology, 2nd ed. (the "Gold Book"). Compiled by A. D. McNaught and A. Wilkinson. Blackwell Scientific Publications, Oxford (1997). Online version (2019-) created by S. J. Chalk. ISBN 0-9678550-9-8. https://doi.org/10.1351/goldbook.
9. H. Klapper and T. Hahn, *Point-group symmetry and physical properties of crystals*, in *International Tables for Crystallography*, Vol. A (International Union of Crystallography, 2006) pp. 804–808.
10. K. Shiota, A. Inui, Y. Hosaka, R. Amano, Y. Ōnuki, M. Hedo, T. Nakama, D. Hirobe, J. Ohe, J. Kishine, H. M. Yamamoto, H. Shishido, and Y. Togawa, Chirality-Induced Spin Polarization over Macroscopic Distances in Chiral Disilicide Crystals, *Phys. Rev. Lett.* **127**, 126602 (2021).
11. A. J. Gellman, Chiral surfaces: accomplishments and challenges, *ACS Nano* **4**, 5 (2010).
12. J. Kübler and C. Felser, Non-collinar antiferromagntes and the anomalous Hall effect, *Europhys. Lett.* **108**, 67001 (2014).
13. S. Nakatsuji, N. Kiyohara, and T. Higo, Large anomalous Hall effect in a non-collinear antiferromagnet at room temperature, *Nature* **527**, 212 (2015).
14. A. K. Nayak, J. E. Fischer, Y. Sun, B. Yan, J. Karel, A. C. Komarek, C. Shekhar, N. Kumar, W. Schnelle, J. Kübler, C. Felser, and S. S. P. Parkin, Large anomalous Hall effect driven by a nonvanishing Berry curvature in the noncolinear antiferromagnet Mn_3Ge, *Sci. Adv.* **2**, e1501870 (2016).
15. B. Bradlyn, J. Cano, Z. Wang, M. G. Vergniory, C. Felser, R. J. Cava, and B. A. Bernevig, Beyond Dirac and Weyl Fermions: Unconventional quasiparticles in conventional crystals, *Science* **353**, aaf5037 (2016).
16. N. B. M. Schröter, S. Stolz, K. Manna. F. de Juan, M. G. Vergniory, J. A. Krieger, D. Pei, T. Schmitt, P. Didin, T. K. Kim, C. Cacho, B. Bradlyn, H. Borrmann, M. Schmidt, R. Widmer, V. N. Strocov, and C. Felser, Observation and control of maximal Chern numbers in a chiral topological semimetal, *Science* **369**, 179 (2020).
17. M. Z. Hasan, G. Chan, I. Belopolski, G. Bian, S.-Y. Xu, and J.-X. Yin, Weyl, Dirac and high-fold chiral Fermions in topological quantum matter, *Nature Reviews Materials* **6**, 784–803 (2021).

18. M.V. Berry, Quantal phase factors accompanying adiabatic changes, *Proc. R. Soc. London A* **392**, 45 (1984).
19. M. Ünzelmann, H. Bentmann, T. Figgemeier, P. Eck, J. N. Neu, B. Geldiyev, F. Diekmann, S. Rohlf, J. Buck, M. Hoesch, M. Kalläne, K. Rossnagel, R. Thomale, T. Siegrist, G. Sangiovanni, D. Di Sante, and F. Reinert, Momentum-space signatures of Berry flux monopoles in a Weyl semimetal, *Nature communications* (2021), DOI: 10.1038/s41467-021-23727-3,
20. C. H. Li, O. M. J. van 't Erve, J. T. Robinson, Y. Liu, L. Li, and B. T. Jonker, Electrical detection of charge-current-induced spin polarization due to spin-momentum locking in Bi_2Se_3, *Nature nanotechnology* **9**, 218 (2014).
21. R. Naaman, Y. Paltiel, and D. H. Waldeck, Chiral induced spin selectivity gives a new twist on spin-control in chemistry, *Acc. Chem. Res.* **53**, 2659 (2020).
22. H. B. Nielsen and M. Ninomiya, The Adler-Bell-Jackiw anomaly and Weyl Fermions in a crystal. *Physics Letters B* **130**, 389 (1983).
23. J. Xiong, S. K. Kushwaha, T. Liang, J. W. Krizan, M. Hirschberger, W. Wang, R. J. Cava, and N. P. Ong, Evidence for the chiral anomaly in the Dirac semimetal Na_3Bi, *Science* **350**, 413 (2015).
24. Q. Li, D. E. Kharzeev, C. Zhang, Y. Huang, I. Pletikosić, A. V. Fedorov, R. D. Zhong, J. A. Schneeloch, G. D. Gu, and T. Valla, Chiral magnetic effect in $ZrTe_5$, *Nature Physics* **12**, 550 (2016).
25. J. Gooth, J. Kübler und C. Felser, Weyl sie exotisch sind, *Physik Journal* **20**, 29 (2021).
26. P. Sessi, F. R. Fan, F. Küster, K. Manna, N. B. M. Schröter, J. R. Ji, S. Stolz, J. A. Krieger, D. Pei, T. K. Kim, P. Dudin, C. Cacho, R. Widmer, H. Borrmann, W. Shi, K. Chang, Y. Sun, C. Felser, and S. S. P. Parkin, Handedness-dependent quasiparticle interference in the two enantiomers of the topological chiral semimetal PdGa, *Nature Communications* **11**, 3507 (2020).
27. F. Tang, Y. Ren, P. Wang, R. Zhong, J. Schneeloch, S. A. Yang, K. Yang, P. A. Lee, G. Gu, Z. Qiao, and L. Zhang, Three-dimensional quantum Hall effect and metal–insulator transition in $ZrTe_5$, *Nature* **569**, 537 (2019).
28. B. I. Halperin, Possible states for a three-dimensional electron gas in a strong magnetic field, *Jpn. J. Appl. Phys.* **26**, 1913 (1987).
29. S. Galeski, X. Zhao, R. Wawrzyńczak, T. Meng, T. Förster, S. Honnali, N. Lamba, T. Ehmcke, A. Markou, W. Zhu, J. Wosnitza, C. Felser, G. F. Chen, and J. Gooth, Unconventional Hall response in the quantum limit of $HfTe_5$, *Nature Communication* **11**, 5926 (2020).
30. H. Weyl, Elektron und gravitation. I, *Z. Phys.* **56**, 330 (1929).
31. B. Yan and C. Felser, Topological materials: Weyl semimetals, *Annual Review in Condensed Matter* **8**, 337–354 (2017).
32. G. E. Volovik, The Universe in a Helium Droplet. Oxford University Press, New York 2003.
33. C. Shekhar, A. K. Nayak, Y. Sun, M. Schmidt, M. Nicklas, I. Leermakers, U. Zeitler, Y. Skourski, J. Wosnitza, W. Schnelle, H. Borrmann, W. Schnelle, J. Grin, C. Felser, and B. Yan, Extremely large magnetoresistance and ultrahigh mobility in the topological Weyl semimetal NbP, *Nature Physics* **11**, 645 (2015).
34. S. L. Adler, Axial-vector vertex in spinor electrodynamics, *Phys. Rev.* **177**, 2426 (1969).
35. J. S. Bell and R. Jackiw, Can really regularized amplitudes be obtained as consistent with their expected symmetry properties? *Nuovo Cim. A* **60**, 47 (1969).

36. A. C. Niemann, J. Gooth, S.-C. Wu, S. Bäßler, P. Sergelius, R. Hühne, B. Rellinghaus, C. Shekhar, V. Süß, M. Schmidt, C. Felser, B. Yan, and K. Nielsch, Chiral magnetoresistance in the Weyl semimetal NbP, *Sci. Rep.* **7**, 43394 (2017).

37. F. Arnold, C. Shekhar, S.-C. Wu, Y. Sun, R. Donizeth dos Reis, N. Kumar, M. Naumann, M. O. Ajeesh, M. Schmidt, A. G. Grushin, J. H. Bardarson, M. Baenitz, D. Sokolov, H. Borrmann, M. Nicklas, C. Felser, E. Hassinger, and B. Yan, Negative magnetoresistance without well-defined chirality in the Weyl semimetal TaP, *Nature Communications* **7**, 11615 (2016).

38. J. Gooth, A. C. Niemann, T. Meng, A. G. Grushin, K. Landsteiner, B. Gotsmann, F. Menges, M. Schmidt, C. Shekhar, V. Süß, R. Hühne, B. Rellingshaus, C. Felser, B. Yan, and K. Nielsch, Experimental signatures of the gravitational anomaly in the Weyl semimetal NbP, *Nature* **547**, 324 (2017).

39. X. Wan, A. M. Turner, A. Vishwanath, and S. Y. Savrasov, Topological semimetal and Fermi-arc surface states in the electronic structure of pyrochlore iridates, *Phys. Rev. B* **83**, 205101 (2011).

40. G. Xu, H. Weng, Z. Wang, X. Dai, and Z. Fang, Chern semimetal and the quantized anomalous Hall effect in $HgCr_2Se_4$. *Phys. Rev. Lett.* **107**, 186806 (2011).

41. B. A. Bernevig, C. Felser, and H. Beidenkopf, Progress and prospects in magnetic topological materials, *Nature* (2021) in press.

42. H. Yang, Y. Sun, Y. Zhang, W.-J. Shi, S. S. P. Parkin, and B. Yan, Topological Weyl semimetals in the chiral antiferromagnetic materials Mn_3Ge and Mn_3Sn, *New J. Phys.* **19**, 015008 (2017).

43. E. Liu, Y. Sun, N. Kumar, L. Muechler, A. Sun, L. Jiao, S.-Y. Yang, D. Liu, A. Liang, Q. Xu, J. Kroder, V. Süß, H. Borrmann, C. Shekhar, Z. Wang, C. Xi, W. Wang, W. Schnelle, S. Wirth, Y. Chen, S. T. B. Goennenwein, and C. Felser Giant anomalous Hall effect in a ferromagnetic Kagomé-lattice semimetal, *Nature Physics* **14**, 1125 (2018).

44. J. Kübler and C. Felser, Weyl Fermions in antiferromagnetic Mn_3Sn and Mn_3Ge, *Europhys. Lett.* **120**, 47002 (2017).

45. Z. Wang, M. G. Vergniory, S. Kushwaha, M. Hirschberger, E. V. Chulkov, A. Ernst, N. P. Ong, R. J. Cava, and B. A. Bernevig, Time-reversal-breaking weyl fermions in magnetic heusler alloys, *Phys. Rev. Lett.* **117**, 236401 (2016).

46. G. Chang, S.-Y. Xu, H.- Zheng. B. Singh, C.-H. Hsu, G. Bian, N. Alidoust, I. Belopolski, D. S. Sanchez, S. Zhang, H. Lin, and M. Z. Hasan, Room-temperature magnetic topological Weyl Fermion and nodal line semimetal states in half-metallic Heusler CoTiX ($X =$Si, Ge, or Sn), *Sci. Rep.* **6**, 38839 (2016).

47. J. Kübler and C. Felser, Berry curvature and the anomalous Hall effect in Heusler compounds, *Phys. Rev. B* **85**, 012405 (2012).

48. J. Kübler and C. Felser, Weyl points in the ferromagnetic Heusler compound Co_2MnAl. *Europhys. Lett.* **114**, 47005 (2016).

49. E. Vilanova Vidal, G. Stryganyuk, H. Schneider, C. Felser, and G. Jakob, Exploring Co_2MnAl Heusler compound for anomalous Hall effect sensors. *Appl. Phys. Lett.* **99**, 132509 (2011).

50. K. Manna, Y. Sun, L. Müchler, J. Kübler, and C. Felser, Heusler, Weyl, and Berry, *Nature Reviews Materials* **3**, 24 (2018).

51. I. Belopolski, K. Manna, D-.S. Sanchez, G. Chang, B. Ernst, J. Yin, S. S. Zhang, T. Cochran, and N. Shumiya, Discovery of topological Weyl lines and drumhead surface states in a room-temperature magnet, *Science* **365**, 1278 (2019).

52. D. F. Liu, A. J. Liang, E. K. Liu, Q. N. Xu, Y. W. Li, C. Chen, D. Pei, W. J. Shi, S. K. Mo, P. Dudin, T. Kim, C. Cacho, G. Li, Y. Sun, L. X. Yang, Z. K. Liu, S. S. P. Parkin, C. Felser, and Y. L. Chen, Magnetic Weyl Semimetal Phase in a Kagomé Crystal, *Science* **365**, 1282 (2019).

53. S. Howard, L. Jiao, Z. Wang, N. Morali, R. Batabyal, P. Kumar-Nag, N. Avraham, H. Beidenkopf, P. Vir, E. Liu, C. Shekhar, C. Felser, T. Hughes, and Vidya Madhavan, Evidence for one-dimensional chiral edge states in a magnetic Weyl semimetal $Co_3Sn_2S_2$, *Nature Communications* **12**, 4269 (2021).

54. G. Chang, B. J. Wieder, F. Schindler, D. S. Sanchez, I. Belopolski, S.-M. Huang, B. Singh, D. Wu, T.-R. Chang, T. Neupert, S.-Y. Xu, H. Lin, and M. Z. Hasan, Topological quantum properties of chiral crystals, *Nature Materials* **17**, 978 (2018).

55. M. Yao, K. Manna, Q. Yang, A. Fedorov, V. Voroshnin, B. V. Schwarze, J. Hornung, S. Chattopadhyay, Z. Sun, S. N. Guin, J. Wosnitza, H. Borrmann, C. Shekhar, N. Kumar, J. Fink, Y. Sun, and C. Felser, Observation of giant spin-split Fermi-arc with maximal Chern number in the chiral topological semimetal PtGa, *Nature Communications* **11**, 2033 (2020).

56. D. Rees, K. Manna, B. Lu, T. Morimoto, H. Borrmann, C. Felser, J. E. Moore, D. H. Torchinsky, and J. Orenstein, Helicity-dependent photocurrents in the chiral Weyl semimetal RhSi, *Sci. Adv.* **6**, eaba0509 (2020).

57. Z. Ni, K. Wang, Y. Zhang, O. Pozo, B. Xu, X. Han, K. Manna, J. Paglione, C. Felser, A. G. Grushin, F. de Juan, E. J. Mele, and L. Wu, Giant topological longitudinal circular photo-galvanic effect in the chiral multifold semimetal CoSi, *Nature Communications* **12**, 154 (2021).

58. D. Rees, B. Lu, Y. Sun, K. Manna, R. Özgür, S. Subedi, H. Borrmann, C. Felser, J. Orenstein, and D. H. Torchinsky, Direct Measurement of Helicoid Surface States in RhSi Using Nonlinear Optics, *Phys. Rev. Lett.* **127**, 157405 (2021).

59. C. Dryzun and D. Avnir, On the abundance of chiral crystals, *Chem. Commun.* **48**, 5874 (2012).

60. H. D. Flack, Chiral and Achiral Crystal Structures, *Helvetica Chimica Acta* **86**, 905 (2003).

61. A. M. Glazer and K. Stadnicka, On the Use of the Term 'Absolute' in Crystallography, *Acta Cryst.* *A***45**, 234 (1989).

62. D. Morikawa, K. Shibata, N. Kanazawa, X. Z. Yu, and Y. Tokura, Crystal chirality, and skyrmion helicity in MnSi and (Fe, Co)Si as determined by transmission electron microscopy, *Phys. Rev. B* **88**, 024408 (2013).

63. P. Sessi, F.-R. Fan, F. Küster, K. Manna, N. B. M. Schröter, J.-R. Ji, S. Stolz, J. A. Krieger, D. Pei, T. K. Kim, P. Dudin, C. Cacho, R. Widmer, H. Borrmann, W. Shi, K. Chang, Y. Sun, C. Felser, and S. S. P. Parkin, Handedness-dependent quasiparticle interference in the two enantiomers of the topological chiral semimetal PdGa, *Nature Communications* **11**, 3507 (2020).

64. D. S. Sanchez, I. Belopolski, T. Cochran, X. Xu, J. Yin, G. Chang, W. Xie, K. Manna, V. Süss, C.-Y. Huang, N. Alidoust, D. Multer, S. S. Zhang, N. Shumiya, X. Wang, G.-Q. Wang, T.-R. Chang, H. Lin, S.-Y. Xu, C. Felser, S. Jia, and M. Z. Hasan, Topological chiral crystals with helicoid-arc quantum states, *Nature* **567**, 500 (2019).

65. D. Takane, Z. Wang, S. Souma, K. Nakayama, T. Nakamura, H. Oinuma, Y. Nakata, H. Iwasawa, C. Cacho, T. Kim, K. Horiba, H. Kumigashira, T. Takahashi, Y. Ando, T. Sato, Observation of Chiral Fermions with a Large Topological Charge and Associated Fermi-Arc Surface States in CoSi, *Phys. Rev. Lett.* **122**, 076402 (2019).

66. J. Gooth, B. Bradlyn, S. Honnali, C. Schindler, N. Kumar, J. Noky, Y. Qi, C. Shekhar, Y. Sun, Z. Wang, B. A. Bernevig, and C. Felser, Evidence for an axionic charge density wave in the Weyl semimetal $(TaSe_4)_2I$, *Nature* **575**, 315 (2019).

67. W. Shi, B. J. Wieder, H. L. Meyerheim, Y. Sun, Y. Zhang, Y. Li, L. Shen, Y. Qi, L. Yang, J. Jena, P. Werner, K. Koepernik, S. Parkin, Y. Chen, C. Felser, B. A. Bernevig, and Z. Wang, A Charge-Density-Wave topological semimetal, *Nature Physics* **17**, 284 (2021).

68. V. A. Davankov, Biological Homochirality on the Earth, or in the Universe? A Selective Review, *Symmetry* **10**, 749 (2018).

69. A. Dorta-Urra and P. Bargueño, Homochirality: A Perspective from Fundamental Physics, *Symmetry* **11**, 661 (2019).

70. L. D. Barron, Symmetry and Chirality: Where Physics Shakes Hands with Chemistry and Biology, *Israel J. Chem.* **61**, 517 (2021).

71. L. Chen and K. Chang, Chiral-Anomaly-Driven Casimir-Lifshitz Torque between Weyl Semimetals, *Phys. Rev. Lett.* **125**, 47402 (2020).

72. T. Stauber, T. Low, and G. Gómez-Santos, Plasmon-Enhanced Near-Field Chirality in Twisted van der Waals Heterostructures, *Nano Letters* **20**, 8711 (2020).

73. J. Jung, M. Polini, A. H. MacDonald, Persistent current states in bilayer graphene, *Phys. Rev. B* **91**, 155423 (2015).

74. F. Wilzek, Problem of Strong P and T Invariance in the Presence of Instantons, *Phys. Rev. Lett.* **40**, 279 (1978).

75. M. Fore, Physicists Have Finally Seen Traces of a Long-Sought Particle. Here' Why That's a Big Deal, https://www.livescience.com/axion-found-in-weyl-semimetal.html

76. Q. Yang, G. Li, K. Manna, F. Fan, C. Felser, and Y. Sun, Topological engineering of Pt-group metal based chiral crystals toward high-efficiency hydrogen evolution catalysts, *Advanced Materials* **2020**, 1908518 (2020).

77. G. Li and C. Felser, Heterogeneous catalysis at the surface of topological materials, *Applied Physics Letters* **116**, 070501 (2020).

78. S. Stolz, M. Danese, M. Di Giovannantonio, J. I. Urgel, Q. Sun, A. Kinikar, M. Bommert, S. Mishra, H. Brune, O. Gröning, D. Passerone, R. Widmer, Asymmetric Elimination Reaction on Chiral Metal Surfaces, *Advanced Materials* **2021**, 2104481 (2021).

79. R. Naaman and D. H. Waldeck, Chiral-induced spin selectivity effect, *J. Phys. Chem. Letters* **3**, 2178 (2012).

80. R. Naaman, Y. Paltiel, and D. H. Waldeck, Chiral Induced Spin Selectivity Gives a New Twist on Spin-Control in Chemistry, *Acc. Chem. Res.* **53**, 2659 (2020).

81. F. Evers, A. Aharony, N. Bar-Gill, O. Entin-Wohlman, P. Hedegård, O. Hod, P. Jelinek, G. Kamieniarz, M. Lemeshko, K. Michaeli, V. Mujica, R. Naaman, Y. Paltiel, S. Refaely-Abramson, O. Tal, J. Thijssen, M. Thoss, J. M. van Ruitenbeek, L. Venkataraman, D. H. Waldeck, B. Yan, and L. Kronik, Theory of Chirality Induced Spin Selectivity: Progress and Challenges, arXiv 2108.09998 (2021).

Asymmetric Autocatalysis and the Elucidation of the Origin of Homochirality

Kenso Soai

Department of Applied Chemistry, Tokyo University of Science,
Kagurazaka, Shinjuku-ku, Tokyo 162-8601, Japan
Research Organization for Nano & Life Innovation, Waseda University,
Wasedatsurumaki-cho, Shinjuku-ku, Tokyo 162-0041, Japan
E-mail: soai@rs.tus.ac.jp
www.rs.kagu.tus.ac.jp/soai

Homochirality of biomolecules such as L-amino acids and D-sugars has been a puzzle for the chemical origin of life. Asymmetric autocatalysis is a reaction in which a chiral product acts as a chiral catalyst for its own production. Pyrimidyl alkanol, quinolyl alkanol and 5-carbamoylpyridyl alkanol were found to act as asymmetric autocatalysts to produce more of itself with amplified enantiomeric excess (ee) in the addition of diisopropylzinc to the corresponding pyrimidine-5-carbaldehyde, quinoline-3-carbaldehyde and pyridine-3-carbaldehyde. Pyrimidyl alkanol with very low ee enhances its ee to >99.5% in asymmetric autocatalysis. Circularly polarized light, quartz, chiral crystals composed of achiral compounds, chiral carbon, nitrogen and oxygen isotopomers act as chiral triggers of asymmetric autocatalysis to afford highly enantioenriched pyrimidyl alkanol. Spontaneous absolute asymmetric synthesis was achieved in the reaction between achiral pyrimidine-5-carbaldehyde and diisopropylzinc in conjunction with asymmetric autocatalysis.

Keywords: Asymmetric autocatalysis, Origin of chirality, Soai reaction, Absolute asymmetric synthesis

1. Introduction

Self-replication and the homochirality of biomolecules such as L-amino acids and D-sugars are two of the characteristic features of life. A broad spectrum of attention has been focused on the origin of homochirality [1] ever since Pasteur discovered molecular dissymmetry in 1848 [2]. In most cases, the enantiomeric excesses induced by the proposed theories of the origins of homochirality of organic compounds have remained to be very low. Thus, an amplification process of low enantiomeric excess (ee) to high ee is required for organic compounds to achieve the homochirality [3].

In 1953, Frank proposed a mathematical mechanism of asymmetric autocatalysis without showing any chemical structure [3k]. However, no experimental realization of asymmetric autocatalysis had been reported until we first reported on the asymmetric autocatalysis of 3-pyridyl alkanol in 1990 [4].

Asymmetric autocatalysis is a reaction in which chiral product serves as chiral catalyst for its own production (Figure 1). The reaction involves a catalytic self-replication of chiral compound., The a priori advantages of asymmetric autocatalysis over usual asymmetric catalysis are as follows: (1) the self-replication is a process that life adopts, the efficiency is high. (2) Chiral product acts as newly formed catalyst, the amount of catalyst increases. In ideal case, no decrease in the catalytic activity occurs. (3) No necessity of the separation of product from catalyst because their structure is the same.

144

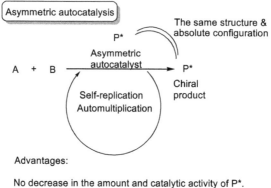

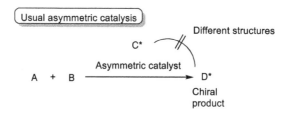

Fig. 1. Principles of asymmetric autocatalysis and usual asymmetric catalysis.

Mukaiyama and Soai *et al.* reported enantioselective addition of alkylmetal reagents such as alkyllithium to aldehydes using diamino alcohols as chiral ligands [5]. The addition of diethylzinc to benzaldehyde was also observed in the presence of diamino alcohol [5]. In the enantioselective addition of dialkylzincs to aldehydes using chiral amino alcohols as catalysts [6], we observed that the absolute configuration of the alcohol moiety of catalyst rather than that of the amine moiety controls the absolute configuration of the product [6a]. We also observed that, in the *N,N*-dibutylnorehedrine catalyzed enantioselective addition of diethylzinc, the reaction with pyridine-3-carbaldehyde (1 h) is faster than benzaldehyde (16 h) [6b]. The *in situ* product, *i.e.*, ethylzinc alkoxide of 3-pyridyl alkanol, was considered to accelerate the reaction.

In this review, we describe the study on asymmetric autocatalysis and on the elucidation of origin of homochirality of organic compounds [7].

2. Asymmetric autocatalysis: The Soai reaction

2.1. *Asymmetric autocatalysis of pyridyl-, pyrimidyl- and quinolyl alkanols with amplification of enantiomeric excess*

Based on the observations discovered in the preceding section, we envisaged asymmetric autocatalysis of a suitable chiral nitrogen containing alcohol in the reaction between nitrogen containing aldehyde and dialkylzinc.

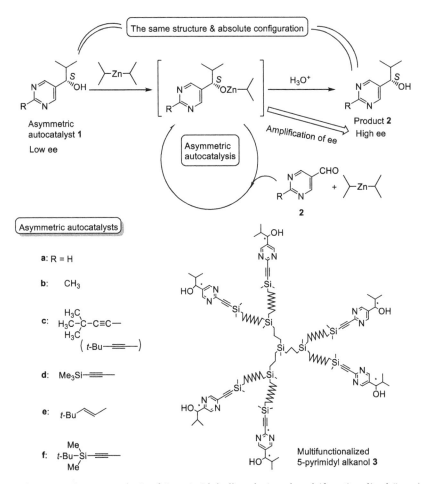

Fig. 2. Asymmetric autocatalysis of 5-pyrimidyl alkanols **1** and multifunctionalized 5-pyrimidyl alkanol **3** with amplification of ee.

The first asymmetric autocatalysis was reported by us in 1990 with pyridyl alkanol in the reaction between pyridine-3-carbaldehyde and dialkylzincs [4]. We kept working on developing more efficient asymmetric autocatalysis.

In 1995, asymmetric autocatalysis with amplification of ee of pyrimidyl alkanol was realized by us in the addition of diisopropyl zinc (i-Pr$_2$Zn) to pyrimidine-5-carbaldehyde (Figure 2) [8a,b]. Using (S)-pyrimidyl alkanol **1a** with low ee (2% ee) as asymmetric autocatalyst, the first reaction afforded the same (S)-pyrimidyl alkanol **1a** with enhanced ee (10% ee) as a mixture of the same structure of product and original catalyst. Taking advantage of asymmetric autocatalysis, given that the structures of the catalyst and the product are the same, the next asymmetric autocatalysis using the catalyst with 10% ee gave 57% ee, 81% ee, then 88% ee, successively. During three consecutive asymmetric autocatalysis reactions, the initial 2% ee was amplified to 88% ee with multiplication of the amount of

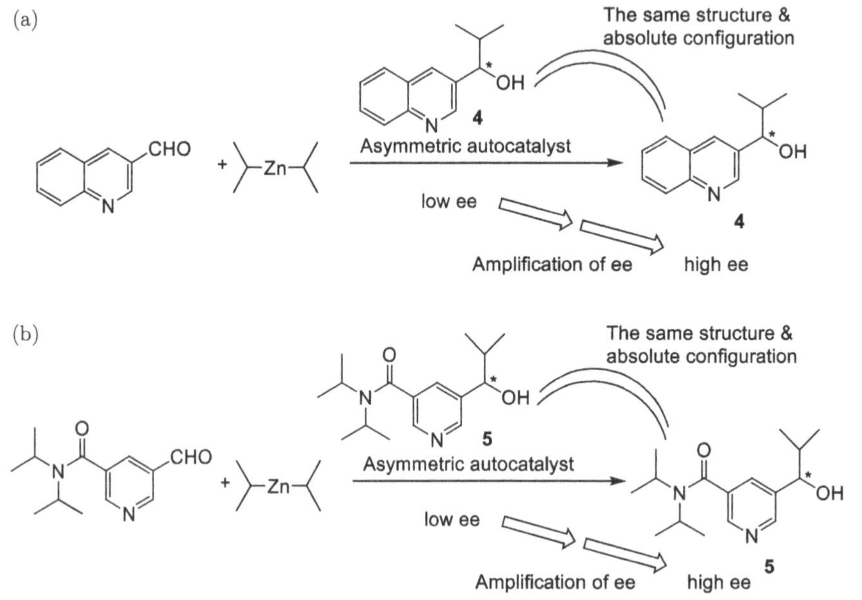

146

Fig. 3. Asymmetric autocatalysis with amplification of ee: (a) 3-quinolyl alkanol **4**, (b) 5-carbamoyl-3-pyridyl alkanol **5**.

enantiomer [8a]. It was also found that multi-functionalized 5-pyrimidyl alkanol **3** act as asymmetric autocatalyst with amplification of ee [8c]. It should be emphasized that this amplification of ee was possible without the intervention of any other chiral factor. The initial low ee of the asymmetric autocatalyst was the origin of chirality. In contrast, in nonautocatalytic asymmetric reactions, amplification of ee is possible only once [3a,b].

Furthermore, 3-quinolyl alkanol **4** [9a] and 5-carbamoylpyridyl alkanol **5** [9b] were also asymmetric autocatalysts with amplification of ee (Figure 3). Mislow referred to the asymmetric autocatalysis with amplification of ee as the Soai reaction [1a].

2.2. *Practically perfect asymmetric autocatalysis with amplification of enantiomeric excess from extremely low to near enantiopurity*

When 2-*t*-butylethynyl-5-pyrimidyl alkanol **1c** was used as asymmetric autocatalyst, highly efficient asymmetric autocatalysis was achieved. (*S*)-2-Pyrimidyl alkanol **1c** with >99.5% ee was used as asymmetric autocatalyst in the reaction between 2-alkynylpyrimidine-5-carbaldehyde **2c** and *i*-Pr₂Zn. (*S*)-Pyrimidyl alkanol **1c** including the product and catalyst with the same absolute configuration with >99.5% ee was formed in >99% yield [8d]. The obtained pyrimidyl alkanol **1c** was used in the next rounds of asymmetric autocatalysis, successively. Even after the tenth round, no deterioration in the enantioselectivity and catalytic activity was observed; the

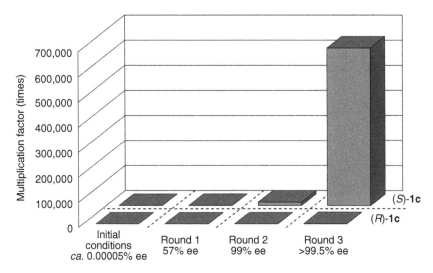

Fig. 4. Amplification of ee from extremely low to >99.5% ee in consecutive asymmetric auto-catalysis of (S)-2-t-butylethynyl-5-pyrimidyl alkanol **1c**.

(S)-pyrimidyl alkanol **1c** exhibited >99.5% ee and the yield was >99%. Isopropy-lzinc alkoxide of pyrimidyl alkanol **1c** is formed *in situ*, and this species acts in asymmetric autocatalysis to produce more of itself, with the same structure and absolute configuration. Thus, the overall process is a practically perfect catalytic self-replication of a chiral compound.

We also examined the amplification of ee in the asymmetric autocatalysis of 2-alkynylpyrimidyl alkanol **1c** with extremely low ee (*ca.* 0.00005%) (Figure 4) [8e]. This ee corresponds to the difference in a small number (several) of molecules between the (S)- and (R)- enantiomers (5 million of each). We found that the first run of asymmetric autocatalysis of (S)-pyrimidyl alkanol **1c** with *ca.* 0.00005% ee amplified ee to 57%, the second run to 99% ee, and then the third run to >99.5% ee. During these three consecutive asymmetric autocatalysis reactions, the amount of the initial slightly major (S)-pyrimidyl alkanol **1c** automultiplied by a factor of *ca.* 630,000 times while automultiplication of the minor (R)-enantiomer **1c** was less than 1,000 times. Thus, a chemical reaction was found in which very low ee is amplified to >99.5% ee by asymmetric autocatalysis.

2.3. *Research on the mechanism of asymmetric autocatalysis: the Soai reaction*

Concerned with non-linear effect in non-autocatalytic asymmetric catalysis, Kagan proposed MLn mechanism [3a] and Noyori proposed the dimer mechanism [3b]. Several groups including us have investigated the mechanism of asymmetric auto-catalysis [10]. Blackmond and Brown suggested the dimeric catalyst model based on the heat flow measurement [10a]. They also suggested the dimeric and tetrameric species from the NMR measurement of the reaction solution [10b,c]. Ercolani [10d] and Gridnev [10e] suggested the structure of catalyst aggregates based on the DFT calculations. Micheau and Buhse [10f-h], Micskei and Palyi [10i], Ribo [10j] and Lente [10k] proposed the mechanistic frameworks and reaction models of asymmetric autocatalysis.

Our study on the relationship between the time, yield and ee of the product by using HPLC suggested the dimeric or higher order aggregated catalytic species [10l,m]. Single crystal X-ray diffraction of isopropylzinc alkoxide of pyrimidyl alkanol revealed the tetrameric or oligomeric crystal structures depending on the excess molar amount of i-Pr$_2$Zn [10n]. Buhse and Micheau proposed the reaction modeling which includes the tetramer or higher order aggregates [10o]. Very recently, NMR and IR analyses by Denmark [10p] and MS analysis by Trapp [10q] of the asymmetric autocatalysis were reported.

3. Elucidation of the origin of homochirality by using asymmetric autocatalysis

Several theories have been proposed as the origin of homochirality. However, in most cases, the enantiomeric imbalances induced by the proposed theories are very low. As described in the preceding section, asymmetric autocatalysis of pyrimidyl alkanol is capable of enhancing very low ee to >99.5% ee [8c]. We believed that the reaction between pyrimidine-5-carbaldehyde and i-Pr$_2$Zn would be triggered by a chiral factor, as the origin of chirality, to afford slightly enantioenriched zinc alkoxide of pyrimidyl alkanol of the corresponding absolute configuration to the origin of chirality. Once asymmetric autocatalysis is initiated, the ee is amplified to become very high ee.

3.1. *Circularly polarized light*

Circularly polarized light (CPL) is a chiral physical factor. CPL has been proposed as the origin of chirality of organic compounds. Indeed, irradiation by CPL of organic compounds such as racemic leucine induces low (mostly <2% ee) enantioenrichments [1d]. We reported that asymmetric autocatalysis is triggered by leucine with low 2% ee to afford pyrimidyl alkanol with high ee with corresponding absolute configurations [11a]. Thus, low ee induced by CPL was correlated with organic compound with high ee in conjunction with asymmetric autocatalysis.

Asymmetric photodecomposition of racemic pyrimidyl alkanol **1c** by irradiation of l-CPL (313 nm) followed by the subsequent consecutive asymmetric autocatalysis with amplification of ee afforded (S)-pyrimidyl alkanol **1c** with >99.5% ee (Figure 5) [11b]. In contrast, irradiation of d-CPL and the following asymmetric autocatalysis afforded (R)-pyrimidyl alkanol **1c** with >99.5% ee. Thus, for the first time, the chirality of the CPL is correlated with the chirality of a highly enantioenriched organic compound. Asymmetric photoequilibrium of a racemic olefin **6**

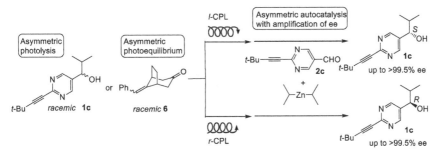

Fig. 5. Asymmetric photolysis and photoequilibrium by *l*- or *r*-circularly polarized light, and correlation of the induced chiral product with highly enantioenriched pyrimidyl alkanol **1c** by asymmetric autocatalysis.

by *l*-CPL irradiation followed by asymmetric autocatalysis led to the formation of (*S*)-pyrimidyl alkanol with >99.5% ee. [11c] Thus, CPL acts as the origin of chirality in conjunction with asymmetric autocatalysis to afford a highly enantioenriched organic compound.

3.2. *Chiral inorganic crystals (minerals)*

Quartz, which occurs commonly in the Earth's crust, is a single crystal of silicon dioxide (SiO_2). The rotation of plane-polarized light, *i.e.*, optical activity, was first observed with quartz. Quartz exhibits enantiomorphism. Quartz has for many years been considered as the origin of chirality of organic compounds [1e]. Although there are several papers that report on the use of quartz as the origin of chirality of organic compounds [12a], no significant asymmetric induction by quartz have yet been reported.

We reasoned that the presence of *d*-quartz would trigger the reaction of pyrimidine-5-carbaldehyde and *i*-Pr$_2$Zn, and that the initially formed isopropylzinc alkoxide of pyrimidyl alkanol with the corresponding absolute configuration to *d*-quartz would act as asymmetric autocatalyst with amplification of ee. Subsequently, (*S*)-pyrimidyl alkanol **1c** with 97% ee was formed (Figure 6) [12b]. In contrast, *l*-quartz afforded (*R*)-pyrimidyl alkanol **1c** with 97% ee. Thus, in conjunction with asymmetric autocatalysis, the chirality of quartz was correlated for the first time with the chirality of an organic compound with very high ee.

Cinnabar is a red-colored chiral -Hg-S-Hg-S- helical crystal of mercury(II) sulfide (α-HgS). It has been used as a pigment for paint, as a raw material for mercury, and as a Chinese traditional medicine. Asymmetric autocatalysis triggered by *P*-HgS afforded (*R*)-pyrimidyl alkanol **1c** with >99.5% ee, while the reaction triggered by *M*-HgS afforded (*S*)-pyrimidyl alkanol **1c** with >99.5% ee [12c]. *d*-Sodium chlorate ($NaClO_3$), a chiral inorganic ionic crystal, [12d] triggered asymmetric autocatalysis to afford (*S*)-pyrimidyl alkanol **1c** with >99.5% ee, while *l*-NaClO$_3$ triggered the formation of (*R*)-pyrimidyl alkanol **1c** with >99.5% ee [12e].

Thus, in conjunction with asymmetric autocatalysis, a correlation between chiral inorganic crystals as the origin of chirality and a chiral organic compound with very high (>99.5%) ee was established, for the first time.

3.3. *Enantiotopic face of achiral inorganic crystal: gypsum*

Gypsum (calcium sulfate dihydrate) is a widely used common mineral. Although gypsum has an achiral crystal structure, it readily exhibits a two-dimensional enantiotopic cleavage (010) and (0-10) face upon slicing. On the enantiotopic (010) face of gypsum, pyrimidine-5-carbaldehyde **2c** was placed and exposed to *i*-Pr₂Zn vapor. The formation of (*R*)-pyrimidyl alkanol **1c** resulted from asymmetric autocatalysis on the enantiotopic (010) face (Figure 6) [12f]. In contrast, the opposite enantiotopic (0-10) face triggered asymmetric autocatalysis to afford (*S*)-pyrimidyl alkanol **1c**.

As we have described, we successfully demonstrated that chiral inorganic crystals and enantiotopic faces of an achiral inorganic crystal act as the origins of chirality to afford, in conjunction with asymmetric autocatalysis with amplification of ee, highly enantioenriched organic compounds.

3.4. *Chiral organic crystals composed of achiral compounds*

There are chiral organic crystals that are known to be composed of achiral molecules [13a]. Although highly stereospecific reactions involving the use of these crystals as reactants have been reported [1i], these crystals have rarely been used as efficient chiral inducers in enantioselective reactions. We examined asymmetric autocatalysis triggered by chiral crystals composed of achiral molecules (Figure 7).

Cytosine is an achiral nucleobase in DNA and RNA. It forms chiral crystals upon crystallization from methanol. Chiral cytosine crystals of [CD(+)310 nm,

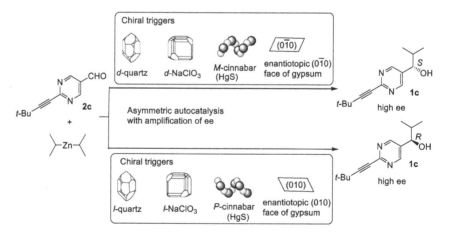

Fig. 6. Chiral inorganic crystals and enantiotopic face of gypsum act as the chiral trigger of asymmetric autocatalysis.

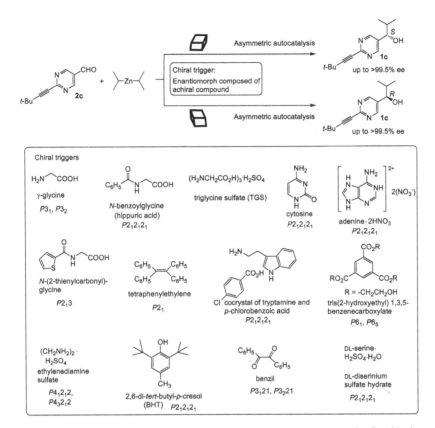

Fig. 7. Asymmetric autocatalysis triggered by chiral crystals composed of achiral organic compounds.

Nujol] act as chiral triggers of the reaction between pyrimidine-5-carbaldehyde **2c** and i-Pr$_2$Zn [13b]. The subsequent asymmetric autocatalysis with amplification of ee afforded (R)-pyrimidyl alkanol **1c** with very high ee. In contrast, cytosine crystals of [CD(-)310 nm, Nujol] afforded (S)-pyrimidyl alkanol **1c**. Thus, achiral cytosine, a nucleobase in DNA and RNA, acts as the origin of chirality in conjunction with asymmetric autocatalysis.

Cytosine, when it is crystallized from water, forms achiral crystals of cytosine monohydrate. It was found that upon dehydration of crystal water (by heating) from one of the enantiotopic faces of achiral cytosine monohydrate, the chiral dehydrated cytosine crystal was formed, the chirality of which corresponds to the enantiotopic face of achiral cytosine monohydrate [13c]. It is noteworthy that dehydration of crystal water of cytosine monohydrate under reduced pressure at room temperature instead of heating from the enantiotopic face of an achiral crystal afforded a chiral dehydrated cytosine crystal with the opposite chirality to that of the heated crystal [13d]. These chiral cytosine crystals formed by dehydration trigger asymmetric autocatalysis. Adenine, another achiral nucleobase, forms chiral

152

crystals of adenine dinitrate. Asymmetric autocatalysis is triggered by the chiral adenine dinitrate crystal [13e]. Thus, chiral crystals of achiral nucleobases such as cytosine and adenine act as the origins of chirality in conjunction with asymmetric autocatalysis.

Among the 20 proteinogenic amino acids on Earth, 19 amino acids exhibit the L-form. Glycine does not possess any asymmetric carbon atom and stands as the only achiral proteinogenic amino acid. The crystal structure of the γ-glycine polymorph is chiral. We correlated the absolute chiral crystal structure with optical rotation [13f] and CD spectra [13g]. We found that the $P3_2$ crystal of γ-glycine acts as the origin of chirality and that, in conjunction with asymmetric autocatalysis, (S)-pyrimidyl alkanol **1c** with up to >99.5% ee was formed [13g]. The $P3_1$ crystal afforded (R)-pyrimidyl alkanol **1c** with up to >99.5% ee. Even achiral glycine, as the chiral γ-polymorph, is capable of becoming the origin of chirality in conjunction with asymmetric autocatalysis. A wide range of chiral crystals composed of achiral compounds such as cocrystals of tryptamine/p-chlorobenzoic acid [13h] have been shown to trigger asymmetric autocatalysis.

3.5. Enantiotopic faces of achiral crystal composed of achiral organic compound

Some achiral organic crystals composed of achiral compounds have enantiotopic faces. 2-(t-Butyldimethylsilylethynyl)pyrimidine-5-carbaldehyde **2f** forms an achiral crystal (P-1). When the vapor of i-Pr$_2$Zn is exposed to the Re face of the crystal, (R)-pyrimidyl alkanol **1f** is formed with high ee [14]. Exposure of i-Pr$_2$Zn to the Si face affords (S)-pyrimidyl alkanol **1f**. Thus, the enantiotopic faces of an achiral crystal work as chiral trigger in asymmetric autocatalysis to afford a highly enantioenriched compound with the corresponding absolute configurations to those of the enantiotopic faces.

3.6. Absolute asymmetric synthesis in conjunction with asymmetric autocatalysis

There has been a common knowledge in organic reactions that, when chiral product C is formed in the reaction between achiral A and achiral B without the presence of any chiral factor, chiral product C always becomes the 1:1 mixture of two enantiomers, *i.e.*, racemate. To obtain enantioenriched product C, it is necessary to use some form of chiral factor such as a chiral catalyst. However, based on the theory of probability, the fluctuations in the ratio of the two enantiomers are almost always present in a given racemic mixture [15a]. Absolute asymmetric synthesis is the reaction in which enantioenriched product is formed from achiral reagents without the intervention of any chiral factor [1a].

As described in the preceding sections, asymmetric autocatalysis of pyrimidyl alkanol is capable of enhancing very low ee (*ca.* 0.00005% ee) to near enantiopurity (>99.5% ee) [8a,e]. Based on these results, we reasoned as follows: (1) when

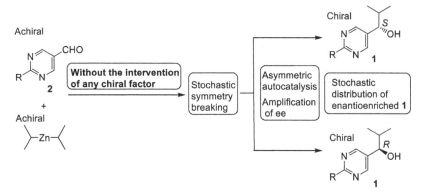

Fig. 8. Spontaneous absolute asymmetric synthesis in conjunction with asymmetric autocatalysis with amplification of ee.

pyrimidine-5-carbaldehyde and i-Pr$_2$Zn are reacted without the intervention of any chiral factor, S or R enantiomeric imbalances occur randomly in the initially formed isopropylzinc alkoxide of pyrimidyl alkanol that would act as asymmetric autocatalyst, and (2) the subsequent asymmetric autocatalysis with amplification of ee affords pyrimidyl alkanol with enhanced detectable ee.

We then examined the reaction between pyrimidine-5-carbaldehyde and i-Pr$_2$Zn without the addition of any chiral material (Figure 8). (S)- or (R)-Pyrimidyl alkanol with detectable ee was formed [15b]. In the 37 reactions that were carried out, in ether and toluene, the formation of (S)- and (R)-pyrimidyl alkanol **1c** was observed 19 and 18 times, respectively (Figure 9) [15c]. The distribution of the S and R enantiomers is stochastic. This result satisfies one of the conditions necessary for spontaneous absolute asymmetric synthesis. Furthermore, in the presence of achiral silica gel [15d] or achiral amine [15e], spontaneous absolute asymmetric synthesis, using pyrimidine-5-carbaldehyde **2c** and i-Pr$_2$Zn, was found to afford enantioenriched pyrimidyl alkanol **1c** with stochastic distribution of enantiomers. The absolute asymmetric synthesis of pyrimidyl alkanol in a stochastic distribution has also been reported [15f]. These results of spontaneous absolute asymmetric synthesis are essentially achieved under homogeneous conditions (excluding the reaction in the presence of insoluble silica gel [15d]).

Spontaneous absolute asymmetric synthesis under heterogeneous solid–vapor phase conditions was then examined. One of the characteristic features of i-Pr$_2$Zn is that it can be distilled. Vapor of i-Pr$_2$Zn/solvent was exposed to powder crystals of pyrimidine-5-carbaldehyde **2c** at 1 atm and room temperature (Figure 10) [15g]. In the 129 reactions that were carried out, the formation of (R)- and (S)-pyrimidyl alkanol **1c** was observed 61 and 58 times, respectively. In the other 10 reactions, the result was <0.5% ee. Powder crystals of pyrimidine-5-carbaldehyde **2c** locate in random orientation in vials, exposing Re or Si enantiotopic faces of the aldehyde **2c** to the approaching vapor of i-Pr$_2$Zn and toluene. On the surface of the aldehyde, at the most suitable position, the initial reaction forms either the (R)- or

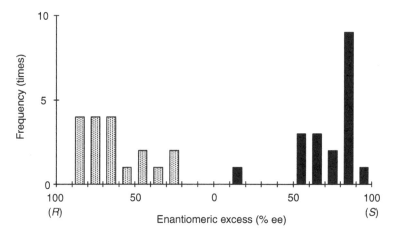

Fig. 9. Absolute asymmetric synthesis of pyrimidyl alkanol **1c** in the reaction of pyrimidine-5-carbaldehyde **2c** and i-Pr$_2$Zn: formation of (R)- and (S)-pyrimidyl alkanol **1c** in 19 times and 18 times, respectively [15c].

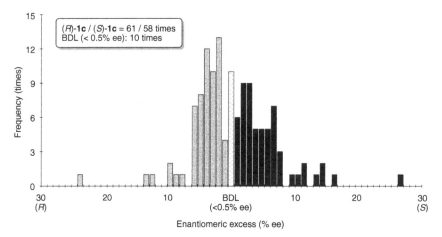

Fig. 10. Absolute asymmetric synthesis of pyrimidyl alkanol **1c** under solid-vapor phase conditions in the reaction of pyrimidine-5-carbaldehyde **2c** and i-Pr$_2$Zn [15g].

(S)-isopropylzinc alkoxide of pyrimidyl alkanol **1c**, then acting as the subsequent asymmetric autocatalyst with amplification of ee.

In summary, asymmetric autocatalysis with amplification of ee is capable of realizing spontaneous asymmetric synthesis.

3.7. Chiral carbon, nitrogen, oxygen and hydrogen isotopomers

There are many apparent achiral organic compounds **7-10** that become chiral upon substitution of hydrogen (^{1}H), carbon (^{12}C), nitrogen (^{14}N), and oxygen (^{16}O) for their isotopes of D (^{2}H), ^{13}C, ^{15}N, and ^{18}O, respectively. According to earlier reports

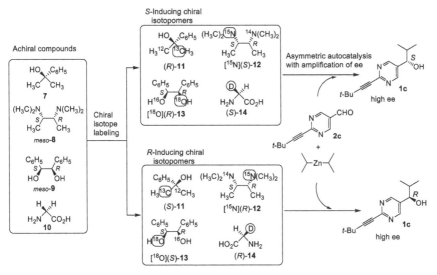

Fig. 11. Asymmetric autocatalysis triggered by chiral carbon (^{13}C/^{12}C), nitrogen (^{15}N/^{14}N), oxygen (^{18}O/^{16}O) and hydrogen (D/H) isotopomers.

on asymmetric synthesis using chiral hydrogen, the D/H isotopomer induces a chiral product with very low ee [16a]. Although the ratio of atomic weight difference of D/H is 100%, that of ^{13}C/^{12}C is only 8%. Therefore, the ratio of the atomic weight difference of other heavier isotope atoms (other than hydrogen) is too small to act as chiral auxiliaries in asymmetric reactions. Thus, the asymmetric induction of detectable ee is unprecedented by using carbon (^{13}C/^{12}C), nitrogen (^{15}N/^{14}N) and oxygen (^{18}O/^{16}O) isotopomers as chiral auxiliaries.

Dimethylphenylmethanol 7, an achiral compound, becomes chiral when one of the carbon atoms of the two methyl groups is labeled with ^{13}C. We found that (R)-dimethylphenylmethanol 11, arising from ^{13}C labeling of one of the methyl groups, triggered the reaction of pyrimidine-5-carbaldehyde 2c and i-Pr$_2$Zn to afford, in conjunction with asymmetric autocatalysis, (S)-pyrimidyl alkanol 1c with high ee (Figure 11) [16b]. In contrast, (S)-alkanol 11 triggered the formation of (R)-pyrimidyl alkanol 1c with high ee. Chiral nitrogen (^{15}N/^{14}N) [16c] 12 and oxygen (^{18}O/^{16}O) [16d] 13 isotopomers were found to trigger asymmetric autocatalysis to afford pyrimidyl alkanol 1c with high ee with the corresponding absolute configurations to those of chiral nitrogen and oxygen isotopomers.

Glycine 10 is an achiral amino acid that does not have an asymmetric carbon atom. When one of the hydrogen atoms of the methylene group is labeled with D, chiral (S)- or (R)-glycine-α-d 14 is formed. (S)-Glycine-α-d 14 triggers asymmetric autocatalysis to afford (S)-pyrimidyl alkanol 1c with high ee, while the opposite (R)-glycine-α-d 14 affords (R)-pyrimidyl alkanol 1c [16e]. Chiral primary alcohols due to deuterium substitution also work as chiral triggers of asymmetric autocatalysis of 1c [16f].

As described, we found that chiral isotopomers of carbon ($^{13}C/^{12}C$) **11**, nitrogen ($^{15}N/^{14}N$) **12** and oxygen ($^{18}O/^{16}O$) **13** are capable of inducing enantiomeric imbalances in the reaction between pyrimidine-5-carbaldehyde **2c** and i-Pr$_2$Zn, and that the subsequent asymmetric autocatalysis with amplification of ee affords pyrimidyl alkanol **1c** of high ee with the corresponding absolute configurations to those of the chiral isotopomers.

4. Discrimination of cryptochirality by asymmetric autocatalysis

Asymmetric autocatalysis is capable of discriminating cryprochiral compound. 5-Ethyl-5-propylundecane, *i.e.*, (n-butyl)ethyl(n-hexyl)(n-propyl)methane, is a cryptochiral compound. It does not exhibit detectable value of optical rotation because the differences in the structures of the four substituents are so small [17a]. By using the compound as a chiral trigger of asymmetric autocatalysis, its chirality was discriminated by analyzing the absolute configuration of the obtained pyrimidyl alkanol **1c** [17b].

Furthermore, unusual reversal of the sense of enantioselectivity was detected by asymmetric autocatalysis. Chiral ($1R,2S$)-N,N-dimethylnorephedrine (DMNE) triggers the formation of (R)-pyrimidyl alkanol **1c** in the asymmetric autocatalysis using pyrimidine-5-carbaldehyde **2c** and i-Pr$_2$Zn. Surprisingly, when a mixture of ($1R,2S$)-DMNE and achiral N,N-dibutylaminoethanol (DBAE) used as chiral trigger, the opposite enantiomer of (S)-pyrimidy alkanol **1c** was formed [17c,d].

5. Conclusions

We observed the first asymmetric autocatalysis of 3-pyrimidyl alkanol in the reaction between pyridine-3-carbaldehyde and dialkyl zincs. Asymmetric autocatalysis with amplification of ee was achieved using 5-pyrimidyl alkanol, 3-quinolyl alkanol, and 5-carbamoyl-3-pyridyl alkanol as asymmetric autocatalysts in the reactions between pyrimidine-5-carbaldehyde, quinoline-3-carbaldehyde, and 5-carbamoylpyridin-3-carbaldehyde, respectively. When 2-alkynyl-5-pyrimidyl alkanol with extremely low ee was used as asymmetric autocatalyst, three consecutive asymmetric autocatalysis reactions enhanced the ee to near enantiopurity (>99.5% ee), with a multiplication factor of *ca.* 630,000. Thus, it is demonstrated that there exists a chemical reaction in which the initial low ee can be enhanced to near enantiopurity.

Asymmetric autocatalysis was employed to examine the proposed theories of the origins of homochirality. Chiral inorganic crystals such as quartz, cinnabar, and sodium chlorate trigger asymmetric autocatalysis to afford pyrimidyl alkanol of high ee with the corresponding absolute configurations to those of chiral crystals. Thus, quartz acts as the origin of homochirality and, for the first time, the chirality of CPL is correlated with a chiral organic compound with high ee. Asymmetric photodecomposition and photoequilibrium by CPL afford, in conjunction with asymmetric autocatalysis, nearly enantiopure enantioenriched pyrimidyl alkanol. Spontaneous absolute asymmetric synthesis, i.e., asymmetric synthesis without the intervention of any chiral factor, is achieved in the reaction between pyrimidine-5-carbaldehyde and i-Pr$_2$Zn, followed by asymmetric autocatalysis. (S)- or (R)-Pyrimidyl alkanol of detectable ee are formed with stochastic distribution of enantiomers. Chiral crystals composed of achiral organic compounds such as cytosine and γ-glycine polymorph act as the origin of homochirality in asymmetric autocatalysis. Chiral carbon (^{13}C/^{12}C), nitrogen (^{15}N/^{14}N) and oxygen (^{18}O/^{16}O) isotopomers were found to induce chirality in conjunction with asymmetric autocatalysis.

Asymmetric autocatalysis, the Soai reaction, involves the catalytic replication of a chiral molecule. Replication and chirality constitute the essence of life. Asymmetric autocatalysis with amplification of enantiomeric excess sheds light on the origin of homochirality. It also sheds light on how replication began at the molecular level, and today it constitutes a branch of systems chemistry [18].

Acknowledgments

This work was financially supported by JSPS KAKENHI 19K05482.

References

1. (a) K. Mislow, Absolute asymmetric synthesis: A commentary, *Collect. Czech. Chem. Commun.* **68**, 849 (2003). (b) A. Guijarro and M. Yus, *The Origin of Chirality in the Molecules of Life* (The Royal Society of Chemistry: Cambridge, UK, 2009). (c) B. L. Feringa and R. A. van Delden, Absolute asymmetric synthesis: The origin, control, and amplification of chirality, *Angew. Chem. Int. Ed.* **38**, 3418 (1999). (d) Y. Inoue, Asymmetric photochemical reactions in solution, *Chem. Rev.* **92**, 741 (1992). (e) R. M. Hazen and D. S. Sholl, Chiral selection on inorganic crystalline surfaces, *Nat. Mater*, **2**, 367 (2003). (f) M. Bolli, R.Micura, and A. Eschenmoser, Pyranosyl-RNA: Chiroselective self-assembly of base sequences by ligative oligomerization of tetranucleotide-2', 3'-cyclophosphates (with a commentary concerning the origin of biomolecular homochirality), *Chem. Biol.* **4**, 309 (1997). (g) J. M. Ribó, J. Crusats, F. Sagués, J. Claret, and R. Rubires, Chiral sign induction by vortices during the formation of mesophases in stirred solutions, *Science* **292**, 2063 (2001). (h) K.-H. Ernst, Molecular chirality at surfaces, *Phys. Status Solidi* **249**, 2057 (2012). (i) I. Weissbuch and M. Lahav, Crystalline architectures as templates of relevance to the origins of homochirality, *Chem. Rev.* **111**, 3236 (2011). (j) A. Lennartson and M. Håkansson, Absolute asymmetric synthesis of five-coordinate complexes, *N.J. Chem.* **39**, 5936 (2015).
2. L. Pasteur, Recherches sur les relations qui peuvent exister entre la forme cristalline, la composition chimique et le sens de la polarisation rotatoire. *Ann. Chim. Phys.* **24**, 442 (1848).
3. (a) T. Satyanarayana, S. Abraham, and H. B. Kagan, Nonlinear effects in asymmetric catalysis. *Angew. Chem. Int. Ed.* **48**, 456 (2009). (b) M. Kitamura, S. Okada, S. Suga, and R. Noyori, Enantioselective addition of dialkylzincs to aldehydes promoted by chiral amino alcohols. Mechanism and nonlinear effect, *J. Am. Chem. Soc.* **111**, 4028 (1989). (c) D. K. Kondepudi and K. Asakura, Chiral autocatalysis, spontaneous symmetry breaking, and stochastic behavior, *Acc. Chem. Res.* **34**, 946 (2001). (d) C. Viedma, Chiral symmetry breaking during crystallization: Complete chiral purity

induced by nonlinear autocatalysis and recycling, *Phys. Rev. Lett.*, **94**, 065504 (2005). (e) V. A. Soloshonok, H. Ueki, M. Yasumoto, S. Mekala, J. S. Hirschi, and D. A. Singleton, Phenomenon of optical self-purification of chiral non-racemic compounds, *J. Am. Chem. Soc.* **129**, 12112 (2007). (f) Y. Hayashi, M. Matsuzawa, J. Yamaguchi, S. Yonehara, Y. Matsumoto, M. Shoji, D. Hashizume, and H. Koshino, Large nonlinear effect observed in the enantiomeric excess of proline in solution and that in the solid state, *Angew. Chem. Int. Ed.* **45**, 4593 (2006). (g) A. Córdova, M. Engqvist, I. Ibrahem, J. Casas, H. Sundén, Plausible origins of homochirality in the amino acid catalyzed neogenesis of carbohydrates, *Chem. Commun.* 2047 (2005). (h) M. M. Green, J. -W. Park, T. Sato, A. Teramoto, S. Lifson, R. L. B. Selinger, and J. V. Selinger, The macromolecular route to chiral amplification, *Angew. Chem. Int. Ed.*, **38**, 3138 (1999). (i) W. L. Noorduin, E. Vlieg, R. M. Kellogg, and B. Kaptein, From Ostwald ripening to single chirality, *Angew. Chem. Int. Ed.* **48**, 9600 (2009). (j) Y. Saito and H. Hyuga, Colloquium: Homochirality: Symmetry breaking in systems driven far from equilibrium, *Rev. Mod. Phys.* **85**, 603 (2013). (k) F. C. Frank, On spontaneous asymmetric synthesis, *Biochim. Biophys. Acta* **11**, 459 (1953). (l) J. Han, O. Kitagawa, A. Wzorek, K. D. Klia, and V. A. Soloshonok, The self-disproportionation of enantiomers (SDE): A menace or an opportunity? *Chem. Sci.* **9**, 1718 (2018). (m) C. Moberg, Recycling in asymmetric catalysis, *Acc. Chem. Res.* **49**, 2736 (2016).

4. K. Soai, S. Niwa, and H. Hori, Asymmetric self-catalytic reaction. Self-production of chiral 1-(3-pyridyl)alkanols as chiral self-catalysts in the enantioselective addition of dialkylzinc reagents to pyridine-3-carbaldehyde, *J. Chem. Soc. Chem. Commun.* 982 (1990).

5. T. Mukaiyama, K. Soai, T. Sato, H. Shimizu, and K. Suzuki, Enantioface-differentiating (asymmetric) addition of alkyllithium and dialkylmagnesium to aldehydes by using (2S, 2'S)-2-hydroxymethyl-1-[(1-alkylpyrrolidin-2-yl)methyl]pyrrolidines as chiral ligands. *J. Am. Chem. Soc.* **101**, 1455 (1979).

6. (a) K. Soai, A. Ookawa, T. Kaba, and K. Ogawa, Catalytic asymmetric induction. Highly enantioselective addition of dialkylzincs to aldehydes using chiral pyrrolidinylmethanols and their metal salts, *J. Am. Chem. Soc.* **109**, 7111 (1987). (b) K. Soai, H. Hori, and S. Niwa, Enantioselective addition of dialkylzincs to pyridinecarbaldehyde in the presence of chiral aminoalcohols: asymmetric synthesis of pyridylalkyl alcohols, *Heterocycles*, **29**, 2065 (1989). (c) K. Soai and S. Niwa, Enantioselective addition of organozinc reagents to aldehydes, *Chem. Rev.* **92**, 833 (1992).

7. (a) K. Soai, T. Shibata, and I. Sato, Enantioselective automultiplication of chiral molecules by asymmetric autocatalysis, *Acc. Chem. Res.* **33**, 382 (2000). (b) K. Soai and T. Kawasaki, Asymmetric autocatalysis with amplification of chirality, *Top. Curr. Chem.* **284**, 1 (2008). (c) K. Soai, T. Kawasaki, and A. Matsumoto, The origins of homochirality examined by using asymmetric autocatalysis, *Chem. Rec.* **14**, 70 (2014). (d) K. Soai, T. Kawasaki, and A. Matsumoto, Asymmetric autocatalysis of pyrimidyl alkanol and its application to the study on the origin of homochirality, *Acc. Chem. Res.* **47**, 3643 (2014). (e) K. Soai, Asymmetric autocatalysis. Chiral symmetry breaking and the origins of homochirality of organic molecules, *Proc. Jpn. Acad. Ser. B* **95**, 89 (2019).

8. (a) K. Soai, T. Shibata, H. Morioka, and K. Choji, Asymmetric autocatalysis and amplification of enantiomeric excess of a chiral molecule, *Nature* **378**, 767 (1995). (b) T. Shibata, H. Morioka, T. Hayase, K. Choji, and K. Soai, Highly enantioselective catalytic asymmetric automultiplication of chiral pyrimidylalcohol, *J. Am. Chem. Soc.* **118**, 471 (1996). (c) T. Kawasaki, M. Nakaoda, Y. Takahashi, Y. Kanto, N. Kuruhara, K. Hosoi, I. Sato, A. Matsumoto, and K. Soai, Self-replication and amplification of

enantiomeric excess of chiral multi-functionalized large molecule by asymmetric auto-catalysis *Angew. Chem. Int. Ed.* **53**, 11199 (2014). (d) T. Shibata, S. Yonekubo, and K. Soai, Practically perfect asymmetric autocatalysis using 2-alkynyl-5-pyrimidylalkanol, *Angew. Chem. Int. Ed.* **38**, 659 (1999). (e) I. Sato, H. Urabe, S. Ishiguro, T. Shibata, and K. Soai, Amplification of chirality from extremely low to greater than 99.5% ee by asymmetric autocatalysis, *Angew. Chem. Int. Ed.* **42**, 315 (2003).

9. (a) T. Shibata, K. Choji, T. Hayase, Y. Aizu, and K. Soai, Asymmetric autocat-alytic reaction of 3-quinolylalkanol with amplification of enantiomeric excess, *Chem. Commun.* 1235 (1996). (b) T. Shibata, H. Morioka, S. Tanji, T. Hayase, Y. Kodaka, and K. Soai, Enantioselective synthesis of chiral 5-carbamoyl-3-pyridyl alcohols by asymmetric autocatalytic reaction, *Tetrahedron Lett.* **37**, 8783 (1996).

10. (a) D. G. Blackmond, C. R. McMillan, S. Ramdeehul, A. Schorm, and J. M. Brown, Origins of asymmetric amplification in autocatalytic alkylzinc additions, *J. Am. Chem. Soc.* **123**, 10103 (2001). (b) M. Quaranta, T. Gehring, B. Odell, J. M. Brown, and D. G. Blackmond, Unusual inverse temperature dependence on reaction rate in the asymmetric autocatalytic alkylation of pyrimidyl aldehydes. *J. Am. Chem. Soc.* **132**, 15104 (2010). (c) T. Gehring, M. Quaranta, B. Odell, D. G. Blackmond, and J. M. Brown, Observation of a transient intermediate in Soai's asymmetric autocatalysis: Insights from ^{1}H NMR turnover in real time, *Angew. Chem. Int. Ed.* **51**, 9539 (2012). (d) L. Schiaffino and G. Ercolani, Unraveling the mechanism of the Soai asymmetric autocatalytic reaction by first-principles calculations: Induction and amplification of chirality by self-assembly of hexamolecular complexes, *Angew. Chem. Int. Ed.* **47**, 6832 (2008). (e) I. D. Gridnev and A. K. Vorobiev, Quantification of sophisticated equilibria in the reaction pool and amplifying catalytic cycle of the Soai reaction, *ACS Catal.* **2**, 2137 (2012). (f) J. C. Micheau, J. M. Cruz, C. Coudret, and T. Buhse, An auto-catalytic cycle model of asymmetric amplification and mirror-symmetry breaking in the Soai reaction, *ChemPhysChem* **11**, 3417 (2010). (g) J. C. Micheau, C. Coudret, J. M. Cruz, and T. Buhse, Amplification of enantiomeric excess, mirror-image symmetry breaking and kinetic proofreading in Soai reaction models with different oligomeric orders, *Phys. Chem. Chem. Phys.* **14**, 13239 (2012). (h) J. -C. Micheau, C. Coudret, J. -M. Cruz, and T. Buhse, Amplification of enantiomeric excess, mirror-image symmetry breaking and kinetic proofreading in Soai reaction models with different oligomeric orders, *Phys. Chem. Chem. Phys.* **214**, 13239 (2012). (i) K. Micskei, G. Rábai, E. Gál, L. Caglioti, and G. Pályi, Oscillatory symmetry breaking in the Soai reaction. *J. Phys. Chem. B* **112**, 9196 (2008). (j) J. Crusats, D. Hochberg, A. Moyano, and J. M. Ribó, Frank model and spontaneous emergence of chirality in closed systems. *Chem. Phys. Chem.* **10**, 2123 (2009). (k) É. Dóka, and G. Lente, Mechanism-based chemical understanding of chiral symmetry breaking in the Soai reaction. A combined probabilistic and deterministic description of chemical reactions, *J. Am. Chem. Soc.* **133**, 17878 (2011). (l) I. Sato, D. Omiya, K. Tsukiyama, Y. Ogi, and K. Soai, Evi-dence of asymmetric autocatalysis in the enantioselective addition of diisopropylzinc to pyrimidine-5-carbaldehyde using chiral pyrimidyl alkanol, *Tetrahedron Asymm.* **12**, 1965 (2001). (m) I. Sato, D. Omiya, H. Igarashi, K. Kato, Y. Ogi, K. Tsukiyama, and K. Soai, Relationship between the time, yield, and enantiomeric excess of asymmetric autocatalysis of chiral 2-alkynyl-5-pyrimidyl alkanol with amplification of enantiomeric excess, *Tetrahedron Asymm.* **14**, 975 (2003). (n) A. Matsumoto, T. Abe, A. Hara, T. Tobita, T. Sasagawa, T. Kawasaki, and K. Soai, Crystal structure of isopropylzinc alkoxide of pyrimidyl alkanol: Mechanistic insights for asymmetric autocatalysis with amplification of enantiomeric excess. *Angew. Chem. Int. Ed.* **54**, 15218 (2015). (o) M. E. Noble-Teran, J. -M. Cruz, J. -C. Micheau, and T. Buhse, A quantification of

the Soai reaction, *ChemCatChem* **10**, 642 (2018). (p) S. V. Athavale, A. Simon, K. N. Houk, and S. E. Denmark, Demystifying the asymmetry-amplifying, autocatalytic behaviour of the Soai reaction through structural, mechanistic and computational studies, *Nature Chem.* **12**, 412 (2020). (q) O. Trapp, S. Lamour, F. Maier, A. F. Siegle, K. Zawatzky, and B. F. Straub, In situ mass spectrometric and kinetic investigations of Soai's asymmetric autocatalysis, *Chem. Eur. J.* **26**, 15871 (2020). (r) D. G. Blackmond, Autocatalytic models for the origin of biological homochirality, *Chem. Rev.*, **120**, 4831 (2020). (s) T. Buhse, J.-M. Cruz, M. E. Noble-Terán, D. Hochberg, J. M. Ribó, J. Crusats, and J.-C. Micheau, Spontaneous Deracemizations, *Chem. Rev.* **121**, 2147 (2021). (t) A. J. Bissette and S. P. Fletcher, Mechanisms of autocatalysis, *Angew. Chem. Int. Ed.* **52**, 12800 (2013).

11. (a) T. Shibata, J. Yamamoto, N. Matsumoto, S. Yonekubo, S. Osanai, and K. Soai, Amplification of a slight enantiomeric imbalance in molecules based on asymmetric autocatalysis. -The first correlation between high enantiomeric enrichment in a chiral molecule and circularly polarized light, *J. Am. Chem. Soc.* **120**, 12157 (1998). (b) T. Kawasaki, M. Sato, S. Ishiguro, T. Saito, Y. Morishita, I. Sato, H. Nishino, Y. Inoue, and K. Soai, Enantioselective synthesis of near enantiopure compound by asymmetric autocatalysis triggered by asymmetric photolysis with circularly polarized light, *J. Am. Chem. Soc.* **127**, 3274 (2005). (c) I. Sato, R. Sugie, Y. Matsueda, Y. Furumura, and K. Soai, Asymmetric synthesis utilizing circularly polarized light mediated by the photoequilibrium of chiral olefins in conjunction with asymmetric autocatalysis. *Angew. Chem. Int. Ed.* **43**, 4490 (2004).

12. (a) W. A. Bonner, P. R. Kavasmaneck, F. S. Martin, and J. J. Flores, Asymmetric adsorption of alanine by quartz, *Science* **186**, 143 (1974). (b) K. Soai, S. Osanai, K. Kadowaki, S. Yonekubo, T. Shibata, and I. Sato, *d*- and *l*-Quartz-promoted highly enantioselective synthesis of a chiral organic compound, *J. Am. Chem. Soc.* **121**, 11235 (1999). (c) H. Shindo, Y. Shirota, K. Niki, T. Kawasaki, K. Suzuki, Y. Araki, A. Matsumoto, and K. Soai, Asymmetric autocatalysis induced by cinnabar: Observation of the enantioselective adsorption of a 5-pyrimidyl alkanol on the crystal surface, *Angew. Chem. Int. Ed.* **52**, 9135 (2013). (d) D. K. Kondepudi, R. J. Kaufman, and N. Singh, Chiral symmetry breaking in sodium chlorate crystallization, *Science* **250**, 975 (1990). (e) I. Sato, K. Kadowaki, and K. Soai, Asymmetric synthesis of an organic compound with high enantiomeric excess induced by inorganic ionic sodium chlorate, *Angew. Chem. Int. Ed.* **39**, 1510–1512 (2000). (f) A. Matsumoto, Y. Kaimori, M. Uchida, H. Omori, T. Kawasaki, and K. Soai, Achiral inorganic gypsum acts as an origin of chirality through its enantiotopic surface in conjunction with asymmetric autocatalysis, *Angew. Chem. Int. Ed.* **56**, 545 (2017).

13. (a) T. Matsuura and H. Koshima, Introduction to chiral crystallization of achiral organic compounds. Spontaneous generation of chirality, *J. Photochem. Photobiol. C Photochem. Rev.* **6**, 7 (2005). (b) T. Kawasaki, K. Suzuki, Y. Hakoda, and K. Soai, Achiral nucleobase cytosine acts as an origin of homochirality of biomolecules in conjunction with asymmetric autocatalysis, *Angew. Chem. Int. Ed.* **47**, 496 (2008). (c) T. Kawasaki, Y. Hakoda, H. Mineki, K. Suzuki, and K. Soai, Generation of absolute controlled crystal chirality by the removal of crystal water from achiral crystal of nucleobase cytosine, *J. Am. Chem. Soc.* **132**, 2874 (2010). (d) H. Mineki, Y. Kaimori, T. Kawasaki, A. Matsumoto, and K. Soai, Enantiodivergent formation of a chiral cytosine crystal by removal of crystal water from an achiral monohydrate crystal under reduced pressure, *Tetrahedron Asymm.* **24**, 1365 (2013). (e) H. Mineki, T. Hanasaki,

A. Matsumoto, T. Kawasaki, and K. Soai, Asymmetric autocatalysis initiated by achiral nucleic acid base adenine: Implications on the origin of homochirality of biomolecule, *Chem. Commun.* **48**, 10538 (2012). (f) K. Ishikawa, M. Tanaka, T. Suzuki, A. Sekine, T. Kawasaki, K. Soai, M. Shiro, M. Lahav, and T. Asahi, Absolute chirality of the γ-polymorph of glycine: Correlation of the absolute structure with the optical rotation, *Chem. Commun.* **48**, 6031 (2012). (g) A. Matsumoto, H. Ozaki, S. Tsuchiya, T. Asahi, M. Lahav, T. Kawasaki, and K. Soai, *Org. Biomol. Chem.* **17**, 4200 (2019). (h) T. Kawasaki, K. Jo, H. Igarashi, I. Sato, M. Nagano, H. Koshima, and K. Soai, Asymmetric amplification using chiral co-crystal formed from achiral organic molecules by asymmetric autocatalysis, *Angew. Chem. Int. Ed.* **44**, 2774 (2005).

14. T. Kawasaki, S. Kamimura, A. Amihara, K. Suzuki, and K. Soai, Enantioselective C–C bond formation as a result of the oriented prochirality of an achiral aldehyde at the single-crystal face upon treatment with a dialkyl zinc vapor, *Angew. Chem. Int. Ed.* **50**, 6796 (2011).

15. (a) W. H. Mills, Some aspects of stereochemistry, *Chem. Ind. (London)* **51**, 750 (1932). (b) K. Soai, T. Shibata Y. Kowata, Production of optically active pyrimidylalkyl alcohol by spontaneous asymmetric synthesis, *Jpn. Kokai Tokkyo Koho*, Patent No. JP1997-268179 (1997). An abstract is readily available as JPH09268179 from the European Patent Office. (c) K. Soai, I. Sato, T. Shibata, S. Komiya, M. Hayashi, Y. Matsueda, H. Imamura, T. Hayase, H. Morioka, H. Tabira, J. Yamamoto, and Y. Kowata, Asymmetric synthesis of pyrimidyl alkanol without adding chiral substances by the addition of diisopropylzinc to pyrimidine-5-carbaldehyde in conjunction with asymmetric autocatalysis, *Tetrahedron: Asymmetry* **14**, 185 (2003). (d) T. Kawasaki, K. Suzuki, M. Shimizu, K. Ishikawa, and K. Soai, Spontaneous absolute asymmetric synthesis in the presence of achiral silica gel in conjunction with asymmetric autocatalysis, *Chirality* **18**, 479 (2006). (e) K. Suzuki, K. Hatase, D. Nishiyama, T. Kawasaki, and K. Soai, Spontaneous absolute asymmetric synthesis promoted by achiral amines in conjunction with asymmetric autocatalysis, *J. Systems Chem.* **1**, 5 (2010). (f) D. A. Singleton, and L. K. Vo, A few molecules can control the enantiomeric outcome. Evidence supporting absolute asymmetric synthesis using the Soai asymmetric autocatalysis, *Org. Lett.* **5**, 4337 (2003). (g) Y. Kaimori, Y. Hiyoshi, T. Kawasaki, A. Matsumoto, and K. Soai, Formation of enantioenriched alkanol with stochastic distribution of enantiomers in the absolute asymmetric synthesis under heterogeneous solid–vapor phase conditions, *Chem. Commun.* **55**, 5223 (2019).

16. (a) A. Horeau, A. Nouaille, and K. Mislow, Secondary deuterium isotope effects in asymmetric syntheses and kinetic resolutions, *J. Am. Chem. Soc.* **87**, 4957 (1965). (b) T. Kawasaki, Y. Matsumura, T. Tsutsumi, K. Suzuki, M. Ito, and K. Soai, Asymmetric autocatalysis triggered by carbon isotope ($^{13}C/^{12}C$) chirality, *Science* **324**, 492 (2009). (c) A. Matsumoto, H. Ozaki, S. Harada, K. Tada, T. Ayugase, H. Ozawa, T. Kawasaki, and K. Soai, Asymmetric induction by nitrogen $^{14}N/^{15}N$ isotopomer in conjunction with asymmetric autocatalysis, *Angew. Chem. Int. Ed.* **55**, 15246 (2016). (d) T. Kawasaki, Y. Okano, E. Suzuki, S. Takano, S. Oji, and K. Soai, Asymmetric autocatalysis: Triggered by chiral isotopomer arising from oxygen isotope substitution, *Angew. Chem. Int. Ed.* **50**, 8131 (2011). (e) T. Kawasaki, M. Shimizu, D. Nishiyama, M. Ito, H. Ozawa, and K. Soai, Asymmetric autocatalysis induced by meteoritic amino acids with hydrogen isotope chirality, *Chem. Commun.* 4396 (2009). (f) I. Sato, D. Omiya, T. Saito, and K. Soai, Highly enantioselective synthesis induced by chiral primary alcohols due to deuterium substitution, *J. Am. Chem. Soc.* **122**, 11739 (2000).

17. (a) H. Wynberg, G. L. Hekkert, J. P. M. Houbiers, and H. W. Bosch, The Optical activity of butylethylhexylpropylmethane, *J. Am. Chem. Soc.* **87**, 2635 (1965). (b) T. Kawasaki, H. Tanaka, T. Tsutsumi, T. Kasahara, I. Sato, and K. Soai, Chiral discrimination of cryptochiral saturated quaternary and tertiary hydrocarbons by asymmetric autocatalysis, *J. Am. Chem. Soc.* **128**, 6032 (2006). (c) F. Lutz, T. Igarashi, T. Kawasaki, and K. Soai, Small amounts of achiral beta-amino alcohols reverse the enantioselectivity of chiral catalysts in cooperative asymmetric autocatalysis, *J. Am. Chem. Soc.* **127**, 12206 (2005). (d) F. Lutz, T. Igarashi, T. Kinoshita, M. Asahina, K. Tsukiyama, T. Kawasaki, and K. Soai, Mechanistic insights in the reversal of enantioselectivity of chiral catalysts by achiral catalysts in asymmetric autocatalysis, *J. Am. Chem. Soc.* **130**, 2956 (2008).
18. G. Ashkenasy, T. M. Hermans, S. Otto, and A. F. Taylor, Systems chemistry, *Chem. Soc. Rev.* **46**, 2543 (2017).

On Chirality, Symmetry, Entropy and Isotopes

R. A. Zubarev[†]

Division of Chemistry I, Department of Medical Biochemistry & Biophysics,
Karolinska Institutet, Stockholm, 17177, Sweden
[†] *E-mail: roman.zubarev@ki.se*
www.ki.se

Stable isotopes of light elements, with the exception of deuterium, have all been considered nearly equal participants in biochemical reactions. This paradigm is now questioned. Does the isotopic composition of biological molecules, such as enzymes, affect significantly their kinetic behavior, and if yes, what is the mechanism of such action? We found that the isotopic effects are indeed far greater than the conventional kinetic isotopic theory predicts, and are in broad agreement with the Isotopic Resonance hypothesis.

Keywords: Enzymes; Kinetics; Isotopes

1. Isotopic symmetry and isotopic resonance

1.1. *What is symmetry?*

The Nobel laureate PW Anderson is credited with postulating that "it is only slightly overstating the case to say that Physics is the study of symmetry." Yet the definition of symmetry is surprisingly elusive for being a central concept of an exact science. The Oxford dictionary defines symmetry as "correct or pleasing proportion of the parts". In other words, one knows symmetry when one sees it. In agreement with that Anderson also said that "by symmetry we mean the existence of different viewpoints from which a system appears the same". But such different viewpoints may not be so easy to find even when they do exist. Also, what happens when from some viewpoint the system looks almost the same but not quite? Are there different kinds of symmetry and different grades of symmetry?

At this point it makes sense to inquire whether any of these questions are practically important. They may be, if symmetry of molecules affects their chemical properties or kinetics of their chemical reactions. Indeed, according to the Jahn-Teller (JT) theorem, spatially symmetric molecules are less thermodynamically stable due to degeneracy of quantum-mechanical states and thus are more active chemically. What if the same is valid for all kinds of symmetry, not only geometric one?

1.2. *What is isotopic resonance?*

More than a decade ago it was found that the average terrestrial isotopic compositions of the four most biologically important elements, namely carbon, hydrogen, nitrogen and oxygen are not random. [1] More specifically, the relationship between their nominal masses (numbers of nucleons), monoisotopic masses (in calculation of which only the most abundant light isotopes are considered) and average isotopic masses is degenerate for a large class of proteins, the most biologically important molecules. Furthermore, it appears that the proportion of the isotopically degenerate (or "resonant") molecules has been much higher in the distant past, when first primitive life appeared on the planet. The "isotopic resonance" (IsoRes) hypothesis formulated around 2009 postulates that the presence of such non-random isotopic compositions was beneficial for early life, as isotopically resonant molecules are more chemically active and thus have faster kinetics.

During the first years of its existence, the IsoRes hypothesis was just a theoretical curiosity. Albeit supported by literature data, [2] it has not been backed up experimentally. Bust since it made testable predictions, in early 2010s the idea to verify these predictions gained support of the Swedish Research Council. The easiest venue for testing was found in microbiology rather than in synthetic chemistry. Indeed, microorganisms, such as the laboratory strains of *Escherichia coli*, are self-organizing bioreactors that can multiply within a day or two thousand-fold in a minimal media composed of water, inorganic salts and a simple carbon source, such as sugar or an amino acid. It is quite easy to manipulate the isotopic composition of such a media, moving the system closer to or farther away from the resonance at will.

1.3. *Testing the isotopic resonance hypothesis*

As often the case, this new tool created new opportunities. Theoretical considerations suggested that there are many isotopic resonances of different strengths, and that the terrestrial IsoRes is not among the strongest ones (the resonance strength is assessed by the fraction of protein molecules that are "on the resonance" at a given isotopic composition of C, H, N and O). Thus other, stronger resonances should give even higher rates of bacterial growth than terrestrial conditions. This prediction of the IsoRes hypothesis was tested in thousands individual bacterial experiments on several unnatural resonances, and by 2015 the IsoRes hypothesis was confirmed at $p << 10^{-6}$ level. [3] Particularly impressive was the significantly faster growth at the "super-resonance" when two out of four elements were enriched with heavy isotopes to a >10% degree and one – to a >6% degree.

There was still uncertainty related to the role in IsoRes of the heavy stable hydrogen isotope deuterium. Indeed, deuterium ^2H is known to exert on biological organisms by far the largest effect due to its large relative difference in mass and physico-chemical properties with the light hydrogen isotope ^1H. The so far tested isotopic resonances were obtained at enrichment or depletion of the elements C, N and O, but not hydrogen. A separate study on the effect of low-level deuterium enrichment on bacterial growth gave results in qualitative agreement with the IsoRes hypothesis, but quantitatively shaky. [4] A new and broader study on the role of deuterium in IsoRes was required.

It was realized that, as hydrogen has two stable isotopes and oxygen has three isotopes, at certain conditions even water becomes isotopically resonant. One such resonant condition occurs when the deuterium content roughly doubles to

250–350 ppm. This resonance was tested first on a simple biochemical system containing the recombinant enzyme luciferase, its substrate luciferin and ATP. [5] The reaction between these reactants produces light at a certain wavelength, which determines the wide use of this reaction in the read-out step of many biochemical assays.

As all chemicals were highly purified and competently handled in manufacturer's laboratory, the measurements were expected to be very precise, but early results turned out to be noisy. It was then realized that, as deuterium enrichment in water was obtained by mixing normal water with heavy water ^2H_2O, achieving homogeneous mixture was a problem. After realizing that several days of heating and stirring were needed to prepare a single mixture, this obstacle was overcome, and the precision of measurements was significantly improved. The results showed a staggering 30% reduction in luciferase activity at the predicted resonance (it should be noted that the IsoRes hypothesis predicts at a resonance only a large change in kinetics but not its direction; having said that, in most systems studied so far the kinetics increased).

As bacteria, especially *E. coli,* are much more robust towards external influences than mammalian cells, the effect of the deuterium content in water was studied on human cell growth. A significant (up to 30%) increase in growth of A549 cells was found at ≈350 ppm deuterium in growth media, in agreement with earlier less quantitative observations. [6] Using sophisticated methods of chemical proteomics, it was found that mitochondria inside the cells are the entities most affected by water deuterium content, which reflects in their production of reactive oxygen species (ROS).

Since the effect of the isotopic resonance conditions on biological systems is now firmly established, supporting the role of IsoRes in life emergence or taking root on our planet, the question arises what implications IsoRes phenomenon has for the present. One IsoRes aspect still concerns the origin of life, as neither Mars nor Venus have conditions close to any strong isotopic resonance for protein molecules. The conclusion that "life on Mars is unlikely" [7] may sound too categorical, but this is exactly what follows from IsoRes studies at Mars-like conditions. Furthermore, exoplanets' isotopic signatures for C, H, N and O (when these can be obtained with sufficient precision) could be used to decide whether or not a given planet has conditions advantageous for protein life forms. It is also not unlikely that chirality itself, the prime feature of biology, emerges naturally as a byproduct of an isotopic resonance.

2. Towards monoisotopic biotechnology

One intriguing prediction of the IsoRes hypothesis is that monoisotopic molecules are the "hottest" and thus monoisotopic enzymes should work faster (or slower, in an unfortunate but less likely case). The problem is in obtaining truly monoisotopic enzymes, as an average protein molecule contains several thousands carbon atoms and hundreds of nitrogen and oxygen atoms, each of which having to be a single isotope. However, we know from the IsoRes studies that the isotopic resonances tend to be quite broad, and thus even significant depletion of the minor (heavy) isotopes of C, H, N and O might still provide significant kinetic boost. This viewpoint gave a platform for testing the monoisotopic IsoRes prediction: as many enzymes can be recombinantly expressed in *E. coli* bacteria grown on minimal media, a nearly monoisotopic growth media should yield nearly monoisotopic, or at least significantly depleted (ultra-light) enzymes.

Such experiments have recently been performed, and the results exceeded the expectations: removing at least half of the natural heavy isotopes produced a significant boost in enzyme kinetics, multiplying the reaction rates by a factor of 2 to 3 depending upon the enzyme. [8] Note that the conventional kinetic isotopic effect in this case is only of the order of 0.1%.

Monoisotopic biotechnology is thus has been born as a field but it is still in infancy. There are many mysteries that are yet to be uncovered, and several important questions remain unanswered. It is clear however that the old paradigm prescribing to ignore the role of stable isotopes in biology or reduce them to negligible manifestations needs to be amended. The isotopic dimension of biology may be as complex and nearly as important as the role of post-translational modifications in protein function.

Acknowledgments

This work was supported by the Swedish Research Council's grant 2017-04303.

References

1. R. A. Zubarev, K. A. Artemenko, A. R. Zubarev, C. Mayrhofer, H. Yang, E. Y. M. Fung, Early life relict feature in peptide mass distribution, *Cent. Eur. J. Biol.* **5**, 190 (2010).
2. R. A. Zubarev, Role of Stable Isotopes in Life - Testing Isotopic Resonance Hypothesis, *Genomics, Proteomics & Bioinformatics* **9**, 15 (2011).
3. X. Xie, R. A. Zubarev, Isotopic Resonance Hypothesis: Experimental Verification by *Escherichia coli* Growth Measurements, *Sci. Rep.* **5**, #9215 (2015).
4. X. Xie, R. A. Zubarev, Effects of low-level deuterium enrichment on bacterial growth, PLOS ONE, **9**, e102071 (2014).
5. S. Rodin, P. Rebellato, A. Lundin, R. A. Zubarev, Isotopic resonance at 370 ppm deuterium negatively affects kinetics of luciferin oxidation by luciferase, *Sci. Rep.* **8**, #16249 (2018).
6. X. Zhang, J. Wang, R. A. Zubarev, Slight deuterium enrichment in water acts as an antioxidant: is deuterium a cell growth regulator? *Mol. & Cell. Proteomics* **19**, 1790 (2020).
7. X. Xie, R. A. Zubarev, On the Effect of Planetary Stable Isotope Compositions on Growth and Survival of Terrestrial Organisms, *PLOS ONE* **12**, e0169296 (2017).
8. X. Zhang, Z. Meng, C. Beusch, H. Gharibi, Q. Cheng, L. di Stefano, J. Wang, A. Saei, A. Vegvari, M. Gaetani, R. A. Zubarev, https://assets.researchsquare.com/files/rs-1103656/v1/3df32040-a0c1-4991-bc0f-4b06ea881147.pdf?c=1639602628

The Chiral-Induced Spin Selectivity Effect

R. Naaman

*Department of Chemical and Biological Physics, Weizmann Institute of Science,
Rehovot 7610001, Israel*
† *E-mail: ron.naaman@weizmann.ac.il*
https://www.weizmann.ac.il/chembiophys/naaman/home

It has been shown that when electrons are transferred through chiral molecules, the efficiency of the transfer depends on the electron's spin and on the handedness of the chiral potential. Hence, chiral molecules serve as spin filters. This effect is termed chiral-induced spin selectivity (CISS). The dependence of the CISS effect on the molecular properties is discussed here as well as some implications of the effect such as in bio-recognition and in the interaction of chiral molecules with ferromagnetic substrates. The role of the CISS effect in various fields is also reviewed.

Keywords: Spin, Electrons, Chirality, Enantio-selectivity, ferromagnetism.

1. Introduction

The chiral symmetry in molecules was recognized in the 19th century by Pasteur, and Lord Kelvin first used this term. This phenomenon was established as an important structural parameter, and it was found in Biology that amino acids and DNA appear almost exclusively with one handedness. This issue of "homo-chirality" was and is still discussed extensively. Present discussion focuses on a question that should precede the question of the origin of "homochirality", namely, why does chirality exist at all? Why does Nature preserve chirality so persistently throughout evolution without a single known case in which mutations converted a chiral system to an achiral one. Maintenance of chirality is especially surprising when one realizes that it requires energy, since in chiral systems the entropy is lower than in racemic mixtures. The 'conventional wisdom' about chirality in biomolecules is that it serves as a structural motif that places chemical functionalities in defined positions and orientations that enable biologically relevant functions. However, as was discovered in the last decade, chiral molecules also have unique electronic and spintronic properties that may explain their important role in Biology. Those properties result from the chiral-induced spin selectivity (CISS) effect [1].

According to the CISS effect, when electrons move within a chiral potential, their motion is spin dependent; therefore, one spin state is preferred. Which spin is preferred depends on the handedness of the potential: for one handedness the spin is aligned parallel to the velocity, whereas for the other, the spin is aligned antiparallel to the velocity.

Extensive theoretical studies were performed in an attempt to reveal the mechanism underlying this effect [2–13]. It was soon realized that indeed, some spin-dependent conduction is obtained, due to the chiral structure. However, since the magnitude of the spin orbit coupling is associated with hydrocarbons, it is impossible to explain the large spin polarization observed experimentally at room temperature. It was suggested that to obtain quantitative agreement between the theory and the experiments, (1) the "spin selectivity" should be replaced with "angular momentum" selectivity and the spin observed in the SOC should be associated with leads [14–16], or (2) an enhancement mechanism should be found that will somehow magnify the effect of the small SOC [17–19]. Recently, models that include spin interaction with polarons [20], vibrations [21, 22], and electron correlation [19] were considered and were able to obtain spin polarization values that are similar to those obtained experimentally.

The recent theoretical and experimental studies seem to indicate that the CISS effect is not a purely electronic effect and that it involves coupling with the nuclei degrees of freedom. In this way, low frequency vibrations that have angular momentum, termed "chiral vibrations/phonons", enhance the effective SOC.

As a result of the CISS effect, when a chiral molecule is charged polarized due to an external electric field, this charge polarization is accompanied by transient spin polarization. It is important to emphasize that the charge polarization process itself involves mixing with an electronic excited state and with vibrations; hence, it can be accompanied by spin polarization, since the quantum state associated with the polarization is not the electronic ground state of the system.

2. What molecular properties determine the magnitude of the CISS effect?

All models associated with the CISS effect result in some optimum ratio between the pitch of the chiral-helical potential and its radius. Typically this optimal ratio approaches unity. However, it is not always obvious from the structure of the molecule what the two parameters that govern the electron motion are. In addition, there must be additional properties that control the magnitude of the spin polarization. Early experimental studies have established that for a given oligomer, such as an oligopeptide or DNA, the spin polarization scales with the length of the oligomer, within the size range investigated, namely, up to about 10 nm. This experimental observation was also reproduced by some of the model calculations.

Several experimental studies revealed the correlation between the optical activity of the molecules studied and the sign and magnitude of the spin polarization [23–26]. For experiments performed with supramolecular wires made from chiral and achiral porphyrines, the correlation between the optical activity and the spin polarization was revealed. As shown in Figure 1a, the molar circular dichroism ($\Delta\varepsilon$) shows a non-linear dependence on the fraction of the chiral molecules in the solution, confirming the chiral amplification in the system. An example of this is the study described in ref. [24], in which we made use of the well-known "Sergeant and Soldiers" principle of chiral amplification [27, 28]. Mixing chiral and achiral building blocks leads to amplification in the net helicity of a supramolecular wire structure. Here, the chiral molecule determines the supramolecular chirality of the helix and the achiral molecules follow the helicity of the chiral one. Sergeant and soldier experiments with chiral amplification can be expressed by the

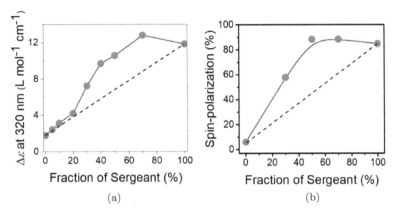

Fig. 1. CD signal and spin polarization. (a) Evolution of the molar circular dichroism ($\Delta\varepsilon$) as a function of a fraction of chiral sergeant in a sergeant and solider experiment involving chiral and achiral molecules, as measured by circular dichroism spectroscopy. The non-linear dependence of $\Delta\varepsilon$ on the fraction of sergeant suggests chiral amplification in the solution-state assembly processes. (b) Spin-polarization at +3 V measured using mc-AFM-based I-V curves at different fractions of chiral molecules in nanofibers made from a mixture of chiral and achiral porphyrines. The same solutions used for CD studies were used for forming the nanofibers by drop-casting on Au/Ni substrates. The solid gray lines in (a) and (b) guide the eye. The dashed black line denotes the expected trend in the data in the absence of chiral amplification (copied with permission from ref. [24]).

spin polarization of electron transport measured via magnetic contact atomic force microscope (mc-AFM) studies on nanofibers drop-casted on Au/Ni substrates from mixtures of chiral and achiral molecules at various compositions. Interestingly, we have found a non-linear increase in spin-polarization in up to 30% of the chiral molecules; this strongly resembles the trend observed in the CD studies. At a fraction of 50% chiral molecules and beyond, the magnitude of spin-polarization is comparable to that observed for nanofibers assembled from enantiomerically pure chiral wires (Figure 1b). The results directly prove that the CISS effect results from the supramolecular chirality and is not just due to the number of stereocenters present in the film.

Hence, the optical activity of molecules seems to be a good predictor of the efficiency of the CISS effect and shows a sign of the preferred spin.

3. Spin-dependent charge reorganization (SDCR) in chiral molecules

When a molecule is exposed to an electric field, or when a molecule interacts with another molecule or with a substrate, charge reorganization takes place. Regarding molecules interacting with other molecules or with a substrate, the charge reorganization is characterized by London's dispersion forces [29]. Typically when calculated, the wavefunction that results is not antisymmetric. It is assumed that the forces have a long range and therefore the electron wave functions of the two interacting species do not significantly overlap. This issue was discussed in several works that provided a path towards a symmetrically correct wave function [30, 31]. However, the role of the electrons' spin was almost always neglected because of two reasons; First, the coupling of the spin to the molecular axis is weak; hence, any spin-exchange term is supposed to average to zero. Second, the exchange terms have a short range and therefore they are not supposed to significantly affect the interaction.

The common concepts described above must be modified when chiral molecules are concerned, especially chiral molecules in a bio-environment, when the molecules are very crowded and therefore, the interactions have a very short range. Owing to the CISS effect, when charge is reorganized in chiral molecules, due to dispersion forces or due to interactions that involve two species with different electro-chemical potentials, this charge reorganization is accompanied by a transient spin polarization. Hence, when a dipole moment is induced in a chiral molecule, each electric pole is associated with a specific spin. Which spin is associated with which pole depends on the handedness of the molecule [32]. Since the spin is strongly coupled, in this case, to the molecular frame, one has to consider the spin exchange interaction. One then realizes that this interaction will be different for the interaction of two molecules with the same handedness or molecules with opposite handedness. Therefore, beyond the spatial effect in the enantioselective interaction, there is an additional electronic term that is enantio-specific. This spin exchange term can reach values of hundreds of meV electrons, when two chiral molecules are in proximity. Recent studies, using atomic force microscopy, indeed support the existence of this interaction and its magnitude [33].

An interesting manifestation of the SDCR is the observation of thermally activated ferromagnetism in chiral metalo-organic crystals. When the magnetic properties of Cu-phenyl-alanine crystals were studied, it was found that these chiral crystals indeed exhibit the CISS effect; however, they also exhibit a very peculiar behavior. For example, at low temperatures, of up to about 50K, these crystals are antiferromagnetic. Upon warming, they exhibit ferromagnetic properties that increase with temperature [34]. It was suggested that this effect results from the interaction between the unpaired electron on the copper ion and the chiral lattice. This interaction increases with temperature due to the anharmonicity of the potential of the copper ion. The motion of the ion towards the chiral lattice induces SDCR in the lattice and as a result, an exchange interaction occurs between the polarized spin on the lattice and the spin on the ion. Figure 2 presents evidence for the CISS effect in 300 nm-thick crystals of Cu-phenylalanine, measured by a contact magnetic atomic force microscope. In addition, the magnetic hysteresis is shown for 4 and 300K, measured by a superconducting quantum interference device (SQUID). The results indicate that hysteresis is enhanced with increasing temperature.

The temperature-activated ferromagnetism is a new form of magnetism that does not result from the exchange interaction among the unpaired electrons, but rather, by interaction of each unpaired electron with the chiral lattice. Here the hysteresis increases with temperature, and it does not display a sharp square-like

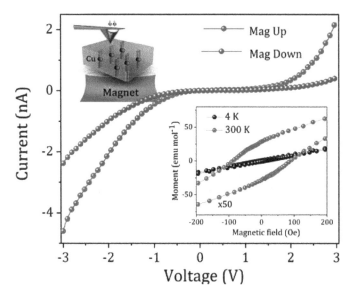

Fig. 2. The CISS effect and the temperature-enhanced magnetism. The image presents current versus voltage curves measured for Cu-phenyl alanine crystal by a magnetic conducting atomic force microscope (see the inserted image) when the magnet is pointing up (red) or down (blue). A clear spin-dependent conduction is observed. In the inserted curves the magnetic moment as a function of an applied field is shown for the same crystals at 4 and 300K, measured by SQUID. The hysteresis increases with increasing temperature. Copied with permission from ref. [34].

shape as a function of the applied magnetic field, but rather, it decreases slowly as a function of the applied field (see the insert in Figure 2).

The SDCR effect is also manifested in the case of interaction between chiral molecules, and it clearly contributes to the enantioselectivity of this interaction [32]. Another result from the SDCR effect is the enantiospecific interaction of chiral molecules with a ferromagnetic substrate, when the substrate is magnetized perpendicular to the surface. This property is utilized to separate enantiomers from racemic mixtures [35, 37, 38].

4. Summary

The CISS effect has a wide range of implications in various disciplines. Next, some of these implications will be briefly described.

When considering Physics and electronic devices, the CISS effect presents chiral molecules as "nano devices" that produce spin-polarized electrons without the need to have a ferromagnet as a spin injector [39]. The advantage here is the ability to miniaturize devices, since the reduction in the size of a regular ferromagnet is limited by the superparamagnetic-ferromagnetic transition. Namely, a small ferromagnet becomes superparamagnetic and therefore cannot function as a spin-selective source. However, chiral molecules or inorganic chiral materials are not limited by this size constraint. In addition, the CISS effect presents the chiral molecules as a new topological material, as was described recently [16].

In Chemistry the CISS effect was shown to allow the enhancement of multiple electron reactions, by controlling the relative orientation of the electron spins. An interesting example is the enhancement observed in the case of water splitting, where in addition, unwanted byproducts could also be eliminated [40, 41]. The CISS effect opened a new venue towards efficient enantioseparation [35] of chiral molecules by magnetic substrates, and as was recently shown, it allows for spin enantio-selective Chemistry [42].

In Biology the CISS effect may provide the answers to several phenomena not completely understood so far. It explains how very long-range electron transfer occurs in Biology and specifically, how it occurs through proteins. Apparently, the coupling between the spin and the electron's linear momentum protects the electron from being back scattered [26]. This property enhances the range of efficient electron transfer [43]. The spin-dependent charge reorganization also explains the long-range allosteric control of protein association [44]. Owing to the SDCR effect, the bio-affinities and the enantioselectivity of biological processes can be better understood and can in principle be calculated better [35]. In addition, the CISS effect increases the efficiency of multiple electron redox reactions that involve oxygen and may explain the efficient respiration process [36]. Hence the CISS effect explains why chirality is preserved so persistently in Nature through evolution.

The short summary provided here indicates the impact the CISS effect has; it is clear that when more studies will be performed, the influence of the effect will be explored even further. It is also expected that technologies, such as spin-dependent OLED, spin-dependent photovoltaic, and hydrogen production will emerge as a result of the CISS phenomenon.

Acknowledgments

The author is very thankful to all his collaborators over the years and especially to Profs. David H. Waldeck and Yossi Paltiel for being essential in developing the research on CISS.

References

1. R. Naaman, Y. Paltiel, D. Waldeck, Chiral molecules, and the electron's spin. *Nature Reviews Chemistry*, **3**, 250–260 (2019).
2. S. Yeganeh, M. A. Ratner, E. Medina, V. Mujica, Chiral electron transport: Scattering through helical potentials. *J. Chem. Phys.*, **131**, 014707 (2009).
3. A.-M. Guo, Q.-f. Sun, Spin-selective transport of electrons in DNA double helix. *Phys. Rev. Lett.*, **108**, 218102 (2012).
4. R. Gutierrez, E. Diaz, C. Gaul, T. Brumme, F. Dominguez Adame, G. Cuniberti, Modeling spin transport in helical fields: derivation of an effective low-dimensional Hamiltonian. *J. Phys. Chem. C*, **117**, 22276–22284 (2013).
5. A.-M. Guo, E. Diaz, C. Gaul, R. Gutierrez, F. Dominguez Adame, G. Cuniberti, Q.-f. Sun, Contact effects in spin transport. *Nano Lett.*, **19**, 5253–5259 (2019).
6. A.-M. Guo, Q.-F. Sun, Spin-dependent electron transport in protein-like single-helical molecules. *Proc. Natl. Acad. Sci. U. S. A.*, **111**, 11658–11662 (2014).

7. H.-N. Wu, Y.-L. Zhu, X. Sun, W.-J. Gong, Spin polarization and spin separation realized in the double-helical molecules. *Phys. E*, **74**, 156–159 (2015).
8. E. Medina, L. A. Gonzalez-Arraga, D. Finkelstein-Shapiro, B. Berche, V. Mujica, Continuum model for chiral induced spin selectivity in helical molecules. *J. Chem. Phys.*, **142**, 194308 (2015).
9. S. Matityahu, Y. Utsumi, A. Aharony, O. Entin-Wohlman, C. A. Balseiro, Spin-dependent transport through a chiral molecule in the presence of spin-orbit interaction and nonunitary effects. *Phys. Rev. B: Condens. Matter Mater. Phys.*, **93**, 075407 (2016).
10. T.-R. Pan, A.-M. Guo, Q.-F. Sun, Spin-polarized electron transport through helicene molecular junctions. *Phys. Rev. B: Condens. Matter Mater. Phys.*, **94**, 235448 (2016).
11. S. Matityahu, A. Aharony, O. Entin-Wohlman, C. A. Balseiro, Spin filtering in all-electrical three-terminal interferometers. *Phys. Rev. B: Condens. Matter Mater. Phys.*, **95**, 085411 (2017).
12. E. Diaz, F. Dominguez-Adame, R. Gutierrez, G. Cuniberti, V. Mujica, Thermal decoherence and disorder effects on chiral induced spin selectivity. *J. Phys. Chem. Lett.*, **9**, 5753–5758 (2018).
13. V. V. Maslyuk, R. Gutierrez, A. Dianat, V. Mujica, G. Cuniberti, Enhanced magnetoresistance in chiral molecular junctions. *J. Phys. Chem. Lett.*, **9**, 5453–5459 (2018).
14. S. S. Skourtis, D. N. Beratan, R. Naaman, A. Nitzan, D. H. Waldeck, Chiral control of electron transmission through molecules. *Phys Rev Lett.*, **101**, 238103 (2008).
15. J. Gersten, K. Kaasbjerg, A. Nitzan, Induced spin filtering in electron transmission through chiral molecular layers adsorbed on metals with strong spin–orbit coupling. *J. Chem. Phys.*, **139**, 114111 (2013).
16. Y. Liu, J. Xiao, J. Koo, B. Yan, Chirality-driven topological electronic structure of DNA-like materials. *Nat. Mat.*, **20**, 638–644 (2021)
17. S. Dalum, P. Hedegård, Theory of chiral induced spin selectivity. *Nano Lett.*, **19**, 5253–5259 (2019).
18. K. Michaeli, R. Naaman, Origin of spin dependent tunneling through chiral molecules. *J. Phys. Chem. C*, **123**, 17043–17048 (2019).
19. J. Fransson, Chirality-Induced Spin Selectivity: The role of electron correlations. *J. Phys. Chem. Lett.*, **10**, 7126–7132 (2019).
20. L. Zhang, Y. Hao, W. Qin, S. Xie, F. Qu, Chiral-induced spin selectivity: A polaron transport model. *Phys. Rev. B*, **102**, 214303 (2020).
21. G.-F. Du, H.-H. Fu, R. Wu, Vibration-enhanced spin-selective transport of electrons in the DNA double helix. *Phys. Rev. B*, **102**, 035431 (2020).
22. J. Fransson, Vibrational origin of exchange splitting and chiral-induced spin selectivity. *Phys. Rev. B*, **102**, 235416 (2020).
23. B. P. Bloom, B. M. Graff, S. Ghosh, D. N. Beratan, D. H. Waldeck, Chirality control of electron transfer in quantum dot assemblies. *J. Am. Chem. Soc.*, **139**, 9038–9043 (2017).
24. C. Kulkarni, A. K. Mondal, T. K. Das, G. Grinbom, F. Tassinari, M. F. J. Mabesoone, E. W. Meijer, R. Naaman, Highly efficient and tunable filtering of electrons' spin by supramolecular chirality of nanofiber-based materials. *Adv. Mat.*, **32**, 1904965 (2020).
25. A. K. Mondal, M. D. Preuss, M. L. Ślęczkowski, T. K. Das, G. Vantomme, E. W. Meijer, R. Naaman, Spin filtering in supramolecular polymers assembled from achiral monomers mediated by chiral solvents. *J. Am. Chem. Soc.*, **143**, 7189–7195 (2021).
26. S. Mishra, A. K. Mondal, S. Pal, T. K. Das, E. Z. B. Smolinsky, G. Siligardi, R. Naaman, Length-dependent electron spin polarization in oligopeptides and DNA. *J. Phys. Chem. C*, **124**, 10776–10782 (2020).

174

27. M. M. Green, M. P. Reidy, R. J. Johnson, G. Darling, D. J. O'Leary, G. Willson, Macromolecular stereochemistry: the out-of-proportion influence of optically active comonomers on the conformational characteristics of polyisocyanates. The sergeants and soldiers experiment. *J. Am. Chem. Soc.*, **111**, 6452 (1989).
28. A. R. A. Palmans, E. W. Meijer, Amplification of chirality in dynamic supramolecular aggregates. *Angew. Chem. Int. Ed.*, **46**, 8948 (2007).
29. F. London, The general theory of molecular forces". *Transactions of the Faraday Society*, **33**, 8–26 (1937).
30. K. Szalewicz, Symmetry-adapted perturbation theory of intermolecular forces. *Wiley Interdiscip Rev. Comput. Mol. Sci.*, **2**, 187–374 (2012).
31. M. Massimi Pauli's Exclusion Principle (Cambridge Univ Press, Cambridge, UK) (2005).
32. A. Kumar, E. Capua, M. K. Kesharwani, J. M. L. Martin, E. Sitbon, D. H. Waldeck, R. Naaman, Chirality-induced spin polarization places symmetry constraints on biomolecular interactions. *PNAS*, **114**, 2474–2478 (2017).
33. Y. Kapon, A. Saha, T. Duanis-Assaf, T. Stuyver, A. Ziv, T. Metzger, S. Yochelis, S. Shaik, R. Naaman, M. Reches, Y. Paltiel, Evidence for new enantiospecific interaction force in chiral biomolecules. *Chem.*, **7**, 1–13 (2021).
34. A. K. Mondal, N. Brown, S. Mishra, P. Makam, D. Wing, S. Gilead, Y. Wiesenfeld, G. Leitus, L. J. W. Shimon, R. Carmieli, D. Ehre, G. Kamieniarz, J. Fransson, O. Hod, L. Kronik, E. Gazit, R. Naaman, Long-range spin-selective transport in chiral metal-organic crystals with temperature-activated magnetization. *ACS Nano*, **14**, 16624–16633 (2020).
35. K. Banerjee-Ghosh, O. Ben Dor, F. Tassinari, E. Capua, S. Yochelis, A. Capua, S.-H. Yang, S. S. P. Parkin, S. Sarkar, L. Kronik, L. T. Baczewski, R. Naaman, Y. Paltiel, Separation of enantiomers by their enantiospecific interaction with achiral magnetic substrates. *Science*, **360**, 1331–1334 (2018).
36. Y. Sang, F. Tassinaria, K. Santra, W. Zhang, C. Fontanesi, B. P. Bloom, D. H. Waldeck, J. Fransson, R. Naaman, Chirality enhances oxygen reduction, *PNAS* **119** e2202650119 (2022).
37. F. Tassinari, J. Steidel, S. Paltiel, C. Fontanesi, M. Lahav, Y. Paltiel, R. Naaman, Enantioseparation by crystallization using magnetic substrates. *Chemical Science*, **10**, 5246–5250 (2019).
38. K. Santra, D. Bhowmick, Q. Zhu, T. Bendikov, R. Naaman, A method for separating chiral enantiomers by enantiospecific interaction with ferromagnetic substrates. *J. Phys. Chem. C*, **125**, 17530–17536 (2021).
39. K. Michaeli, V. Varade, R. Naaman, D. Waldeck, A new approach towards spintronics-spintronics with no magnets. *J. of Physics: Condensed Matter*, **29**, 103002 (2017).
40. W. Mtangi, V. Kiran, C. Fontanesi, R. Naaman, The role of the electron spin polarization in water splitting. *J. Phys. Chem. Lett.*, **6**, 4916–4922 (2015).
41. W. Mtangi, F. Tassinari, K. Vankayala, A. V. Jentzsch, B. Adelizzi, A. R. A. Palmans, C. Fontanesi, E. W. Meijer, R. Naaman, Control of electrons' spin eliminates hydrogen peroxide formation during water splitting. *JACS*, **139**, 2794–2798 (2017).
42. R. Naaman, Y. Paltiel, D. H. Waldeck, Chiral induced spin selectivity gives a new twist on spin-control in chemistry. *Acc. Chem. Res.*, **53**, 2659–2667 (2020).
43. S. Mishra, S. Pirbadian, A. K. Mondal, M. Y. El-Naggar, R. Naaman, Spin-dependent electron transport through bacterial cell surface multiheme electron conduits. *J. Am. Chem. Soc.*, **141**, 19198–19202 (2019).
44. K. Banerjee-Ghosh, S. Ghosh, H. Mazal, I. Riven, G. Haran, R. Naaman, Long-range charge reorganization as an allosteric control signal in proteins. *J. Am. Chem. Soc.*, **142**, 20456–20462 (2020).

Ultrafast Chiral Dynamics and Geometric Fields in Chiral Molecules

Olga Smirnova*

Max-Born-Institut, Max-Born-Str. 2A, 12489 Berlin, Germany, and Technische Universität Berlin, Straße des 17. Juni 135, 10623 Berlin, Germany
* E-mail: smirnova@mbi-berlin.de

This paper has been stimulated by the many discussions and questions during the symposium. Its goal is to provide a broad picture of the emerging families of new, highly efficient methods for sensing molecular chirality. Their common foundation is the use of dynamics: rotational, vibrational, or electronic, to sense molecular structures; this paper focuses on the fastest among those – the electronic response. The new, highly efficient methods are broadly based on the idea that it is the molecule itself that is ideally suited for sensing its own structure, and that charge dynamics coherently excited inside the molecule enables it to do so. We briefly review the two complementary roads one can take starting from this general premise: the so-called "chiral observer" and "chiral reagent". The observer uses achiral light and achiral detectors to trigger and measure dynamics, which becomes chiral thanks to the structure of the molecule. The chiral measurement is possible because achiral light and achiral detectors jointly form a chiral reference frame, which can be used to resolve enantio-sensitive vectorial observables such as induced polarization or electron current. The reagent uses light with polarization shaped into a 3D chiral structure at any point in space. This shape can be tuned to aid or hinder the dynamics developing in the molecule. We also identify a new object that underlies chiral dynamics – a chiral geometric field associated with it. We point out that the chiral geometric field underlies several classes of enantio-sensitive observables in photo-ionization of chiral molecules, including a completely new class.

Keywords: Ultrafast chirality, Synthetic chiral light, geometric magnetism.

1. Introduction

There is no denying that chirality is important at all scales, from elementary particles to the cosmos, and at varied levels, from corkscrews and fragrances to handshakes – both between people and between molecules in our bodies. Chiral structures facilitate molecular recognition, enable and direct chemical processes. Perhaps, chirality can also be viewed as a basic element of information processing at the molecular level: a single bit, left-handed (0) or right-handed (1), and a two-bit gate – match or no match – when two chiral structures meet. In the information processing context, invariance and robustness of the geometricallly protected bit – the molecular handedness – makes it an excellent information carrier. How is this property is encoded in the interaction of chiral molecules with light?

A chiral object is generally needed to recognize another chiral object. Chiral nature of a helix drawn by a circularly polarized light wave as it propagates in space offers just such an opportunity. Static helical structures can thus be identified by the difference in the interaction (e.g. absorption) of, say, left-handed light with DNA, RNA, some proteins, etc. A complementary effect is optical activity – the left– or right– rotation of the polarization vector of a linearly polarized light wave as it propagates in a chiral medium. The strength of these interactions benefits from a match between the spatial scale of the light wave and the size of the molecule. For infrared and visible light, the several orders of magnitude mismatch between its micron-scale helix and angstrom-scale structures of small chiral molecules makes chiro-optical recognition very challenging. Formally, chiro-optical recognition by circularly polarized light relies on the interplay of electric-dipole and magnetic dipole interactions; the latter are lost in the electric dipole approximation. Large efforts have been made in improving the sensitivity of chiro-optical techniques that rely on magnetic interactions (e.g. [1–14]), with Raman optical activity (ROA) [13] being one striking example. Yet, the enantio-sensitive effects in e.g. ROA remain small, at 10^{-3} level, with chiral high harmonic generation (e.g. [6, 8]) pushing the sensitivity to a few percent level but at the expense of using rather intense driving light.

Even more challenging is chiro-optical sensing of ultrafast chiral dynamics: for conventional chiro-optical methods, the time-resolved signal can be just barely above the noise level (e.g. [2–6]). Yet, dynamical response provides an independent and complementary view at the physical mechanisms underlying chiral function. It is essential for understanding the impact of chiral structures on such ultrafast electron dynamics as ring currents [15, 16], the fields they generate inside molecules [17, 18], ultrafast charge migration [19–28], or the nonlinear electronic response [29].

The search for better approaches has encompassed diverse fields from nonlinear optics [30] to photo-electron spectroscopy [31, 32]. It resulted in a revolution eschewing the need for magnetic interactions. It has led to the discovery of many extremely efficient enantio-sensitive methods addressing rotational [33, 34], vibrational [35] and electronic degrees of freedom [31, 32, 36–46] and giving access to both chiral structures and chiral dynamics. These new methods can be broadly divided into two complementary classes, based on either a (tacitly present) chiral experimental set-up [47] or on replacing the spatial light helix with a temporal chiral structure [48–52, 54].

In both cases, one relies on the chiral dynamics induced by a chiral molecule itself, taking advantage of the fact that the molecule knows best how to twist the motion of its own electrons. One can try to measure the handedness of this twist by detecting the properties of the electronic current, e.g. via photo-electron spectroscopy. Alternatively, one can try to aide or counteract the twist by controlling the temporal evolution of the electric field of the lightwave, which drives the dynamics and can either compete or collaborate with the molecule. In this case, the electric field has to draw a chiral structure during its oscillations, leading to the concept of the synthetic chiral light [48–52].

Below, we describe some of the key ideas behind these methods and outline the pathways ahead, leading to extremely efficient enantio-sensitive signals from optically thin targets. Deeper analysis of these concepts can be found in our recent perspective paper [53].

We also address the question about the interplay between geometry and physics, pointing out a new, uniquely chiral property triggered by ultrafast electron dynamics in randomly oriented molecules – chiral geometric fields [55, 56]. This field is generated by helical currents excited in chiral molecules. It can be felt by processes occurring inside the molecule, making them enantio-sensitive. The chiral geometric field bears distinct signatures of molecular chirality: the molecular handedness is mapped onto its shape and strength.

In solids, the importance of geometric fields is well documented. It leads to the concepts of Berry curvature, topological phases, geometric magnetism, and provides a framework for understanding electronic response [57]. This gives us reasons to expect that chiral geometric field should play a prominent role in the electronic response, and also enable new enantio-sensitive phenomena. It may provide a bridge, along which chiral and topological phenomena established in condensed phase could be imported into chiral molecular gases, and the concepts found in molecular gases can be tested in condensed matter systems (see e.g. a review by C. Felser and J. Gooth in this book).

2. Chiral observer and chiral reagent

To check if a macroscopic object is chiral or not, one could compare the object with its mirror image. Following the definition of a chiral object, one then has to see if the two mirror twins (enantiomers) can be superimposed. Below we will refer to this approach as "chiral observer".

Another approach involves some interaction with a reference, i.e. well controlled, chiral object, here referred to as a "chiral reagent." This approach relies on differences in the outcomes of the interaction of the chiral reagent with the two enantiomers of the same chiral object. Circularly polarized light is one example of such chiral reagent, with the handedness of its spatial helix conveniently controlled by a (non-chiral) observer. While the difference in the absorption of circularly polarized light (absorption circular dichroism, CD) remains very popular (e.g. [4, 5, 58]) among all-optical methods, it generally suffers from weak signals unless one decreases light's wavelength by going to X-rays, to match the angstrom-scale molecule [59, 60], or uses very intense fields to increase the magnetic-dipole transitions [6, 8, 10–12, 61] and looks at highly nonlinear effects. Recent spectacular effort on broadband ultrafast circular dichroism spectroscopy in the deep ultraviolet in combination with transient absorption and anisotropy measurements of spin-crossover dynamics in Fe(II) complexes [62] notwithstanding, both routes can enhance chiral dichroism from 0.01% to a few percent level. Further improvements require one to abandon weak magnetic interactions and rely exclusively on the electric dipole transitions.

At the fundamental level, the chiral observer and chiral reagent look at different observables associated with chiral dynamics. For optical methods, these dynamics are induced by light and inevitably start with electronic response: electronic current and electronic polarization. What makes the electronic dynamics chiral? The answer to this question arises when we realize that the basic property of chiral systems we rely on in our daily life, such as a hairdryer or a corkscrew, is their ability to convert planar circular motion into a linear motion, or linear displacement, orthogonal to the plane of the circle (and vice versa). The planar circular motion is characterized by its angular momentum, a pseudovector $\vec{L}$ orthogonal to the plane. The linear motion or displacement is characterized by a vector, e.g. J. Whether J is parallel or antiparallel to L, i.e. the sign of the scalar product $h = \vec{J} \cdot \vec{L}$, tells us if the induced dynamics is left-handed or right-handed. As $\vec{L}$ is a pseudo-vector and $\vec{J}$ is a vector, h is a pseudoscalar. Pseudo-scalars change sign upon mirror reflection and are ideally suited to discriminate opposite enantiomers. Since a pseudovector can be generated by taking a vector product of two vectors, e.g. $\vec{L} = [\vec{r} \times \vec{p}]$, the chiral observable can also be given by a triple product of three vectors.

Ultimately, a chiral reagent allows one to measure different signal intensities, e.g. different absorption of left-handed or right-handed circularly polarized light for the same chiral molecule. Signal intensities are scalars. To make a scalar, the pseudoscalar associated with the handedness of the molecule has to combine with the pseudocalar associated with the handedness of light to generate a product of two pseudoscalars – a scalar measured by a detector.

In contrast, a chiral observer measures vectors, e.g. the direction of the vector J relative to the initial circular motion the observer has induced. The papers [31, 32, 63] discovered the first representative of this family, known as the photo-electron circular dichroism, PECD. A circularly polarized light ejects a photo-electron that, like a nut rotating on a bolt, is directed forward or backward with respect to the plane of rotation, depending on the molecular handedness. The same happens upon electronic excitation by a circularly polarized light into a superposition of bound states [35], see Fig. 1, leading to the phenomenon known as photo-excitation circular dichroism, PXCD.

One does not have to stop at the vectorial observables. Depending on the complexity of the initially induced motion, which can be driven by two orthogonally polarized fields with different frequencies, e.g. ω and 2ω [44–46], the observer may need (or want) to measure quantities more complex than simple vectors, that is, tensors. The hierarchy continues *ad infinitum* as the tensor rank increases, providing more and more detailed insight into the coupling of the dynamics triggered by the light field into dynamics imposed by the molecule and thus allowing one to reach better understanding of the molecular structure giving rise to such motion (a rigorous description of hierarchy of chiral observables can be found in our recent perspective paper [53].

The concept of chiral observer embodies a powerful principle of detecting chirality. It allows one to use several non-chiral objects to form a single chiral object used to detect a vectorial (or tensorial) chiral response. For example, in the case of PECD, the chiral object is formed by the plane of rotation of the circularly polarized light and the axis of the photoelectron detector orthogonal to this plane, with the signs "forward" and "backward" attached to it. Freed from the need to rely on the weak magnetic interactions, the observer gains freedom to build the chiral

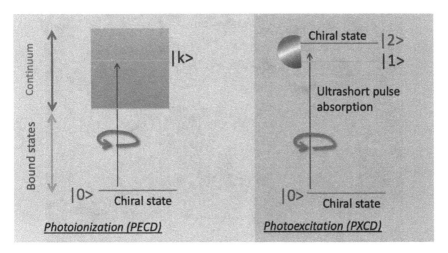

Fig. 1. Schematic representation of the difference between photo-ionization (left) and photo-excitation (right) circular dichroism. In the latter case, coherent excitation of several bound states by circularly polarized light generate oscillating bound-state polarization and current, which acquire components orthogonal to the polarization plane of the pump pulse.

setup from a variety of components available in their experiment, as if from the Lego blocks, taking maximum advantage of available tools and experimental specifics.

However, in spite of such flexibility, the chiral observer cannot detect enantio-sensitive *scalar* observables. Is it possible to modify the chiral reagent, also freeing it from the need to rely on weak magnetic interactions? The answer is "yes". Just like the chiral observer, the "improved" chiral reagent takes advantage of the ultrafast chiral dynamics dictated by the molecule itself, realizing that the molecule is best suited to direct its own chiral dynamics.

3. Synthetic chiral light: polarization "helix" in time

The "improved" chiral reagent replaces inefficient helix of optical light in space with a highly efficient helix (or any other chiral figure) in time, see Fig. 2. We shall refer to it as "synthetic chiral light" [48] to distinguish it from standard chiral light, i.e. the standard light helix in space. Note that the earlier work by [49], is already using three colour synthetic fields, although these fields are not globally chiral [64].

What does the "helix" in time mean? It means that if we follow the tip of the electric field vector of our light, this tip should draw a chiral Lissajous figure during the light oscillation period, at any given point in space. In contrast to the conventional helix of circularly polarized light, which is non-local (it only reveals its chiral structure when moving from one point in space to another), the handedness h of this Lissajous figure is local. It can be defined already in the electric dipole approximation and exists at every point in space. In principle, it could also be space-dependent, changing as a function of $\vec{r}$: $h = h(\vec{r})$. We shall introduce the pseudoscalar h characterizing the handedness of such temporal "helix" below.

180

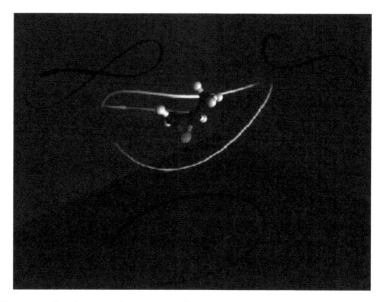

Fig. 2.   Synthetic chiral light. The Lissajous figure drawn by the tip of the electric field vector during one laser period is three-dimensional, chiral, and arises already in the dipole approximation.

Such fields can excite molecular dynamics: rotational, vibrational, or electronic, depending on the light frequencies. Changing the handedness of the Lissajous figure will either aid or counteract the natural chiral dynamics developed by the molecule and determined by the molecular handedness. This interplay changes the strength of the excited dynamics and hence leads to enantio-sensitive changes of such scalar observables as populations of molecular electronic, vibronic or rotational states [34, 67, 68], or the total intensity of the nonlinear optical response [48].

Since the pseudoscalar h – the handedness of the Lissajous figure of the synthetic chiral light – does not include the magnetic field, it has to be constructed from at least three non-coplanar electric field vectors. Thus, this light has to have 3D electric fields (in the dipole approximation), a property that arises already in tightly focused beams or inside photonic structures, thanks to the emergence of the longitudinal field component. Such longitudinal components also naturally arise in vortex beams, making structured light an attractive candidate for inducing and controlling enantio-sensitive dynamics and associated (e.g. optical) response [69]. We shall see below that the field also has to be multi-color, since the three electric field vectors have to be distinguishable. Therefore, such light can sense chirality of matter only via non-linear interactions, as it has to encode all three field components into the light-induced transitions.

To control total enantio-sensitive response, e.g. absorption, in the entire sample, the synthetic chiral light should maintain its handedness globally, across the whole interaction region: $\int h(\vec{r})d\vec{r} \neq 0$. If the light field is not globally chiral, $\int h(\vec{r})d\vec{r} = 0$, higher multipoles of lights' handedness will start to play key role. One example is the enantio-sensitive direction of coherent light emitted by the molecular sample as a result of non-linear processes such as harmonic generation [70] or free induction decay [69]. Three-dimensional sculpting of multi-color light beams appears as a natural next step for the rapidly developing field of structured light, compelled by the rich opportunities arising from the interaction of such light with chiral matter.

How can we characterize the handedness of light's Lissajous figure? An intuitive approach is to take three snapshots of the electric field vector at three instants of time t_1, t_2, t_3 and construct a triple product of these three vectors $\vec{F}(t_1) \cdot [\vec{F}(t_2) \times \vec{F}(t_3)]$. If the result is non-zero, the segment of the figure traced by the field was chiral. To make sure that the entire Lissajous figure is chiral, one should average the triple product over time:

$$H^{(3)}(\tau_1, \tau_2) = \int_0^T dt \vec{F}(t) \cdot [\vec{F}(t + \tau_1) \times \vec{F}(t + \tau_2)], \tag{1}$$

We see that the natural definition of handedness of the synthetic chiral light is given by the third-order $H^{(3)}(\tau_1, \tau_2)$ chiral field correlation function. In the frequency domain, the chiral correlation function $h^{(3)}$ is simply a triple product of the field strengths at three frequency components: ω_1, ω_2, and $-\omega_3 = -(\omega_1 + \omega_2)$ [48]

$$h^{(3)}(-\omega_3, \omega_1, \omega_2) = \vec{F}_{\omega_3}^* \cdot [\vec{F}_{\omega_2} \times \vec{F}_{\omega_1}] \tag{2}$$

The equality $\omega_3 = (\omega_1 + \omega_2)$, together with complex conjugation, arises because we require that the Lissajous curve is closed and that its handedness does not average to zero. The triple product is non-zero only if the three frequency components are non-coplanar, i.e. the non-zero $h^{(3)}$ requires 3 fields along 3 orthogonal axes.

Importantly, the chiral correlation function $h^{(3)}$ is not just a convenient formal construct describing the geometry of the Lissajous figure. It arises naturally when we consider the interaction of this light with matter. Consider the simplest and commonly used model of a chiral quantum system (see e.g. [49, 65]) a V-type 3-level system with states $|0, 1, 2\rangle$ which have dipole couplings $\vec{d}_{10}, \vec{d}_{21}$, and $\vec{d}_{20}$, where the triple product $\vec{d}_{20} \cdot [\vec{d}_{10} \times \vec{d}_{21}] \neq 0$, see Fig. 3. The system is randomly oriented in space. Let us select our z-axis along the field at ω_3 and set $\omega_3 = E_2 - E_0$. The other two fields are polarized in the x-y plane.

The function $h^{(3)}$ describes the interference of two pathways in absorption. The first (achiral) pathway is associated with linear response at the frequency ω_3. It is not sensitive to chirality, and leads to polarization at $\omega_3 = \omega_1 + \omega_2$ along z. The second (chiral) pathway corresponds to the second-order process that records the molecular handedness: the medium absorbs one ω_1 photon and one ω_2 photon from the two fields polarized in the $x - y$ plane, generating polarization at the frequency ω_3 along z. Two-photon absorption in such system would not lead to induced dipole connecting initial and final states. It will interfere with the one-photon polarization induced by the field ω_3. This polarization is unique to chiral media, is out of phase in the media of opposite handedness, and its interference with the one-photon transition will be either constructive or destructive, changing the

182

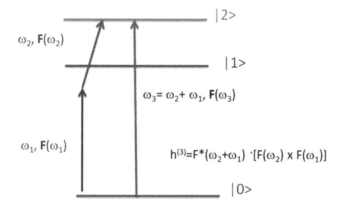

Enantio Absorption =Im{ Molecular pseudoscalar · Light pseudoscalar}		
	Molecular pseudoscalar	**Light pseudoscalar**
Standard chiral field	m·d	helicity
Synthetic chiral field	$[d_1 \times d_2] \cdot d_3$	$h^{(3)}$

Fig. 3. **Enantio-sensitivity in absorption.** A three-level system $|0,1,2\rangle$ is driven by the three-color field with frequencies ω_1, ω_2 polarized in the $x-y$ plane and the field with frequency $\omega_3 = \omega_1 + \omega_2$ polarized along z. Lack of selection rules in a chiral molecule allows for dipole couplings between states $|0\rangle, |1\rangle$, $|0\rangle, |2\rangle$, and $|1\rangle, |2\rangle$, with the two-photon transition at $\omega_1 + \omega_2$ driven along z in randomly oriented chiral media, and interfering with the one-photon transition driven along z at ω_3.

absorption. The correlation function $h^{(3)}$ naturally encodes the "field" part of this interference, while the material response is encoded into the corresponding nonlinear susceptibility proportional to the triple product of the three dipoles [66]. The correlation function $h^{(3)}$ substitutes optical chirality, the latter characterizes the strength of enantio-sensitive absorption in linear response [71] (beyond the dipole approximation). Thus, the chiral light correlation function characterises the strength of enantio-sensitive light-matter interaction in the electric dipole approximation.

For a two-colour field, such as the one in Fig. 2, $h^{(3)} = 0$ simply because the field does not contain three frequencies. It means that nonlinear 3-photon processes driven by this field are not enantio-sensitive, but it does not necessarily mean that the field is achiral, only that its handedness will manifest in the higher order response. For the field in Fig. 2, the lowest-order non-zero chiral correlation function is $h^{(5)}$ [48]

$$h^{(5)}(-2\omega, -\omega, \omega, \omega, \omega) = \vec{F}_{2\omega}^* \cdot [\vec{F}_\omega^* \times \vec{F}_\omega][\vec{F}_\omega \cdot \vec{F}_\omega]. \tag{3}$$

It quantifies the lowest-order enantio-sensitive response of isotropic chiral media to this light, which again arises from the interference of two pathways. The first, achiral, pathway is associated with linear response at the frequency 2ω, which leads to polarization at 2ω along z. In the second, chiral, pathway, the medium absorbs three ω photons from the major field component and emits one ω photon into the minor ellipticity component, also generating polarization at frequency 2ω along z (see [48]), orthogonal to the polarization plane of the ω-field. This combination records the direction of rotation of the driving field in the x-y plane. Just as before, the second pathway exists only in chiral media, and the induced polarization is out of phase in media of opposite handedness. Interference between these two pathways enables enantio-sensitive absorption and emission at the frequency 2ω and the possibility of achieving enantio-sensitive populations of excited electronic states. The enantio-sensitive contributions to these observables can be written as a product of two pseudoscalars [48]: $h^{(5)}$ characterizing field's handedness and a molecular pseudoscalar involving first- and fourth-order susceptibilities.

The possibility of structuring local properties of light in space [72], including both its intensity and phase [73], creates unique opportunities for imaging [74] and manipulating [75] properties of matter. Likewise, structuring light's chirality [71, 76–79] could open new efficient routes for enantio-sensitive imaging and control of chiral matter. One example of the new type of structured chiral light is the *chirality polarized* light [70], where the local handedness $h(\vec{r})$ is distributed in space. For a chiral molecule, interaction with such light will result in different strength of the enantio-sensitive response as a function of the molecular position in space, hereby creating a grating. This grating will be opposite for the molecules of opposite handedness, leading e.g. to the enantio-sensitive deflection of the parametric emission, such as high harmonic emission [70] or the free induction decay [69].

4. From photoelectron circular dichroism to chiral geometric fields

Photoelectron circular dichroism, PECD relies on alternative detection principle — the principle of chiral observer [53]. It was predicted by Ritchie [31], Cherepkov [63] and Powis [32]. The first measurement was done by Bowering *et al.* [36]. Recently it was observed in the multiphoton [37, 38] and strong-field ionization regimes [80] and extended to industrial applications [40].

In PECD, the circularly polarized field

$$\vec{E}(t) = \vec{E}_\omega e^{-i\omega t} + \text{c.c.} \tag{4}$$

ionizes an isotropic sample of chiral molecules. Initial circular rotation imposed on the photo-electron by the laser field couples to the motion orthogonal to the polarization plane imposed by the molecular "corkscrew". A photoelectron current $\vec{j}$ is generated orthogonal to the plane, i.e.

$$\vec{j} = g\vec{L} \equiv g\vec{E}_\omega^* \times \vec{E}_\omega, \tag{5}$$

where L is the spin angular momentum associated with the incident light. Since $\vec{j}$ is a vector and $\vec{L}$ is a pseudo-vector, the coefficient g must be a pseudo-scalar.

184

Indeed, it is [47]

$$g \equiv \frac{1}{6} \int d\Omega_k (\vec{d}^*_{\vec{k},i} \times \vec{d}_{\vec{k},i}) \cdot \vec{k}, \tag{6}$$

where $\int d\Omega_k$ indicates integration over all directions of the photoelectron momentum $\vec{k}$, for fixed magnitude k, and $\vec{d}_{\vec{k},i} \equiv \langle \vec{k}|\vec{d}|i\rangle$ is the transition dipole matrix element between the initial state $|i\rangle$ and the continuum state $|\vec{k}\rangle$. Due to the purely electric dipole nature of PECD, the enantio-sensitive signal reaches tens of percent of the total photoionization signal, both for one-photon and few-photon ionization [36, 37, 41, 81].

A circularly polarized pump pulse can also be used to excite coherent superposition of electronic (or vibronic) states, generating a non-stationary state and the corresponding induced dipole. One can show [35] that, after the end of the pump pulse and for a pair of excited states, $|1\rangle$ and $|2\rangle$, the expectation value of the induced dipole $\langle \vec{d}\rangle$ in the direction perpendicular to the polarization plane oscillates as:

$$\langle \vec{d}\rangle(\omega_{21})(t) = g\vec{L}\sin\omega_{21}t \tag{7}$$

where t is the time measured from the center of the pump pulse, $\omega_{21} \equiv \omega_2 - \omega_1$,

$$g \equiv \frac{1}{6}(\vec{d}_{1,0} \times \vec{d}_{2,0}) \cdot \vec{d}_{2,1} \tag{8}$$

and

$$\vec{L} \equiv \vec{E}^*(\omega_{1,0}) \times \vec{E}(\omega_{2,0}). \tag{9}$$

That is, the excited dipole does indeed convert the initial rotation excited by the circularly polarized pump into the motion orthogonal to this plane. The phase of these oscillations is determined not only by $\vec{L}$, but also by the sign of the molecular pseudoscalar g. Thus, oscillations excited in opposite enantiomers will have opposite phases. This is the essence of photo-excitation circular dichroism (PXCD), introduced recently by Beaulieu et al. [35].

Excitation of chiral currents in molecules, whether oscillating like in PXCD or unidirectional, like in PECD, can be linked to the concept of the geometric magnetism introduced by M. Berry [82]. Geometric magnetism plays particularly important role in solids, where the non-zero Berry curvature is linked to the anomalous electron velocity, the Hall effect, and to the related topological phenomena [57]. Our current work, reported at this symposium shows that a similar geometric field, now associated with chiral currents or chiral polarization, also shows up in photoionizaton of chiral molecules by circularly polarized fields. The chiral arrangements of the nuclei, which "twists" both currents and polarization, plays fundamental role: the nonzero handedness of the currents and fields ensures that they do not vanish upon averaging over the random molecular orientations and thus survive in the molecular frame. As a result, they can enable physical phenomena. Indeed, geometric fields do underlie several classes of chiral observables in photoionization [55].

Their first manifestation is the so-called propensity field arising in PECD [83]:

$$\vec{B}(\vec{k}) = -i[\vec{d}_{\vec{k}g}^* \times \vec{d}_{\vec{k}g}] = -i\frac{[\vec{p}_{\vec{k},g}^* \times \vec{p}_{\vec{k},g}]}{(E - E_g)^2}. \tag{10}$$

Naturally, since $\vec{B}(\vec{k})$ is related to photo-ionization, it is defined in the momentum space $(\vec{k})$. The length-gauge photoionization dipoles $\vec{d}_{g\vec{k}}$ describe the corresponding bound-free transitions to the continuum states with momenta $\vec{k}$ in the molecular frame; $\vec{p}_{g\vec{k}}$ describe the same matrix elements in the velocity gauge. The velocity gauge expression for $\vec{B}(\vec{k})$ is formally identical to the Berry curvature in a two-band solid, here with the parabolic conduction band and flat valence band [55, 56, 83]. The expressions for the PECD current Eqs. (5, 6) show that the proportionality coefficient g between the laser intensity and the PECD current can be written as a scalar product of the field $\vec{B}(\vec{k})$ and the photoelectron momentum $\vec{k}$, averaged over all directions of $\vec{k}$,

$$g \propto \int d\Omega_k \vec{B}(\vec{k}) \cdot \vec{k}, \tag{11}$$

Similar relationship arises in the photogalvanic effect in solids [84], with the Berry curvature for the solid replacing the field $\vec{B}(\vec{k})$ here. Thus, g is the chiral molecule analog of the anomalous photo-induced conductivity in solids.

The $\vec{k}$-resolved geometric field quantifies the PECD for a given direction of the photo-electron in the molecular frame. Importantly, it is already averaged over all orientations of the molecule with respect to the laboratory frame, i.e. with respect to the direction of the ionizing field.

The local (i.e. $\vec{k}$-resolved) photoionization circular dichroism can exist even in achiral molecules, if it does not have the rotational symmetry around the light propagation axis. However, only in chiral molecules the "twist" imparted by the scattering of the electron on the nuclei survives full rotational averaging for a randomly oriented molecule, both over the molecular orientation relative to the light field and the photo-electron momentum relative to the molecular frame.

The direction of $\vec{B}(\vec{k})$ defines the propagation direction of the circularly polarized light pulse, for which CD will maximise for a given $\vec{k}$. The length $|\vec{B}(\vec{k})|$ defines the difference in the photoionization yields for co- and counter-rotating fields propagating in direction of $\vec{B}(\vec{k})$,

$$\vec{B}(\vec{k}) \cdot \vec{e}_B = |\vec{d}_{g\vec{k}}^+|^2 - |\vec{d}_{g\vec{k}}^-|^2, \quad \vec{e}_B \equiv \frac{\vec{B}(\vec{k})}{|\vec{B}(\vec{k})|}. \tag{12}$$

The geometric field arising in one-photon ionization is but a special case of the geometric fields driven by chiral electron dynamics and arising after rotational averaging [55]. The generalization emerges as soon as we consider photoionization not from a single stationary state, but from a coherent superposition of states. Consider a simple model of two coherently populated states with equal initial amplitudes, independent of the molecular orientation. The superposition is prepared at $t = 0$

and evolves in time as $|j\rangle + e^{-i\phi_{ij}}|i\rangle$, where $\phi_{ij} = \omega_{ij}t$, and $\omega_{ij} \equiv \omega_i - \omega_j$ is the transition frequency between the states. Using the same definition for the geometric field as before, but now applying it to the photoionization of this superposition by a sufficiently short, circularly polarized probe pulse, we see that the geometric field also encodes the coherence between the excited states:

$$\vec{B}_{ij}(\vec{k}, \phi_{ij}) = -\frac{1}{2}i\left[\vec{d}^*_{\vec{k}i} \times \vec{d}_{\vec{k}j}\right]e^{i\phi_{ij}} + \text{c.c.} \equiv \vec{Q}_{ij}(\vec{k})\cos\phi_{ij} + \vec{P}_{ij}(\vec{k})\sin\phi_{ij}, \quad (13)$$

where the displacement $\vec{Q}_{ij}(\vec{k})$ and current $\vec{P}_{ij}(\vec{k})$ quadratures are

$$\vec{Q}_{ij}(\vec{k}) \equiv -\Re\left\{i\left[\vec{d}^*_{\vec{k}i} \times \vec{d}_{\vec{k}j}\right]\right\}, \quad (14)$$

$$\vec{P}_{ij}(\vec{k}) \equiv \Im\left\{i\left[\vec{d}^*_{\vec{k}i} \times \vec{d}_{\vec{k}j}\right]\right\}. \quad (15)$$

For $i = j = g$, $\phi_{ij} = 0$ and Eq. (13) reduces to Eq. (10). The generalization to any number of states in the superposition is obvious.

The $\vec{Q}(\vec{k})$ quadrature is present already at $t = 0$, when the initial superposition state is purely real-valued and has not developed any current yet. Its origin is the displacement of the electron density (around the direction of $\vec{Q}(\vec{k})$). The (diagonal) Q-quadrature manifests already in photoionization from a single state and gives rise to the PECD current, which is proportional to the first moment of the $\vec{B}(\vec{k})$ field – its net radial component

$$B_{\text{radial}} = \int d\Omega_k \vec{B}(\vec{k}) \cdot \vec{k}/|k|, \quad (16)$$

The unit vector $\vec{k}/|k|$ is normal to the surface of the sphere with the radius k, describing the energy shell determined by the energy conservation law in photoionization. The net radial component of the geometric field expresses its flux through this sphere. Importantly, one can rigorously prove [55] that, while the Q-quadrature contributes to the net radial field, its average over all directions of $\vec{k}$ is equal to zero:

$$\vec{Q}_0 = \int d\Omega_k \vec{Q}(\vec{k}) = 0 \quad (17)$$

We shall refer to enantio-sensitive observables associated with the first moment of the geometric vector field, its net radial component, as Class II observables.

The $\vec{P}^{L,R}(\vec{k})$ quadrature provides a very important addition to the geometric field. Its contribution maximizes when the phase between the two states is $\pi/2$, i.e. when the superposition carries maximum current. Its physical origin, therefore, is a bound state current. Notably, one can rigorously show [55] that, in contrast to Q-quadrature, the P-quadrature does contribute to the zero moment of the vector field, the net geometric field given by

$$\vec{B}_0 = \vec{P}_0 = \int d\Omega_k \vec{P}(\vec{k}) \neq 0 \quad (18)$$

Thus, current-carrying superpositions should generate new observables in photoionization, associated with $\vec{B}_0$. We shall refer to them as Class I enantio-sensitive observables. What could be the members of this class?

The net geometric field is a pseudo-vector. Thus, an enantio-sensitive observable, which carries a pesudo-scalar reflecting the handedness of the molecule, needs another vector to generate a pseudo-scalar via a scalar product with the pseudo-vector of the geometric field. Such vector could be, for example, a permanent dipole of a chiral molecular ion, generated upon photoionization. In fact, it could be any other polar vector $\vec{e}_M$ associated with nuclear arrangements in the molecular frame of the cation. The non-zero scalar product $\vec{B}_0 \cdot \vec{e}_M$ would then imply orientation of this polar vector relative to the net geometric field generated by the currents. Consequently, we should obtain enantio-sensitive orientation of the molecular frame vector in the laboratory frame, with the orientation dictated by the geometric arrangement of the pump and probe pulses involved in the preparation of the bound-state current, followed by the ionization step. One simple option is to use co-propagating pulses, e.g. linear pump and circular probe, which leads to orientation of the molecular ion relative to the polarization plane [55].

Overall, the geometric field gives rise to three classes of enantio-sensitive observables [55]. While Class I observables are associated with the net geometric field, i.e. zero moment of the vector field, and Class II observables are associated with the first (radial) moment of the vector field, the Class III observables are associated with the higher-order tensorial moments of the vector field, associated with its decomposition into the vector spherical harmonics. The new enantio-sensitive observables of Class I have been completely overlooked so far. Since they only appear if a current is excited prior to photoionization, these observables are the messengers of charge-directed reactivity [20–24]: the chemical reactivity driven by ultrafast electron dynamics. In this case, the reactivity is enantio-sensitive and the dynamics is chiral. The first member of Class I observables is molecular circular dichroism in photoionization [55]. Observables of Class II and III include the PECD (and time-dependent PECD) current and an infinite array of its multipolar versions, the latter arising e.g. when a combination of of orthogonally polarized ω and 2ω fields are used in photoionization [44, 46]. Most of these observables have not been known so far.

5. Conclusions and outlook

It is tempting to try to use the lessons learned from ultrafast chiral dynamics to speculate on possible connections between chiral molecular gases and topological effects in condensed matter systems.

We have described the concept of dynamical chirality emerging in non-linear interaction of light with randomly oriented chiral molecules. This concept is relevant for both – the light and the electrons, or, in general, matter. Dynamical chirality plays a role similar to the typical "structural" chirality, but is encoded not in the spatial structure of the light fields or the positions of the nuclei in a molecule, but rather in the local Lissajous figure of the light field or the electron dynamics inside the molecular frame. In both cases, for the synthetic chiral light and for electrons driven by such light, the chiral dynamics evolves on the sub-laser-cycle time-scale. For electrons, sub-cycle sensitivity arises due to the non-linear nature of light-molecule interaction. For matter, the handedness is encoded in the polarization induced in the medium. For light, it is encoded in the temporal evolution of the light polarization vector.

From this general perspective, one can speculate that there are at least two potentially interesting connections between ultrafast chiral electron dynamics in molecular gases and the condensed matter systems.

The first connection is related to the discovery of the geometric field in molecular photoionization. The second is related to the concept of synthetic chiral light.

The geometric field in molecular photoionization or photoexcitation is a property of electrons, emerging as a result of rotational averaging over random molecular orientations relative to the driving light field.

The similarity between the geometric field in photoionization of randomly oriented chiral molecules and the Berry curvature characterizing topological properties of materials and their anomalous electrical, optical or thermoelectrical response (see paper by C. Felser in this book) suggests that the "twins" of these anomalous effects may exist in chiral molecular gases.

Particularly interesting may be the analogy between chiral molecular gases and Weyl semimetals. Weyl semimetals contain pairs of band crossings, at which electrons behave as the so-called Weyl fermions: massless particles characterized by specific helicity $\xi = +1$, or $\xi = -1$, quantified by the projection of their spin on the direction of their linear momentum. These pairs correspond to opposite linear momenta and same spin. The Berry curvature – the important property of electrons in these materials – characterizes e.g. the conductivity in circular photo-galvanic effect [84] and also leads to strong anomalous thermal conductivity (see paper by C. Felser in this issue).

The possible analogy stems from the fact that ultrafast electron dynamics excited in chiral molecules also creates "chiral electrons", but their chirality is not related to their helicity (invariant only for relativistic electrons), but is due to the fact that their temporal dynamics – their trajectory – is chiral. The handedness of the molecule dictates the handedness of the chiral currents excited by achiral fields. In randomly oriented molecules such currents do not cancel, but lead to a persistent chiral current in the molecular frame. In contrast, in randomly oriented achiral molecules such currents, excited by non-chiral laser fields, would (and do) cancel.

The first sign of such connection is the analogy between the circular photogalvanic effect in solids and PECD [55]. Whether other phenomena driven by the Berry curvature in solids find analogues in molecular systems remains to be seen. One could, for example, imagine that the net Berry curvature in chiral molecular gas could also lead to the anomalous thermoelectric effect in such gases upon application of circularly polarized field.

Chiral electron dynamics can also be excited using synthetic chiral light. In this case, the chirality of induced polarization (which encodes chiral electron trajectory) is determined not only by the chirality of the molecule, but also by the chirality of light's Lissajous figure. The ability to alter the chirality of electrons in molecular medium is somewhat similar to the effect of chiral anomaly, which emerges in Weyl semimetals when applied external field is chiral (e.g. collinear electric and magnetic fields). Chirality of Weyl fermions can be altered by influencing their spin (via a magnetic field) and linear momentum (via an electric field). Here an interesting possible connection comes from the opportunity to apply synthetic chiral light to Weyl semimetals. Since the interaction is chiral, it should also break the balance between the electrons with opposite handedness, however the mechanism of chiral symmetry breaking will be associated with the electron orbital momentum rather than their spin.

Finally, it would be interesting to consider whether the net geometric field in molecular photoionization can affect the electron spin in photoionization or photoexcitation of chiral molecules (see the paper by R. Naaman in this book). Since the presence of the net geometric field indicates enantio-sensitive orientation of molecules upon ionization [55], it indicates that the net field corresponds to net orbital angular momentum of the electron upon ionization, which may be transferred to spin via spin-orbit interaction in the initial or final state.

Conflicts of interest

There are no conflicts to declare.

Acknowledgements

It is my great pleasure to acknowledge tremendous contributions of Dr. D. Ayuso, Dr. A. Ordonez and Prof. P. Decleva to the development of the ideas and concepts reviewed in this paper. I would like to acknowledge fruitful discussions with Dr. Misha Ivanov. Funding by Deutsche Forschungsgemeinschaft (SM 292/5-2) and are gratefully acknowledged. Funded by the European Union (ERC, ULISSES, 101054696). Views and opinions expressed are however those of the author(s) only and do not necessarily reflect those of the European Union or the European Research Council. Neither the European Union nor the granting authority can be held responsible for them.

References

1. J. R. Rouxel, A. Rajabi and S. Mukamel, Chiral four-wave mixing signals with circularly polarized x-ray pulses, *Journal of Chemical Theory and Computation* **16**, 5784 (2020), PMID: 32786909.
2. L. Ye, J. R. Rouxel, S. Asban, B. Rösner and S. Mukamel, Probing molecular chirality by orbital-angular-momentum-carrying x-ray pulses, *Journal of Chemical Theory and Computation* **15**, 4180 (2019), PMID: 31125229.
3. J. R. Rouxel, Y. Zhang and S. Mukamel, X-ray raman optical activity of chiral molecules, *Chem. Sci.* **10**, 898 (2019).

190

4. H. Rhee, Y.-G. June, J.-S. Lee, K.-K. Lee, J.-H. Ha, Z. H. Kim, S.-J. Jeon and M. Cho, Femtosecond characterization of vibrational optical activity of chiral molecules, *Nature* **458**, 310 (Mar 2009).
5. M. Oppermann, J. Spekowius, B. Bauer, R. Pfister, M. Chergui and J. Helbing, Broadband ultraviolet cd spectroscopy of ultrafast peptide backbone conformational dynamics, *The Journal of Physical Chemistry Letters* **10**, 2700 (2019).
6. R. Cireasa, A. E. Boguslavskiy, B. Pons, M. C. H. Wong, D. Descamps, S. Petit, H. Ruf, N. Thiré, A. Ferré, J. Suarez, J. Higuet, B. E. Schmidt, A. F. Alharbi, F. Légaré, V. Blanchet, B. Fabre, S. Patchkovskii, O. Smirnova, Y. Mairesse and V. R. Bhardwaj, Probing molecular chirality on a sub-femtosecond timescale, *Nature Physics* **11**, 654 (Aug 2015).
7. M. Cho, Drive round the twist, *Nature Physics* **11**, 621 (Aug 2015).
8. D. Baykusheva and H. J. Wörner, Chiral discrimination through bielliptical high-harmonic spectroscopy, *Phys. Rev. X* **8**, p. 031060 (Sep 2018).
9. D. Baykusheva, D. Zindel, V. Svoboda, E. Bommeli, M. Ochsner, A. Tehlar and H. J. Wörner, Real-time probing of chirality during a chemical reaction, *Proceedings of the National Academy of Sciences* **116**, 23923 (2019).
10. D. Ayuso, P. Decleva, S. Patchkovskii and O. Smirnova, Strong-field control and enhancement of chiral response in bi-elliptical high-order harmonic generation: an analytical model, *Journal of Physics B: Atomic, Molecular and Optical Physics* **51**, p. 124002 (may 2018).
11. D. Ayuso, P. Decleva, S. Patchkovskii and O. Smirnova, Chiral dichroism in bi-elliptical high-order harmonic generation, *Journal of Physics B: Atomic, Molecular and Optical Physics* **51**, p. 06LT01 (feb 2018).
12. O. Smirnova, Y. Mairesse and S. Patchkovskii, Opportunities for chiral discrimination using high harmonic generation in tailored laser fields, *Journal of Physics B: Atomic, Molecular and Optical Physics* **48**, p. 234005 (oct 2015).
13. L. D. Barron,* L. Hecht, I. H. McColl and E. W. Blanch, Raman optical activity comes of age, *Molecular Physics* **102**, 731 (2004).
14. L. D. Barron, *Molecular light scattering and optical activity* (Cambridge University Press, 2009).
15. I. Barth and J. Manz, Periodic electron circulation induced by circularly polarized laser pulses: Quantum model simulations for mg porphyrin, *Angewandte Chemie International Edition* **45**, 2962 (2006).
16. T. Bredtmann and J. Manz, Electronic bond-to-bond fluxes in pericyclic reactions: Synchronous or asynchronous?, *Angewandte Chemie International Edition* **50**, 12652 (2011).
17. I. Barth and J. Manz, Electric ring currents in atomic orbitals and magnetic fields induced by short intense circularly polarized π laser pulses, *Phys. Rev. A* **75**, p. 012510 (Jan 2007).
18. K.-J. Yuan and A. D. Bandrauk, Attosecond-magnetic-field-pulse generation by intense few-cycle circularly polarized uv laser pulses, *Phys. Rev. A* **88**, p. 013417 (Jul 2013).
19. L. Cederbaum and J. Zobeley, Ultrafast charge migration by electron correlation, *Chemical Physics Letters* **307**, 205 (1999).
20. J. Breidbach and L. S. Cederbaum, Migration of holes: Formalism, mechanisms, and illustrative applications, *The Journal of Chemical Physics* **118**, 3983 (2003).
21. A. I. Kuleff and L. S. Cederbaum, Ultrafast correlation-driven electron dynamics, *Journal of Physics B: Atomic, Molecular and Optical Physics* **47**, p. 124002 (jun 2014).

22. S. Lünnemann, A. I. Kuleff and L. S. Cederbaum, Ultrafast charge migration in 2-phenylethyl-n,n-dimethylamine, *Chemical Physics Letters* **450**, 232 (2008).

23. F. Remacle, R. D. Levine, E. W. Schlag and R. Weinkauf, Electronic control of site selective reactivity: a model combining charge migration and dissociation, *The Journal of Physical Chemistry A* **103**, 10149 (1999).

24. F. Remacle and R. D. Levine, An electronic time scale in chemistry, *Proceedings of the National Academy of Sciences* **103**, 6793 (2006).

25. F. Calegari, D. Ayuso, A. Trabattoni, L. Belshaw, S. De Camillis, S. Anumula, F. Frassetto, L. Poletto, A. Palacios, P. Decleva, J. B. Greenwood, F. Martín and M. Nisoli, Ultrafast electron dynamics in phenylalanine initiated by attosecond pulses, *Science* **346**, 336 (2014).

26. F. Calegari, G. Sansone, S. Stagira, C. Vozzi and M. Nisoli, Advances in attosecond science, *Journal of Physics B: Atomic, Molecular and Optical Physics* **49**, p. 062001 (feb 2016).

27. M. Nisoli, P. Decleva, F. Calegari, A. Palacios and F. Martín, Attosecond electron dynamics in molecules, *Chemical Reviews* **117**, 10760 (2017), PMID: 28488433.

28. D. Ayuso, A. Palacios, P. Decleva and F. Martín, Ultrafast charge dynamics in glycine induced by attosecond pulses, *Phys. Chem. Chem. Phys.* **19**, 19767 (2017).

29. S. Mukamel, *Principles of nonlinear optical spectroscopy* Oxford Series in Optical and Imaging Sciences, Oxford Series in Optical and Imaging Sciences (Oxford University Press, New York, 1995).

30. J. A. Giordmaine, Nonlinear optical properties of liquids, *Physical Review* **138**, A1599–A1606 (1965). https://link.aps.org/doi/10.1103/PhysRev.138.A1599.

31. B. Ritchie, Theory of the angular distribution of photoelectrons ejected from optically active molecules and molecular negative ions, *Phys. Rev. A* **13**, 1411 (Apr 1976).

32. I. Powis, Photoelectron circular dichroism of the randomly oriented chiral molecules glyceraldehyde and lactic acid, *The Journal of Chemical Physics* **112**, 301 (2000).

33. D. Patterson and J. M. Doyle, Sensitive Chiral Analysis via Microwave Three-Wave Mixing, *Physical Review Letters* **111**, p. 023008 (July 2013).

34. S. Eibenberger, J. Doyle and D. Patterson, Enantiomer-specific state transfer of chiral molecules, *Phys. Rev. Lett.* **118**, p. 123002 (Mar 2017).

35. S. Beaulieu, A. Comby, D. Descamps, B. Fabre, G. A. Garcia, R. Géneaux, A. G. Harvey, F. Légaré, Z. Mašín, L. Nahon, A. F. Ordonez, S. Petit, B. Pons, Y. Mairesse, O. Smirnova and V. Blanchet, Photoexcitation circular dichroism in chiral molecules, *Nature Physics* **14**, 484 (May 2018).

36. N. Böwering, T. Lischke, B. Schmidtke, N. Müller, T. Khalil and U. Heinzmann, Asymmetry in photoelectron emission from chiral molecules induced by circularly polarized light, *Phys. Rev. Lett.* **86**, 1187 (Feb 2001).

37. C. Lux, M. Wollenhaupt, T. Bolze, Q. Liang, J. Köhler, C. Sarpe and T. Baumert, Circular dichroism in the photoelectron angular distributions of camphor and fenchone from multiphoton ionization with femtosecond laser pulses, *Angewandte Chemie International Edition* **51**, 5001 (2012).

38. C. S. Lehmann, N. B. Ram, I. Powis and M. H. M. Janssen, Imaging photoelectron circular dichroism of chiral molecules by femtosecond multiphoton coincidence detection, *The Journal of Chemical Physics* **139**, p. 234307 (2013).

39. C. S. Lehmann, N. B. Ram, I. Powis and M. H. M. Janssen, Imaging photoelectron circular dichroism of chiral molecules by femtosecond multiphoton coincidence detection, *The Journal of Chemical Physics* **139**, p. 234307 (2013).

40. M. H. M. Janssen and I. Powis, Detecting chirality in molecules by imaging photoelectron circular dichroism, *Phys. Chem. Chem. Phys.* **16**, 856 (2014).

41. L. Nahon, G. A. Garcia and I. Powis, Valence shell one-photon photoelectron circular dichroism in chiral systems, *Journal of Electron Spectroscopy and Related Phenomena* **204**, 322 (2015).
42. A. Comby, S. Beaulieu, M. Boggio-Pasqua, D. Descamps, F. Légaré, L. Nahon, S. Petit, B. Pons, B. Fabre, Y. Mairesse and V. Blanchet, Relaxation dynamics in photoexcited chiral molecules studied by time-resolved photoelectron circular dichroism: Toward chiral femtochemistry, *The Journal of Physical Chemistry Letters* **7**, 4514 (2016), PMID: 27786493.
43. A. Kastner, C. Lux, T. Ring, S. Züllighoven, C. Sarpe, A. Senftleben and T. Baumert, Enantiomeric excess sensitivity to below one percent by using femtosecond photoelectron circular dichroism, *ChemPhysChem* **17**, 1119 (2016).
44. P. V. Demekhin, A. N. Artemyev, A. Kastner and T. Baumert, Photoelectron circular dichroism with two overlapping laser pulses of carrier frequencies ω and 2ω linearly polarized in two mutually orthogonal directions, *Phys. Rev. Lett.* **121**, p. 253201 (Dec 2018).
45. R. E. Goetz, C. P. Koch and L. Greenman, Quantum control of photoelectron circular dichroism, *Phys. Rev. Lett.* **122**, p. 013204 (Jan 2019).
46. S. Rozen, A. Comby, E. Bloch, S. Beauvarlet, D. Descamps, B. Fabre, S. Petit, V. Blanchet, B. Pons, N. Dudovich and Y. Mairesse, Controlling subcycle optical chirality in the photoionization of chiral molecules, *Phys. Rev. X* **9**, p. 031004 (Jul 2019).
47. A. F. Ordonez and O. Smirnova, Generalized perspective on chiral measurements without magnetic interactions, *Physical Review A* **98**, p. 063428 (December 2018).
48. D. Ayuso, O. Neufeld, A. F. Ordonez, P. Decleva, G. Lerner, O. Cohen, M. Ivanov and O. Smirnova, Synthetic chiral light for efficient control of chiral light-matter interaction, *Nature Photonics* **13**, 866 (2019).
49. P. Král, I. Thanopulos, M. Shapiro and D. Cohen, Two-step enantio-selective optical switch. *Physical review letters*, **90**(3), p. 033001 (2003).
50. C. Pérez, A. L. Steber, S. R. Domingos, A. Krin, D. Schmitz and M. Schnell, Coherent Enantiomer-Selective Population Enrichment Using Tailored Microwave Fields. *Angewandte Chemie International Edition*, **56**(41), pp. 12512–12517 (2017).
51. J. Lee, J. Bischoff, A. O. Hernandez-Castillo, B. Sartakov, G. Meijer and S. Eibenberger-Arias, Quantitative study of enantiomer-specific state transfer. *Physical Review Letters*, **128**(17), p.173001 (2022).
52. K. Schwennicke and J. Yuen-Zhou, Enantioselective Topological Frequency Conversion. *The Journal of Physical Chemistry Letters*, **13**(10), pp. 2434–2441 (2022).
53. D. Ayuso, A. F. Ordonez and O. Smirnova, Ultrafast chirality: the road to efficient chiral measurements. *Phys. Chem. Chem. Phys.* **24**, pp. 26962–26991 (2022).
54. O. Neufeld, D. Ayuso, P. Decleva, M. Y. Ivanov, O. Smirnova and O. Cohen, Ultrasensitive chiral spectroscopy by dynamical symmetry breaking in high harmonic generation, *Phys. Rev. X* **9**, p. 031002 (Jul 2019).
55. A. F. Ordonez, D. Ayuso, P. Decleva and O. Smirnova, Geometric fields and new enantio-sensitive observables in photoionization of chiral molecules, *arXiv:2106.14264 [physics]* (December 2021), arXiv: 2106.14264.
56. A. F. Ordonez and O. Smirnova, Propensity rules in photoelectron circular dichroism in chiral molecules. I. Chiral hydrogen, *Physical Review A* **99**, p. 043416 (April 2019).
57. R. Resta, Macroscopic polarization in crystalline dielectrics: the geometric phase approach, *Reviews of Modern Physics* **66**, 899 (July 1994).
58. N. Berova, P. L. Polavarapu, K. Nakanishi and R. W. Woody, *Comprehensive Chiroptical Spectroscopy* (Wiley, 2013).

59. Y. Zhang, J. R. Rouxel, J. Autschbach, N. Govind and S. Mukamel, X-ray circular dichroism signals: a unique probe of local molecular chirality, *Chem. Sci.* **8**, 5969 (2017).

60. J. R. Rouxel, M. Kowalewski and S. Mukamel, Photoinduced molecular chirality probed by ultrafast resonant x-ray spectroscopy, *Structural Dynamics* **4**, p. 044006 (2017).

61. Y. Harada, E. Haraguchi, K. Kaneshima and T. Sekikawa, Circular dichroism in high-order harmonic generation from chiral molecules, *Phys. Rev. A* **98**, p. 021401 (Aug 2018).

62. M. Oppermann, F. Zinna, J. Lacour and M. Chergui, Chiral control of spin-crossover dynamics in Fe(II) complexes, *Nature Chemistry*, pp. 1–7 (2020), https://www.nature.com/articles/s41557-022-00933-0.

63. N. A. Cherepkov, Angular distribution and spin polarisation of photoelectrons ejected from oriented molecules, *Journal of Physics B: Atomic and Molecular Physics* **14**, p. L623 (1981).

64. Ordóñez Lasso and Andrés Felipe, Chiral measurements in the electric-dipole approximation, Technische Universität Berlin, 2020, Doctoral Thesis, Berlin, 10.14279/depositonce-9677, http://dx.doi.org/10.14279/depositonce-9677.

65. M. Leibscher, E. Pozzoli, C. Pérez, M. Schnell, M. Sigalotti, U. Boscain and C. P. Koch, Full quantum control of enantiomer-selective state transfer in chiral molecules despite degeneracy, *Communications Physics* **5**(1), pp. 1–16 (2022).

66. P. Fischer, D. S. Wiersma, R. Righini, B. Champagne and A. D. Buckingham, *Phys. Rev. Lett.* **85**, p. 4253 (2000).

67. D. Gerbasi, P. Brumer, I. Thanopulos, P. Král and M. Shapiro, Theory of the two step enantiomeric purification of 1,3 dimethylallene, *The Journal of Chemical Physics* **120**, 11557 (2004).

68. A. Yachmenev, J. Onvlee, E. Zak, A. Owens and J. Küpper, Field-induced diastereomers for chiral separation, *Phys. Rev. Lett.* **123**, p. 243202 (Dec 2019).

69. M. Khokhlova, E. Pisanty, S. Patchkovskii, O. Smirnova and M. Ivanov, Enantiosensitive steering of free-induction decay, *arXiv preprint arXiv:2109.15302* (2021).

70. D. Ayuso, A. Ordonez, P. Decleva, M. Ivanov and O. Smirnova, Polarization of chirality (2020).

71. Y. Tang and A. E. Cohen, Optical chirality and its interaction with matter, *Phys. Rev. Lett.* **104**, p. 163901 (Apr 2010).

72. H. Rubinsztein-Dunlop, A. Forbes, M. V. Berry, M. R. Dennis, D. L. Andrews, M. Mansuripur, C. Denz, C. Alpmann, P. Banzer, T. Bauer, E. Karimi, L. Marrucci, M. Padgett, M. Ritsch-Marte, N. M. Litchinitser, N. P. Bigelow, C. Rosales-Guzmán, A. Belmonte, J. P. Torres, T. W. Neely, M. Baker, R. Gordon, A. B. Stilgoe, J. Romero, A. G. White, R. Fickler, A. E. Willner, G. Xie, B. McMorran and A. M. Weiner, Roadmap on structured light, *Journal of Optics* **19**, p. 013001 (nov 2016).

73. E. Pisanty, G. J. Machado, V. Vicuña-Hernández, A. Picón, A. Celi, J. P. Torres and M. Lewenstein, Knotting fractional-order knots with the polarization state of light, *Nature Photonics* **13**, 569 (Aug 2019).

74. S. W. Hell, Nanoscopy with Focused Light (Nobel Lecture), *Angewandte Chemie International Edition* **54**, 8054 (2015), _eprint: https://onlinelibrary.wiley.com/doi/pdf/10.1002/anie.201504181.

75. M. Padgett and R. Bowman, Tweezers with a twist, *Nature Photonics* **5**, 343 (Jun 2011).

76. F. Patti, R. Saija, P. Denti, G. Pellegrini, P. Biagioni, M. A. Iatì and O. M. Maragò, Chiral optical tweezers for optically active particles in the t-matrix formalism, *Scientific Reports* **9**, p. 29 (Jan 2019).

77. M. Li, S. Yan, Y. Zhang, P. Zhang and B. Yao, Enantioselective optical trapping of chiral nanoparticles by tightly focused vector beams, *J. Opt. Soc. Am. B* **36**, 2099 (Aug 2019).

78. D. S. Bradshaw and D. L. Andrews, Laser optical separation of chiral molecules, *Opt. Lett.* **40**, 677 (Feb 2015).

79. R. P. Cameron, A. M. Yao and S. M. Barnett, Diffraction gratings for chiral molecules and their applications, *The Journal of Physical Chemistry A* **118**, 3472 (2014), PMID: 24655409.

80. S. Beaulieu, A. Ferré, R. Géneaux, R. Canonge, D. Descamps, B. Fabre, N. Fedorov, F. Légaré, S. Petit, T. Ruchon *et al.*, Universality of photoelectron circular dichroism in the photoionization of chiral molecules, *New Journal of Physics* **18**, p. 102002 (2016).

81. L. Nahon, G. A. Garcia, C. J. Harding, E. Mikajlo and I. Powis, Determination of chiral asymmetries in the valence photoionization of camphor enantiomers by photoelectron imaging using tunable circularly polarized light, *The Journal of Chemical Physics* **125**, p. 114309 (September 2006).

82. M. V. Berry, Quantal phase factors accompanying adiabatic changes, *Proceedings of the Royal Society of London. A. Mathematical and Physical Sciences* **392**, 45 (1984).

83. A. F. Ordonez and O. Smirnova, Propensity rules in photoelectron circular dichroism in chiral molecules. ii. general picture, *Physical Review A* **99**, p. 043417 (2019).

84. F. de Juan, A. G. Grushin, T. Morimoto and J. E. Moore, Quantized circular photogalvanic effect in Weyl semimetals, *Nature Communications* **8**, p. 15995 (July 2017).

High Sensitivity Chiral Detection in the Gas Phase via Microwave Spectroscopy and the Possible Frontier of Ultracold Chiral Molecules

John M. Doyle,[1,2] Zack D. Lasner,[1,2] Benjamin L. Augenbraun[1,2]

[1] *Harvard-MIT Center for Ultracold Atoms, Cambridge, MA 02138, USA*
[2] *Department of Physics, Harvard University, Cambridge, MA 02138, USA*
jdoyle@g.harvard.edu

Laser-cooling and trapping simple molecules and controlling them at the level of individual quantum states are now established methods in atomic, molecular and optical physics. The frontier of quantum-state-controlled molecules has now moved to polyatomic molecules, including linear, asymmetric top, and chiral varieties. Compared to atoms and diatomic molecules, this molecular complexity offers new quantum resources with distinct advantages for wide-ranging applications, e.g. quantum simulation, precision measurement, and quantum chemistry. Remarkably, it appears that the dramatic increase in structural complexity that comes with polyatomic molecules requires only a modest increase in experimental complexity compared to work with ultracold diatomic molecules. Here we discuss spectroscopic identification of chiral molecules with high sensitivity and specificity in a cold ($\sim$5 K) buffer gas environment, and more recent results on the laser cooling complex polyatomic molecules. Together, these efforts present a road map to full quantum control of ultracold chiral molecules. Other future prospects for ultracold samples of complex molecules are also described.

1. Introduction

The increased structural complexity of molecules, relative to atoms, offers unique quantum resources that are impractical or impossible to attain in atomic systems. While experiments with neutral [1–4] and ionic [5–7] atoms have long exhibited a high degree of control at the single-quantum-state level, more recently ultracold neutral diatomic molecules [8–11] have been proven as viable quantum information platforms [12, 13]. A key advantage in these systems is the large electric dipole moments in molecules, which enable strong, tunable, anisotropic dipole-dipole interactions between molecules. These platforms also possess high-fidelity detection of single particles and provide scalable access to the microscopic details of interesting many-body quantum phenomena.

Quantum-state-controlled diatomic molecules, even in thermal ensembles at temperatures of several kelvin, have also had an important role to play in studies of fundamental physics. Over the last decade, all state-of-the-art electron electric dipole moment (eEDM) searches have been conducted with diatomic molecules [14–17]. In these experiments, the large effective electric fields inside molecules polarized by modest laboratory fields as low as $\sim$100 V/cm can enhance the effect of the eEDM by up to nine orders of magnitude compared with the strength of the eEDM interaction for an isolated electron in the presence of only the laboratory field. These experiments with diatomic molecules may be improved further, for example by laser cooling and trapping molecules for longer interaction times [18].

Polyatomic molecules may ultimately offer even greater sensitivity to the eEDM and other beyond-the-Standard-Model (BSM) physics [19, 20]. While diatomic molecules with specific electronic structures can be easily polarized in laboratory fields below 1 kV/cm, nearly all polyatomic molecules have such a capability in the appropriate states. For example, in the simplest polyatomic molecules like YbOH, which possess three atoms in a linear configuration, these states are excitations of the bending vibrational mode. Such states have lifetimes of ∼1 second. In more complex polyatomic molecules, on the other hand, such as symmetric top molecules including $YbOCH_3$, the easily polarized states are extraordinarily long-lived rotational excitations and the coherence time of any measurement would be limited only by engineering constraints.

Already, molecules with these structures including SrOH [21], CaOH [22], YbOH [23], and $CaOCH_3$ [24] have been laser-cooled to ultracold temperatures in one dimension, and the triatomic molecule CaOH has been magneto-optically trapped [25]. The same experimental tools that have enabled such successes can also be applied to chiral molecules such as CaOCHDT, which naturally occur in two configurations with opposite "handedness." For example, in CaOCHDT, the HDT can be arranged either clockwise or counter-clockwise from the vantage point of the C atom. Molecules with opposite chirality have nearly identical energy levels, but are expected to exhibit small energy differences owing to the parity-violating weak interaction. Thus precision spectroscopy of chiral molecules can probe symmetry violations and test the Standard Model [26].

In addition, a large number of biologically relevant molecules exhibit chirality and exist in two mirror-image (or enantiomeric) configurations. The biological functionality of a molecule often depends critically on its handedness. Tools to identify the enantiomeric composition of a molecule can be useful to biochemistry (including potentially pharmacology) and to understanding the origin of biochemical homochirality.

In this manuscript, we discuss our experimental results detecting chiral molecules with high sensitivity and specificity, using microwave spectroscopy in a cold buffer gas environment. Then, we describe progress in laser cooling and trapping molecules of increasing complexity, and prospects for rapid photon cycling (RPC) and laser cooling of large, asymmetric molecules including chiral species. Finally, we provide an outlook on the scientific goals that could be achieved as RPC and laser cooling are applied to more complex and larger polyatomic molecules.

2. Enantiomer-selective detection and state preparation of chiral molecules

Large, asymmetric molecules commonly come in two mirror images, or enantiomers. Aside from extraordinarily weak parity-violating interactions, in the absence of external fields the energies of opposite enantiomeric configurations are identical. However, the interactions of these molecules with external electric fields can be distinguished. In particular, chiral molecules generally possess three unequal permanent dipole moments, μ_a, μ_b, and μ_c along axes designated a, b, and c. Opposite

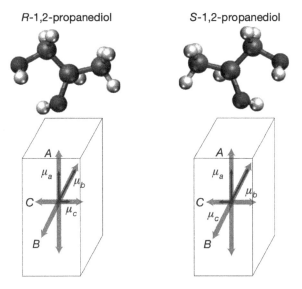

Fig. 1. 1,2-propanediol, an example of a chiral molecule, in both enantiomeric configurations. The products of molecule-frame electric dipole moments have opposite sign, independent of the choice of coordinates. Figure reproduced from [27].

enantiomers possess equal values of $|\mu_a|$, $|\mu_b|$, and $|\mu_c|$, but opposite values of the signed quantity $\mu_a\mu_b\mu_c$. An example chiral molecule is shown for both enantiomers in Fig. 1 along with its electric dipole moments.

An oscillating electric field, for example arising from microwave radiation, can couple to one of the electric dipole moments of a molecule to exert torque and change its rotational state. The transitions driven via the a-, b-, and c-axis dipole moments are known as a-type, b-type, and c-type transitions. If all three types of transitions are driven, then the overall phase of the state evolution can reveal the sign of the product $\mu_a\mu_b\mu_c$, and thus unambiguously identify the enantiomeric composition of a sample of chiral molecules. Here we discuss one implementation of such a technique, known as three-wave-mixing, developed in our lab.

The relevant rotational structure is shown for the example molecule of $1,2$-propanediol in Fig. 2. Both c-type and a-type transitions are driven by microwave or radio-frequency radiation, resulting in a superposition of three rotational states. The time-evolution of this state is characterized by oscillations of the molecular dipole moments, resulting in radiation at the frequencies of the three energy splittings. The phase of the radiation associated with the undriven transition (b-type) depends on the product of dipole moments $\mu_a\mu_b\mu_c$. Therefore, the radiation associated with this transition is perfectly out of phase for left- and right-handed molecules. This out-of-phase radiation can be seen experimentally in Fig. 3.

The double-resonance inherent in the measurement protocol described above also decongests the three-wave mixing spectrum and makes it a highly *specific* detector

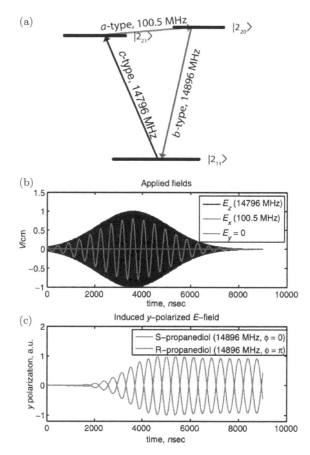

Fig. 2. (a) The rotational levels of 1,2-propanediol used for detection of enantiomeric composition. (b) Applied electric fields, which result in a superposition of all three rotational states. (c) Simulation of $\hat{y}$-polarized radiation, which occurs at the resonance frequency of the b-type transition (which is not directly driven). Figure reproduced from [28].

of chiral molecules. Furthermore, in an equal mixture of enantiomers the three-wave mixing signal exactly vanishes (see Fig. 3), making it highly sensitive to small imbalances in enantiomers.

A crucial experimental requirement that has not yet been discussed is the availability of a cold ($\sim$10 K or less) sample of molecules. In the experiments discussed so far, room-temperature molecules were flowed into a copper cell cooled to $\sim$5 K by a pulse tube cryocooler. The cell is filled with helium buffer gas, which thermalizes with the cell walls and rapidly cools the large, chiral molecules via collisions. In this way, the chiral molecules rotationally cool and the population per quantum state is increased by several orders of magnitude, leading to high spectroscopic sensitivity.

Beyond merely detecting the presence of a chiral molecule and identifying its enantiomeric composition, it is also possible to use three-wave mixing to selectively

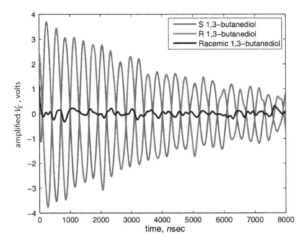

Fig. 3. Measured radiation from three-wave mixing of the chiral molecule 1,3-butanediol. Left- and right-handed enantiomers radiate on the undriven transition perfectly out of phase, while the radiation for an equal (or "racemic") mixture of enantiomers cancels. Figure reproduced from [28].

prepare one enantiomer in a particular quantum state [29]. In particular, we have demonstrated that a chiral molecule can be driven to a given state via two paths. For example, in the structure of Fig. 2(a), the state $|2_{20}\rangle$ can be populated from $|2_{11}\rangle$ directly with a b-type transition, or alternatively via the intermediate state $|2_{21}\rangle$ with both a c-type and a-type transition. If the driving fields are chosen carefully and both paths are driven simultaneously, then the quantum amplitudes to populate a target state will either constructively or destructively interfere depending on the enantiomer.

3. Why laser cool polyatomic molecules?

The methods described in the previous section operate at temperatures of several kelvin, already sufficient to reduce the number of occupied quantum states by several orders of magnitude in large molecules. Further cooling to ultracold temperatures below 1 mK, however, enables qualitatively different experiments. At these lower temperatures, it is feasible to hold molecules in a conservative potential such as an optical dipole trap (or optical lattice) or magnetic trap [8, 30], enabling long coherent times and high densities. Here we provide a brief overview of the applications that are enabled by such traps of polyatomic molecules at ultracold temperatures.

Optical traps, which can only be loaded from ultracold temperatures of $\sim$10 μK, offer several advantages over cryogenic platforms or molecular beams. One key feature is the possibility of confining molecules to sub-micron size scales, allowing them to be brought close enough to interact strongly in a controlled manner. Such strong, controlled interactions are essential to quantum computation and simulation experiments. Another advantage of optical traps, in contrast to magneto-optical or

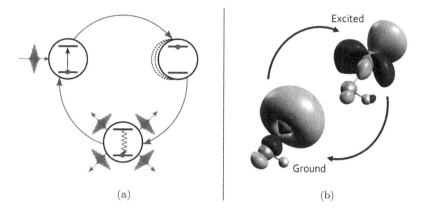

Fig. 4. (a) Schematic diagram of optical cycling in an ideal two-level system: directional photon absorption, momentum change due to photon absorption, and isotropic spontaneous emission leading to decay to the initial quantum level. Complex structures (of either atoms or molecules) can interrupt this cycle, e.g., by the addition of multiple decay pathways for spontaneous emission. (b) Molecular orbitals involved in a molecular optical cycling scheme, computed for the example of calcium hydrosulfide (CaSH). Note that the orbitals are highly localized on the atomic Ca optical cycling center, and that these orbitals somewhat resemble an atom-like s to p transition.

high-field electric and magnetic traps, is that the trapping light field can be weakly perturbing and may not contribute systematic shifts in precision spectroscopy.

The inherent limitation of optical traps is that their trap depths, in practice, are no larger than about 1 mK. To reach this temperature regime, the standard approach in atomic and molecular experiments is to laser cool samples using rapid cycling of $\sim 10^3$ photons (see Fig. 4). RPC has been demonstrated for highly symmetric atoms and molecules in which the inherent structural symmetry leads to electric dipole selection rules that aid the task of scattering approximately 10^4 photons, enough to realize deep laser cooling. It is important to note that RPC also enables high-fidelity quantum state preparation and readout, important for both quantum information applications and spectroscopy. So, achieving RPC provides full quantum control over both the internal and external states of the molecule, providing a near-perfect platform for a variety of quantum science endeavors.

While the advantages described above apply to all laser-coolable atoms and molecules, *polyatomic* molecules (those containing three or more atoms) offer qualitatively distinct vibrational and rotational motions that enable new opportunities in physics, chemistry, and quantum technology. For instance, all polyatomic molecules have long-lived states with body-frame angular momentum arising from nuclear motion. These states offer energy level structures with Debye-scale Stark shifts at low applied electric fields, as well as extremely field-insensitive molecular orientation states, including those with near-zero lab-frame dipole moment. An example of this structure is displayed in Fig. 5, where the calculation is performed for rotational states of the bent triatomic molecule CaSH. Also shown is the ground rotational state in the diatomic molecule CaF, which is negligibly polarized at the

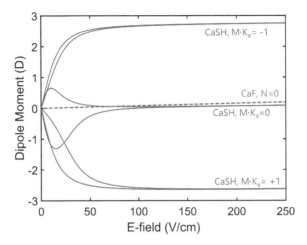

Fig. 5. Lab-frame dipole moments of CaSH $K_a = 1$ states (red solid lines) and the CaF $N = 0$ state (blue dashed line) in applied electric fields between $E = 0$ and 250 V/cm.

low electric fields shown. Polyatomic molecules also often possess nearly degenerate rovibrational levels with enhanced sensitivity to new physics such as ultralight bosonic dark matter [31]. Further, all chiral molecules are necessarily polyatomic. Thus studies of biochemical homochirality or parity-violating spectroscopic splittings between enantiomers require the use of polyatomic molecules.

However, the rich internal structure of polyatomic molecules also makes the task of laser cooling molecules initially appear quite challenging. In particular, the many vibrational and rotational states present in polyatomic molecules constitute a vast reservoir of potential "loss channels" that could interrupt the laser cooling process (the repeated excitation and spontaneous decay process of Fig. 4). In a typical molecule, electron excitation followed by spontaneous photon emission might cause dozens (or more) of vibrational and rotational states to become populated—too many to feasibly control with lasers in the laboratory (at least with current laser technology). However, molecules with a particular structure can be found that violate this trend and which populate only a few vibrational states following electronic excitation, thus allowing rapid photon cycling with a practical number of repumping lasers.

4. Identifying laser-coolable molecules

The essential requirement for rapid photon cycling (and by extension laser cooling) is the existence of an excited state that decays preferentially to a manageable number of ground states (ideally one), with unplugged leaks to all other levels below about 1 part in 10^4. The simplest atoms with this electronic structure are alkali metal atoms with low-lying $^2S \rightarrow {}^2P$ transitions (the D1 and D2 lines). In the case of molecules, certain classes with $^2\Sigma^+$ ground states have been found to be favorable for laser cooling due to their single valence electron in a metal-centered $s\sigma$ orbital [32, 33].

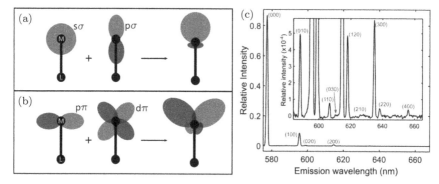

Fig. 6. Illustration of orbital mixing between (a) $s\sigma$ and $p\sigma$ orbitals to generate an $\tilde{X}\,^2\Sigma^+$ state and (b) $p\pi$ and $d\pi$ orbitals to generate an $\tilde{A}\,^2\Pi$ state. (c) Experimental DLIF trace recorded for YbOH molecules following excitation to the $\tilde{A}\,^2\Pi_{1/2}(v=0, J=1/2, p=+)$ level. Inset shows the detail near the baseline, demonstrating that relative intensity sensitivity around 10^{-5} is achieved. Labels above each peak indicate the ground state vibrational level that the decay populates.

These species can be thought of as the molecular analogs of alkali atoms in the sense that they have a single optically active electron with essentially "atom-like" transitions; see Fig. 6(a).

Electronic closure can be ensured by exciting to low-lying electronically excited states that have no levels between them and the absolute ground state. Each time a molecule is excited and spontaneously emits a photon, it is still possible that the vibrational and/or rotational state will change. Fortunately, closed rotational cycling schemes can be found for essentially all molecular structures of interest, relying on a combination of parity and angular momentum selection rules [33–35].

Even without any electronic or rotational loss channels, the absence of angular momentum associated with the stretching modes of vibrational motion will lead to losses to excited vibrational levels. Vibrational losses are strongly suppressed in molecules for which the ground- and excited-state potential energy surfaces resemble one another, as they do for alkaline-earth metal atoms monovalently and ionically bonded to an electronegative ligand (CaF, CaOH, CaOCH$_3$, etc.). As shown in Fig. 6(a–b), electrostatic repulsion between the metal-centered valence electron and the negatively charged ligand leads to orbital hybridization that excludes valence electron density from the bonding region and thus to decoupling between electronic and vibrational excitations [32]. Importantly, this bonding behavior remains true regardless of the alkaline-earth (or alkaline-earth-like) metal atom chosen (e.g., M = Ca, Sr, Ba, Yb) and for a wide range of electronegative ligands (e.g., -OH, -OCH$_3$, -SH, -NH$_2$, etc.).

The intensities of vibronic transitions are governed not by strict selection rules, but by the values of so-called Franck-Condon factors (FCFs), which characterize the overlap between ground and excited vibrational wave functions. The possible decay pathways and their relative strengths can be experimentally measured by driving molecules to a particular excited state and measuring the relative intensities of the

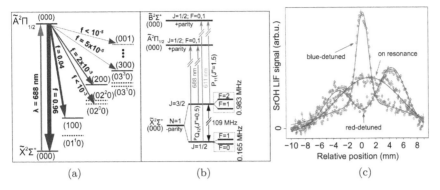

Fig. 7. Laser cooling of the SrOH molecule. (a) Vibrational branching ratios relevant for the SrOH $\tilde{A}\,^2\Pi_{1/2}$ state. In the experiment, the (100) and (02^00) levels were repumped. (b) Rotational substructure of each vibronic band pumped. Two spin-rotation features are addressed. Either the $\tilde{A}$ or $\tilde{B}$ state is targeted for laser cooling. (c) Transverse laser cooling of an SrOH molecular beam from 50 mK (green curve) to 750 μK (blue curve). The dramatic reduction of transverse spatial extent is indicative of cooling. Also shown (red curve) is the effect of laser heating, which increases the molecular velocities in the transverse direction and therefore causes accumulation off of the molecular beam axis.

various wavelengths at which spontaneous emission occurs, e.g. using dispersed laser-induced fluorescence (DLIF). The pioneering use of DLIF measurements to identify laser coolable molecules were carried out by the Steimle group [36–38]. Recently, we have improved the intensity sensitivity achievable in these measurements to about 10^{-5}, allowing us to characterize all loss channels that must be repumped for a full cooling and trapping experiment [39]. Figure 6(c) shows a measurement determining the FCFs for the $\tilde{X} - \tilde{A}$ band of YbOH [39]. Because of the highly diagonal FCFs, fewer than a dozen vibrational decay channels need to be repumped to ensure "closure" to the level of 1 part in 10^4. These highly diagonal FCFs are characteristic of a large class of molecules comprised of alkaline-earth atoms ionically bonded to halogen-like ligands, as was originally noted by Di Rosa [33].

5. Experimental progress

The first polyatomic molecule to be directly laser cooled was SrOH [21]. In these transverse cooling experiments, the transverse temperature of an SrOH molecular beam, produced in a cryogenic buffer gas source, was reduced from 50 mK to 750 μK in one dimension using only 220 scattered photons per molecule. Downstream of the region where molecules interact with the cooling lasers, the spatial profile of the molecular beam is imaged on an electron multiplying charge-coupled device (EMCCD) camera. When molecules have been cooled, they expand less during their time of flight and occupy a smaller region of space. By adjusting laser detunings, molecules can also be laser heated, resulting in a broader spatial distribution downstream of the heating location (Fig. 7(c)). Following the original work on

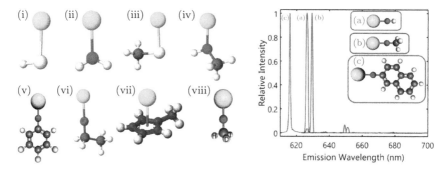

Fig. 8. (Left) Classes of asymmetric top molecules that appear amenable to photon cycling and laser cooling include: (i) MSH, (ii) MNH_2, (iii) $MSCH_3$, (iv) $MNHCH_3$, (v) MOC_6H_5 [including substitutions around the carbon ring], (vi) MOC_2H_5, (vii) $MC_5H_4CH_3$, and (viii) MOCHDT, where M is an alkaline-earth metal atom. (Right) Comparison of DLIF spectra recorded for (a) CaOH, (b) $CaOCH_3$, and (c) $CaOC_{10}H_7$. The behavior of Ca as an optical cycling center is clearly demonstrated by the fact that the number of vibrational-state-changing decays increases only slightly as the number of vibrational modes increases by over an order of magnitude.

SrOH, the linear molecules YbOH and CaOH were also laser cooled in one dimension [22, 23].

Very recently, the linear triatomic molecule CaOH has been loaded into a 3D magneto-optical trap and cooled to 100 μK, below the Doppler limit [25]. This was achieved by optically repumping 11 vibrational and rotational loss channels identified from DLIF measurements [39], enabling an average of 12000 photon scatters per molecule before loss to rovibronic dark states occurred. Notably, only ∼2000 photons were required for magneto-optical trapping of slow molecules, and well under 1000 photons were required for sub-Doppler cooling, indicating that such techniques remain efficient for trapped polyatomic molecules.

Symmetric top molecules represent a natural next step in the "symmetry descent" from linear molecules. These molecules introduce rotational modes with angular momentum about the internuclear axis ($K > 0$ states), but transitions can be chosen to preserve many of the selection rules that were relied upon to construct closed optical cycling transitions of simpler molecules [38, 40]. We have laser-cooled the symmetric top molecule $CaOCH_3$ to approximately 1 mK via rapid photon cycling from states with both $K = 0$ and $K = 1$ [24]. In the case of laser cooling from $K = 0$, the rotational selection rules are equivalent to those found in linear triatomic molecules, while for $K = 1$ a single rotational repumping laser is required for each populated vibrational state. This result establishes that laser cooling remains feasible even for the rotational structures found only in nonlinear molecules.

Whereas the two principal moments of inertia perpendicular to the symmetry axis in $CaOCH_3$ have equal values, in the closely related chiral molecule CaOCHDT all three moments of inertia have distinct values. Species with this property are known as asymmetric top molecules (ATMs) and are the most generic type of molecule. With the exception of rare special cases, all chiral molecules are ATMs.

ATMs lack the symmetries that led to certain selection rules that aided the photon cycling process for linear and symmetric top molecules, but nonetheless our group has shown theoretically that they can be controlled using slight modifications to the standard techniques [34]. In fact, we have found that an extremely broad group of molecules is amenable to these photon cycling techniques, including the structures shown in Fig. 8(a). There is thus no fundamental limitation to laser cooling and trapping chiral molecules such as CaOCHDT. While DLIF measurements of chiral molecules have yet to be performed with a sensitivity sufficient to identify all required repumping lasers, it appears that the vibrational branching of well-chosen molecules will be experimentally manageable.

In addition to the possibility of enantiomeric symmetry, ultracold ATMs would offer several qualitative features useful for a broad range of science [34]. For example, ATMs generally have three permanent dipole moments that could be controlled independently, a feature not present in higher-symmetry species. Furthermore, low-lying states with very long radiative lifetimes and large dipole moments promise strong molecule-molecule coupling in quantum simulation schemes, potentially enabling quantum gates at least an order of magnitude faster than could be achieved with laser-coolable diatomic molecules. Because the parity-doubled states come from rotational motions with typical energies $0.1 - 1$ cm^{-1} above the ground state, these useful "science" states have lifetimes orders of magnitude longer than those present in the bending vibrational modes of linear polyatomic molecules. This enhances the feasible coherence time for both precision measurement and quantum science applications to many tens of seconds.

6. Conclusion

In summary, we have described methods to spectroscopically identify enantiomers of cold ($\sim$5 K) chiral molecules with high sensitivity and specificity. We argue that further cooling of complex polyatomic molecules to the ultracold regime offers many new opportunities for quantum simulation, precision measurement, and studies of chiral molecular physics. Recent experimental results have shown that much of the toolbox developed for ultracold atoms can be applied to polyatomic molecules with equal success. Species with increasingly complex structures have now been laser cooled, including a molecule, CaOCH$_3$, possessing a dozen vibrational modes. Similarly, the polyatomic molecule CaOH has now been magneto-optically trapped and sub-Doppler cooled. The complications in extending such techniques to fully asymmetric molecules, for example CaOCHDT (a chiral isotopologue of CaOCH$_3$), are theoretically understood and appear experimentally tractable.

References

1. W. S. Bakr, A. Peng, M. E. Tai, R. Ma, J. Simon, J. I. Gillen, S. Foelling, L. Pollet, and M. Greiner, Science **329**, 547 (2010).

2. J. F. Sherson, C. Weitenberg, M. Endres, M. Cheneau, I. Bloch, and S. Kuhr, Nature **467**, 68 (2010).
3. H. Labuhn, D. Barredo, S. Ravets, S. D. Léséleuc, T. Macrì, T. Lahaye, and A. Browaeys, Nature **534**, 7609 (2016).
4. H. Bernien, S. Schwartz, A. Keesling, H. Levine, A. Omran, H. Pichler, S. Choi, A. S. Zibrov, M. Endres, M. Greiner, V. Vuletic, and M. D. Lukin, Nature **551**, 7682 (2017).
5. J. G. Bohnet, B. C. Sawyer, J. W. Britton, M. L. Wall, A. M. Rey, M. Foss-Feig, and J. J. Bollinger, Science **352**, 1297 (2016).
6. J. Zhang, G. Pagano, P. W. Hess, A. Kyprianidis, P. Becker, H. Kaplan, A. V. Gorshkov, Z.-X. Gong, and C. Monroe, Nature **551**, 7682 (2017).
7. C. Monroe, W. C. Campbell, L.-M. Duan, Z.-X. Gong, A. V. Gorshkov, P. W. Hess, R. Islam, K. Kim, N. M. Linke, G. Pagano, P. Richerme, C. Senko, and N. Y. Yao, Rev. Mod. Phys **93**, 025001 (2021).
8. L. Anderegg, L. W. Cheuk, Y. Bao, S. Burchesky, W. Ketterle, K.-K. Ni, and J. M. Doyle, Science **365**, 1156 (2019).
9. L. W. Cheuk, L. Anderegg, Y. Bao, S. Burchesky, S. S. Yu, W. Ketterle, K.-K. Ni, and J. M. Doyle, Phys. Rev. Lett. **125**, 043401 (2020).
10. S. Burchesky, L. Anderegg, Y. Bao, S. S. Yu, E. Chae, W. Ketterle, K.-K. Ni, and J. M. Doyle, Phys. Rev. Lett. **127**, 12320 (2021).
11. W. B. Cairncross, J. T. Zhang, L. R. B. Picard, Y. Yu, K. Wang, and K.-K. Ni., Phys. Rev. Lett. **126**, 123402 (2021).
12. A. M. Kaufman and K.-K. Ni., Nature Physics **17**, 1324–1333 (2021).
13. L. Henriet, L. Beguin, A. Signoles, T. Lahaye, A. Browaeys, G.-O. Reymond, and C. Jurczak, Quantum **4**, 327 (2020).
14. ACME Collaboration, Nature **562**, 355 (2018).
15. W. B. Cairncross, D. N. Gresh, M. Grau, K. C. Cossel, T. S. Roussy, Y. Ni, Y. Zhou, J. Ye, and E. A. Cornell, Phys. Rev. Lett. **119**, 153001 (2017).
16. J. Baron, W. C. Campbell, D. DeMille, J. M. Doyle, G. Gabrielse, Y. V. Gurevich, P. W. Hess, N. R. Hut- zler, E. Kirilov, I. Kozyryev, B. R. O'Leary, C. D. Panda, M. F. Parsons, E. S. Petrik, B. Spaun, A. C. Vutha, and A. D. West, Science **343**, 269 (2014).
17. J. J. Hudson, D. M. Kara, I. J. Smallman, B. E. Sauer, M. R. Tarbutt, and E. A. Hinds, Nature **473**, 493 (2011).
18. N. J. Fitch, J. Lim, E. A. Hinds, B. E. Sauer, and M. R. Tarbutt, Quantum Science and Technology **6**, 014006 (2021).
19. I. Kozyryev and N. R. Hutzler, Phys. Rev. Lett. **119**, 133002 (2017).
20. N. R. Hutzler, Quantum Sci. Technol. **5**, 044011 (2020).
21. I. Kozyryev, L. Baum, K. Matsuda, B. L. Augenbraun, L. Anderegg, A. P. Sedlack, and J. M. Doyle, Phys. Rev. Lett. **118**, 173201 (2017).
22. L. Baum, N. B. Vilas, C. Hallas, B. L. Augenbraun, S. Raval, D. Mitra, and J. M. Doyle, Phys. Rev. Lett. **124**, 133201 (2020).
23. B. L. Augenbraun, Z. D. Lasner, A. Frenett, H. Sawaoka, C. Miller, T. C. Steimle, and J. M. Doyle, New J. Phys. **22**, 022003 (2020).
24. D. Mitra, N. B. Vilas, C. Hallas, L. Anderegg, B. L. Augenbraun, L. Baum, C. Miller, S. Raval, and J. M. Doyle, Science **369**, 1366 (2020).
25. N. B. Vilas, C. Hallas, L. Anderegg, P. Robichaud, A. Winnicki, D. Mitra, and J. M. Doyle, (2021) arXiv:2112.08349.
26. M. Quack, J. Stohner, and M. Willeke, Annual Review of Physical Chemistry **59**, 741 (2008).

27. D. Patterson, M. Schnell, and J. M. Doyle, Nature **497**, 475 (2013).
28. D. Patterson and J. M. Doyle, Physical Review Letters **111**, 023008 (2013).
29. S. Eibenberger, J. Doyle, and D. Patterson, Physical Review Letters **118**, 123002 (2017).
30. H. J. Williams, L. Caldwell, N. J. Fitch, S. Truppe, J. Rodewald, E. A. Hinds, B. E. Sauer, and M. R. Tar-butt, Phys. Rev. Lett. **120**, 163201 (2018).
31. I. Kozyryev, Z. Lasner, and J. M. Doyle, Phys. Rev. A **103**, 043313 (2021).
32. A. M. Ellis, Int. Rev. Phys. Chem. **20**, 551 (2001).
33. M. D. D. Rosa, Eur. Phys. J. D **31**, 395 (2004).
34. B. L. Augenbraun, J. M. Doyle, T. Zelevinsky, and I. Kozyryev, Phys. Rev. X **10**, 031022 (2020).
35. I. Kozyryev, L. Baum, K. Matsuda, and J. M. Doyle, ChemPhysChem **17**, 3641 (2016).
36. X. Zhuang, A. Le, T. C. Steimle, N. E. Bulleid, I. J. Smallman, R. J. Hendricks, S. M. Skoff, J. J. Hudson, B. E. Sauer, E. A. Hinds, and M. R. Tarbutt, Phys. Chem. Chem. Phys. **13**, 19013 (2011).
37. D.-T. Nguyen, T. C. Steimle, I. Kozyryev, M. Huang, and A. B. McCoy, J. Mol. Spec. **347**, 7 (2018).
38. I. Kozyryev, T. C. Steimle, P. Yu, D.-T. Nguyen, and J. M. Doyle, New J. Phys. **21**, 052002 (2019).
39. C. Zhang, B. L. Augenbraun, Z. D. Lasner, N. B. Vilas, J. M. Doyle, and L. Cheng, J. Chem. Phys. **155**, 091101 (2021).
40. T. A. Isaev and R. Berger, Phys. Rev. Lett. **116**, 063006 (2016).

Parity Violation in Chiral Molecules:
From Theory towards Spectroscopic Experiment
and the Evolution of Biomolecular Homochirality[a)]

M. Quack*, G. Seyfang and G. Wichmann

Physical Chemistry, ETH Zürich,
CH-8093 Zurich, Switzerland
* E-mail: Martin@Quack.CH
www.ir.ETHz.CH

Molecular chirality is related to the symmetry with respect to space inversion, which was assumed to be fundamental in the theory of chemical bonding and stereochemistry. The symmetry leads to a conservation law for the quantum number parity as a constant of the motion. We know today that there is in fact a slight asymmetry, which leads to a non-conservation of parity or parity violation. We start with an introductory discussion of three fundamental questions on symmetry, relating physics to molecular quantum dynamics and stereochemistry: (*i*) To what extent are the fundamental symmetries and conservation laws of physics and their violations reflected in molecular quantum dynamics and spectroscopy, in general? (*ii*) How important is parity violation - the violation of space inversion symmetry - for the quantum dynamics and spectroscopy of chiral molecules, in particular? (*iii*) How important is parity violation for biomolecular homochirality, i.e. the quasi exclusive preference of L-amino acids and D-sugars in the biopolymers of life (proteins and DNA)? The observation of biomolecular homochirality can be considered to be a quasi-fossil of the evolution of life, the interpretation of which has been an open question for more than a century, with numerous related hypotheses, but no definitive answers. We shall briefly discuss the current status and the relation to the other two questions. The discovery of parity violation led to important developments of physics in the 20th century and is understood within the standard model of particle physics, SMPP. For molecular stereochemistry it leads to the surprising prediction of a small energy difference D of the ground state energies of the enantiomers of chiral molecules, corresponding to a small reaction enthalpy for the stereomutation between the R and S isomers(enantiomers). With exact parity conservation this reaction enthalpy would be exactly zero by symmetry. Theory predicts D to be in the sub-femto eV range, typically, depending on the molecule (about D = 200 ± 50 aeV for ClSSCl or CHFClBr, corresponding to a reaction enthalpy of about 20 ± 5 pJ/mol). We have outlined four decades ago, how this small energy difference D might by measured by spectroscopic experiments, and recent progress indicates that experiments might be successful in the near future. We report here about the development of the quantitative theory for predicting D and we then discuss the development and current status of our experiments including alternatives pursued in other groups and the possible consequences for our understanding of molecular and biomolecular chirality as well as the design of molecular quantum switches for a possible future quantum technology and possible tests of CPT symmetry.

Keywords: chiral molecules, parity violation, symmetry, high resolution spectroscopy, biomolecular homochirality

[a)] Dedicated to the memory of Richard R. Ernst (14 August 1933 - 4 June 2021 (NP Chemistry 1991) after lecture presented at Nobel Symposium 167 on chiral matter, Stockholm, Sweden 28 June - 1 July 2021)

1. Introduction

Ryoji Noyori started his Nobel prize lecture [1] with the sentence: 'Chirality (handedness, left or right) is an intrinsic universal feature of various levels of matter', and this sentence can, indeed, serve as a perfect motto for the Nobel Symposium 167 on 'Chiral Matter', in general terms. However, the terminology 'chiral' ('handed', from the ancient Greek $\chi\varepsilon\iota\varrho$, cheir, for hand) was introduced in the context of the structure and chemistry of chiral molecules [2, 3] replacing the earlier terminology 'dissymmetric', which was used by Pasteur, the discoverer of molecular chirality [4–7].

Molecular chirality is of fundamental importance in stereochemistry and has, indeed, been crucial in the development of our understanding of the foundations of physical-chemical stereochemistry [8]. Advances in the understanding and uses of chiral molecules are reflected by numerous Nobel prizes over more than a century beginning with the first to van't Hoff (1901, a founder of the field of stereochemistry while his research on chemical dynamics and osmotic pressure was emphasized in the prize citation) shortly thereafter to Emil Fischer (1902, emphasizing also his work on organic stereochemistry), and Alfred Werner (1913) [4, 9]. More recently we may mention V. Prelog in 1975 [10], W. S. Knowles, R. Noyori, and K. B. Sharpless in 2001 [1, 11, 12] or the most recent one in 2021 [13, 14] as selected examples. Also molecular motors contain an important 'chiral' aspect [15–17]. Chiral molecules are standard textbook [18] and examination topics in chemistry testing knowledge of chemical nomenclature [19] and at the same time they are of crucial importance in chemical and pharmaceutical industry (see Refs. 20, 21 for example), see also the lecture by Sarah Price at this symposium.

At the same time chiral molecules have a deep connection to the foundations of physics through symmetries and conservation laws [22–24]. Symmetries are very general underlying ordering principles in the systematic approach to understanding nature or even external reality as such, which is assumed to exist as a premiss in science. Starting from observed facts (in experiments or otherwise by 'observations', for instance in astronomy) one has to organize these in well ordered mental 'pictures', which may be called models, hypotheses or theories at various levels of understanding (see Fig. 1). It turns out that symmetries and their violation by asymmetries provide a further underlying structure of these and can even be related to the fundamental question of 'observability' of certain basic facts of theoretical structures as we shall see in Sec. 2. In this sense, symmetries and asymmetries have a very special role for the structure of theory. Indeed, the discovery of the violation of space inversion symmetry, one of the fundamental symmetries of physics, or 'parity violation' in nuclear and elementary particle physics [25–30] not only led to the development of the current 'Standard Model of Particle Physics' (SMPP) [31–37], but also to an interesting interaction between high energy physics, molecular physics, chemistry, and also biochemistry and biology [38, 39]. This interaction results in the following at first perhaps surprising statements [40].

Assuming that there is an external reality, whatever that
may be: How do we understand it?

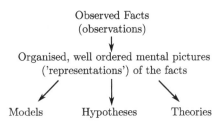

Fig. 1. Organising the facts of reality by models, hypotheses and theories, with symmetries and asymmetries as common fundamental properties (after Ref. 44).

(1) The fundamentally new physics arising from parity violation and the consequent electroweak theory in the SMPP leads to the prediction of fundamental new effects in the dynamics of chiral molecules and thus in the realm of chemistry.

(2) Possible experiments on parity violation in chiral molecules open a new and very special window to looking at the fundamental symmetries of nature and certain aspects of the standard model of high energy physics, and thus molecular physics might contribute to our understanding of the fundamental laws of physics.

(3) Parity violation in chiral molecules provides a unique and very special connection to the early evolution of life as we know it (through the 'homochirality' in the biopolymers of life, which select L-amino acids and D-sugars with absolute preference). This has possibly (but not necessarily) important consequences for the early evolution of life.

Indeed, going beyond parity violation and the standard model, molecular chirality may provide a fresh look at time reversal symmetry and its violation and even the nature of time [41–43].

In the present contribution to this symposium we shall start in Sec. 2 with a conceptual discussion of the use of symmetries in understanding molecular quantum dynamics in general and for chiral molecules in particular. We shall then describe the long road from the idea for an experimental concept to the development of a quantitative theory of parity violation in chiral molecules (Sec. 3) and current developments in spectroscopic experiments on chiral molecules and parity violation (Sec. 4). We shall conclude with a very brief summary of considerations concerning the evolution of biomolecular homochirality (Sec. 5). We restrict attention in this short article on the fundamental concepts and the most important developments as they lead to the current status in the field. We refer to extensive reviews and books in the past for further background and more complete references [22, 23, 38–40, 44–51].

We conclude this brief introduction with a historical remark concerning an important statement by van't Hoff [52] who clearly expressed the consequences for the energetic and thermodynamic equivalence of the enantiomers arising from symmetry (Fig. 2). Considering the equilibrium in the reaction between R and S (a 'stereomutation reaction') he writes:

$$S(L) \rightleftharpoons R(D) \tag{1}$$

and states that the equilibrium constant must be exactly 1 by symmetry

$$K = [R]/[S] = 1 \tag{2}$$

$$\ln K = -\Delta_R S^\ominus(T)/(RT) = 0 \quad \text{all } T \tag{3}$$

We show in Fig. 2 the original citation with its translation and have rewritten here van't Hoff's equations in modern notation. Because of the 'exact symmetry', as he writes, the ground state energies of the enantiomer (E_R^o and E_S^o) and all reaction energies $\Delta_R H^\ominus$, $\Delta_R G^\ominus$ and entropies ($\Delta_R S^\ominus$) must be zero (see [45] for further discussion).

We anticipate here an important change which arose from developments in high energy physics of the second half of the 20th century. We know now that there is no such exact symmetry and in actual fact there is a slight 'parity violating' energy difference $D = \Delta_{pv}E$ between the ground states of the enantiomers, say

$$\Delta_{pv}E = E_R^o - E_S^o = \Delta_R H_0^\ominus / N_A \tag{4}$$

and small non-zero absolute values also for all the thermodynamic quantities mentioned above [53] ($\Delta_R G^\ominus$, $\Delta_R H^\ominus$, $\Delta_R S^\ominus$, etc.). This is shown schematically in

J. H. van't Hoff 'La Chimie dans l'espace' **1887**

S $\rightleftharpoons$ R

'... Un tel équilibre dépend du travail (E) que la transformation peut produire, travail, qui doit être égal à zéro dans le cas en question, vue la symétrie mécanique parfaite des deux isomères...'

'Such an equilibrium depends upon the work (E) which the transformation can generate, work which must be equal to zero in this case, in view of the perfect mechanical symmetry of the two isomers...'

$\ln K = -\frac{\Delta_R G^\ominus}{RT} = 0$

$\Delta_R S^\ominus$ and $\Delta_R H^\ominus$ are exactly zero at all T.

$E_S^0 = E_R^0$

But today $\Delta_R H_0^\ominus \neq 0$ due to parity violation!

Very small, but how small or how big exactly?

Fig. 2. The text from the original publication [52] of van't Hoff stating that the mirror symmetry between the enantiomers S and R of a chiral molecule has consequences for energies and equilibria, rewritten as equations in modern notation replacing 'work E' by the Gibbs free energy $\Delta_R G^\ominus$ (after Ref. 45).

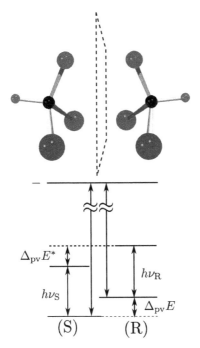

Fig. 3. Energy level scheme for the two enantiomers of CHFClBr including parity violation (see Refs. 39, 47). Following Refs. 54, 55 (S) is calculated to be more stable than (R) by $\Delta_{\mathrm{pv}}E \simeq 236\,\mathrm{aeV}$ corresponding to about $\Delta_{\mathrm{R}}H^{\ominus} \simeq 23 \cdot 10^{-12}\,\mathrm{J/mol}$ ($23\,\mathrm{pJ/mol}$).

Fig. 3 for the example of CHFClBr which will be discussed in the following sections in detail, and which had already used by van't Hoff as a prototype chiral molecule [8].

2. Symmetries, symmetry violations and approximate constants of the motion

The time evolution of a molecular system can be described by the time dependent Schrödinger equation [56–59]:

$$ \mathrm{i}\frac{h}{2\pi}\frac{\partial\Psi(q,t)}{\partial t} = \hat{H}\Psi(q,t) \tag{5} $$

with the Hamilton operator $\hat{H}$

$$ \hat{H} = \hat{T} + \hat{V}(q) \tag{6} $$

as sum of kinetic energy operator $\hat{T}$ and potential energy $\hat{V}(q)$, where q is considered to represent the complete set of space and spin coordinates applicable for the molecule under consideration and is simply one coordinate for a one dimensional model problem. For an isolated molecule the solution of Eq. (5) can be written as

$$\Psi(q,t) = \sum_{k=1}^{\infty} c_k \varphi_k(q) \exp(-2\pi i E_k t/h) \tag{7}$$

with time independent generally complex coefficients c_k. Here the $\varphi_k(q)$ and E_k are obtained from the solution of the time independent Schrödinger equation

$$\hat{H}\varphi_k(q) = E_k\varphi_k(q) \tag{8}$$

From the time dependent wave functions $\Psi(q,t)$ and also the time independent wave function $\varphi_k(q)$ as special case, one can obtain relevant time dependent observable quantities, for example the probability density $P(q,t)$ for the molecular 'structure':

$$P(q,t) = |\Psi(q,t)|^2 \tag{9}$$

$$p_k(q) = |\varphi_k(q)|^2 \tag{10}$$

The time evolution of a molecular (or any other 'microscopic') system according to the Schrödinger equation (Eq. (5)) can also be written in a general abstract way by means of the time evolution operator $\hat{U}(t,t_0)$ [50, 51, 60]:

$$\Psi(q,t) = \hat{U}(t,t_0)\Psi(q,t_0) \tag{11}$$

This operator transforms the wave function $\Psi(q,t_0)$ at the initial time t_0 to the wave function $\Psi(q,t)$ at time t. $\hat{U}$ satisfies a differential equation analogous to Eq. (5) and for an isolated system with a time independent $\hat{H}$ it is given by an exponential function of $\hat{H}$:

$$\hat{U}(t,t_0) = \exp\left[-2\pi i\hat{H}(t-t_0)/h\right] \tag{12}$$

For solutions with a more general time-dependent $\hat{H}$ see [50, 51, 60–62]. Another representation of time dependence in quantum dynamics makes use of the Heisenberg equations of motion [63, 64] for the operator $\hat{Q}(t)$ related to some observable Q, with the solution

$$\hat{Q}(t) = \hat{U}^\dagger(t,t_0)\hat{Q}(t_0)\hat{U}(t,t_0) \tag{13}$$

Instead of asking about the time dependence of observables, of which there are many, one might ask the opposite question, whether in a complex time evolving system there are observables which remain constant in time, the 'constants of the motion'. These are all the observables C_j for which the corresponding operators $\hat{C}_j$ commute with the Hamiltonian $\hat{H}$

$$\hat{H}\hat{C}_j = \hat{C}_j\hat{H} \tag{14}$$

Making use of the solution of the Heisenberg equations of motion given by Eq. (13) the time independence of $\hat{C}_j$ can be proven in one line, because $\hat{U}$ being a function of $\hat{H}$, Eq. (12), it commutes with the $\hat{C}_j$ as well and it is also unitary $\hat{U}^\dagger \hat{U} = 1$, thus:

$$\hat{C}_j = \hat{U}^\dagger(t, t_0) \hat{C}_j(t_0) \hat{U}(t, t_0) = \hat{U}^\dagger(t, t_0) \hat{U}(t, t_0) \hat{C}_j(t_0) = \hat{C}_j(t_0) \quad (15)$$

The operators $\hat{C}_j$ form a group G, the symmetry group of the Hamiltonian [23, 50, 60]. One can furthermore show that for a statistical ensemble described by the density operator $\hat{\varrho}$ also the expectation values $\langle \hat{C}_j(t) \rangle = \mathrm{tr}(\hat{\varrho} \hat{C}_j)$ are time independent. Also, if $\Psi(q, t)$ is an eigenfunction $\zeta_n(t)$ of $\hat{C}_j$ with eigenvalue C_{jn} one has

$$\langle \hat{C}_j(t) \rangle = \langle \zeta_n(t) | \hat{C}_j(t_0) | \zeta_n(t) \rangle = C_{jn} \quad (16)$$

The eigenvalues C_{jn} are 'good quantum numbers' not changing in time. While this allows one to identify some simple structures, which do not change with time even in very complex time-dependent systems, one can proceed further by separating the Hamiltonian into contributions of different magnitude ($\hat{H}_0 \gg \hat{H}_1 \gg \hat{H}_2$ etc.)

$$\hat{H} = \hat{H}_0 + \hat{H}_1 + \hat{H}_2 + ... \quad (17)$$

Then the energy levels of a molecule, say, would be described to a good approximation by $\hat{H}_0$, and when adding $\hat{H}_1$, this would change the energy only a little, being a small 'perturbation'. The symmetry group G of the Hamiltonian may now depend on which terms are retained in Eq. (17), with groups $G_0(\hat{H}_0)$, $G_1(\hat{H}_0 + \hat{H}_1)$, $G_2(\hat{H}_0 + \hat{H}_1 + \hat{H}_2)$, etc.. Then one can have a symmetry or constant of the motion $\hat{C}_{0j}$ belonging to $G_0(\hat{H}_0)$ which may not appear in $G_1(\hat{H}_0 + \hat{H}_1)$. Thus, if one observes a change in time for this observable corresponding to $\hat{C}_{0j}$, this cannot be due to $\hat{H}_0$, as $\hat{C}_{0j}$ would be exactly time independent if only $\hat{H}_0$ is considered. The change of $\hat{C}_{0j}$ in time must entirely arise from $\hat{H}_1$ (or $\hat{H}_2$, if any). At the same time, the different magnitudes of the contributions in $\hat{H}$ introduce a natural hierarchy of time scales, $\hat{H}_0$ leading to some possibly very fast changes for some observables but no change in $\hat{C}_{0j}$, the small $\hat{H}_1$ introducing a slow change of $\hat{C}_{0j}$, and so forth. This allows one to isolate very small contributions in the Hamiltonian arising from $\hat{H}_1$ independent of any large uncertainties, which may occur in the very large $\hat{H}_0$. To use a common picture: One can weigh the 'captain' directly without having to measure a difference in the weight of 'ship+captain' and 'ship alone', where the uncertainty in the latter measurement would be much larger than any possible weight of a captain. For the example of parity violation we shall have differences on the order of magnitude of typical electronic energies of molecules (say, on the order of 1 eV) compared to parity violating energies on the order of 100 aeV to 1 feV, more than 15 orders of magnitude smaller. One may compare this with the weight of a large ship (for example Queen Elisabeth 2) with about 50'000 tons displacement and a captain with 50 to 100 kg, less than 6 orders of magnitude difference.

This concept can be made use of for theory, where the uncertainties in solutions of the Schrödinger equation may be due to theoretical or just numerical

Control of symmetry of
initial state with an approximate
constant of the motion

Time dependence of a certain
approximate constant of the motion
('symmetry breaking' of a symmetry property)

Test of fundamental symmetries
in nature (such as P, T, CP, CPT)
by testing for the time dependence of
the corresponding (perhaps approximate)
constants of the motion

Fig. 4. Scheme for control of symmetries and time evolution in molecular dynamics (modified after Ref. [65], see also Refs. [47, 66]).

uncertainties, typically much larger than a fraction of 10^{-15}: In quantum chemical calculations an uncertainty of 1 meV (in $\hat{H}_0$) would be considered quite acceptable, but is huge compared to 1 feV. The concept can also be made use of in the design of experiments, where a relative experimental uncertainty of less than 10^{-15} may be very difficult to achieve otherwise.

In a time dependent experiment one would follow the scheme of Fig. 4. One prepares a state of a given symmetry corresponding to a 'good quantum number' with respect to $\hat{H}_0$, and then follows the time dependence of this quantum number to observe an effect exclusively due to $\hat{H}_1$ etc. This scheme can be used by means of various kinds of approximate symmetries such as nuclear spin symmetry [67, 68] or one can consider some specific symmetries from simplified molecular models for intramolecular vibrational energy flow in polyatomic molecules, for instance, and we have made extensive use of it in the past (see [62, 67–70] and our recent review [47], for example). Table 1 summarises some results from our work.

However, the approach can also be used for testing fundamental symmetries of physics. These are given by the invariances of the Hamiltonian under [23]

(1) Any translation in space
(2) Any translation in time
(3) Any rotation in space
(4) The reflection of all coordinates of all particles on the center of mass of the system ('P')
(5) 'Time reversal' (reversal of momenta and spins) ('T')

Table 1. Time scales for intramolecular primary processes as successive 'symmetry breakings' for different approximate constants of the motion (modified after Refs. 23, 66, 75, see also Ref. 47).

Approximate constant of the motion	Symmetry breaking process (selected references)	Time scale
Quantum numbers of separate harmonic oscillators (for harmonic approximation in polyatomic molecules)	Selective vibrational stretch-bend Fermi resonance in R$_3$CH [62, 76–83]	10-200 fs
	Ordinary, weakly selective anharmonic couplings [62, 76, 83–86]	500 fs – 10 ps
Quantum numbers of adiabatically separable anharmonic oscillators	'Vibrationally non-adiabatic' couplings R–C $\equiv$ C–H, HF-HF [62, 66, 75, 76, 80, 87–92] Δl coupling in C$_{3v}$ symmetric tops R$_3$CH [70]	10 ps-1 ns
Structural identity for structures separated by high BO barriers	Tunneling processes [46, 93–97, 117]	< 1 ps to very long
Nuclear spin symmetry (separable nuclear spin-rotation-vibration states)	Violation of nuclear spin symmetry (rotation-vibration nuclear spin coupling) [23, 67, 98, 99]	1 ns – 1 s
Parity (space inversion symmetry)	Parity violation [23, 38, 54, 67, 100–105]	1 ms – 1 ks (theory only)
Time reversal symmetry T	T-violation in chiral and achiral molecules? [23, 42, 43, 62, 75, 87, 106–109]	Molecular time scale not known (neither quantitative theory nor experiment) but known in SMPP
CPT Symmetry	Hypothetical CPT violation [23, 43, 87, 106, 107]	So far not found in any part of physics

(6) Any permutation of identical particles (electrons or nuclei for the case of molecules)

(7) Charge conjugation - the replacement of all particles by the corresponding antiparticles ('C')

The symmetries are related to the corresponding conservation laws (for momentum, energy, angular momentum, parity,...) as has been discussed early on in quantum mechanics [71, 72] and in an interesting summary by Pauli [73] for the three discrete symmetry operations C, P, T, which one may all consider as some kind of generalised 'mirror symmetry'. Fig. 5 illustrates this for the case of the parity symmetry ('P' one uses also the symbol E* for the operation of space inversion in molecular physics [74]).

218

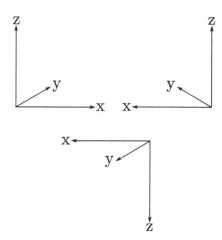

Fig. 5. 'Right handed' Cartesian coordinate system (upper left), mirror image (upper right) and inverted (lower) coordinate systems being 'left handed'.

The inversion of the right handed coordinate system in the upper left part (as defined by the 'right hand rule' convention) by a mirror leads to the left handed coordinate system on the upper right side. The space inversion or parity operator on the right handed coordinate system changes $(x \rightarrow -x, \ y \rightarrow -y, \ z \rightarrow -z)$ into a left handed system as shown in the lower part. This can be transformed to the left handed 'mirror' image on the upper right by a rotation of 180° around the x axis, which corresponds to one of the symmetry operations under (3), the combination being obviously also a symmetry operation for the Hamiltonian, leading to a related constant of the motion, the space inversion P being the elementary operation. A further interesting aspect is the notion that each exact symmetry and conservation law leads to a fundamental 'non-observable' property of nature. This can be nicely understood with the statement by Einstein [110] (original in German, as shown in Fig. 6, as translated in [23]):

'There are thus two types of Cartesian coordinate systems, which are called 'right-handed' and 'left-handed' systems. The difference between the two is familiar to every physicist and engineer. It is interesting that an absolute geometric definition of the right or left handedness is impossible, only the relationship of opposition between the two can be defined'.

For chiral molecules this implies that even if we can determine the 'absolute' configuration of a molecule in comparison to a macroscopic model in the laboratory, we cannot specify for either of the two whether they occur in a space with a left handed or right handed coordinate system. One way to illustrate this property of 'non-observability' is the so called 'Ozma' problem [41, 111]: If there were perfect space inversion symmetry, we would not be able to communicate to a distant civilization with a coded message (without sending a real chiral 'model'), whether we are made of L-amino acids or of D-amino acids, for instance. This is another way of phrasing Einstein's statement. Van't Hoff and Einstein assumed space inversion

'Non-observability' of 'left-right' structure
'Nichtbeobachtbarkeit' der absoluten 'Rechts-
oder Linkshändigkeit' des Raumes
bei perfekter Spiegelsymmetrie

*Es gibt also zweierlei kartesische
Koordinatensysteme, welche man als
'Rechtssysteme' und 'Linkssysteme' bezeichnet.
Der Unterschied zwischen beiden ist jedem
Physiker und Ingenieur geläufig.*

**Interessant ist, dass man Rechtssysteme bzw. Linkssysteme an
sich nicht geometrisch definieren kann,** wohl aber die
Gegensätzlichkeit beider Systeme.
Albert Einstein (ML, 1922)

Non-observability due to a symmetry, but today we know, that the symmetry
is violated, handedness of space can be 'observed absolutely' and
communicated in an absolute sense (by experiments in nuclear and high
energy physics). One alternative experimental approach uses high resolution
spectroscopy of chiral molecules and parity violation by the electroweak
interaction.

Fig. 6.   On the non-observability of absolute handedness with perfect mirror symmetry and parity
conservation (citation from Ref. 110 after Ref. 23, 48).

symmetry to be universally valid and thus the 'handedness' of space would be 'non-
observable'. This was the common assumption until 1956/57, when parity violation
was proposed and observed. Electroweak parity violation makes absolute handedness observable and removes a constant of motion and 'good quantum number'
(parity). We have for the 'electromagnetic Hamiltonian' $\hat{H}_{em}$

$$\hat{H}_{em}\hat{P} = \hat{P}\hat{H}_{em} \tag{18}$$

with common eigenfunctions of energy and parity, where the eigenvalues of parity
are $+1$ or -1 according to

$$\hat{P}\psi_k = (+1)\psi_k \tag{19}$$

$$\hat{P}\psi_j = (-1)\psi_j \tag{20}$$

The eigenfunctions of the Hamiltonian either are symmetric under space inversion (positive parity $+1$) or antisymmetric under inversion (negative parity -1).
With electroweak parity violation due to the weak nuclear force we have $\hat{H}_{em}+\hat{H}_{weak}$

$$(\hat{H}_{em} + \hat{H}_{weak})\hat{P} \neq \hat{P}(\hat{H}_{em} + \hat{H}_{weak}) \tag{21}$$

One consequence is now that the ground state eigenfunctions of chiral molecules are
localized and have no well defined parity as shown in the right hand part of Fig. 7
and the two states have different energies. Given this, the 'handedness' of space

220

Symmetry Breaking and Symmetry Violation

spontanous: classical → quantum

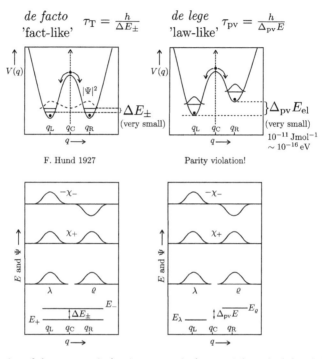

Fig. 7. Illustration of the symmetric (parity conserving) potential on the left side following Hund 1927 for a chiral molecule with energy eigenstates of positive and negative parity separated by a tunneling splitting $\Delta E_\pm$ and for an effective potential with parity violation on the right side leading to two localized ground states at the chiral structures separated by the parity violating energy difference $\Delta_{pv} E$ (after [46], see also [100, 112]).

becomes 'observable' and we can communicate to a distant civilization of what type of amino acids we are made of predominantly (for instance the more stable ones, if that is established experimentally, see below). While the energy differences are truly minute, due to 'weakness' of $\hat{H}_{weak}$, we can nevertheless compute them significantly and devise significant experiments, because of the different symmetries of $\hat{H}_{em}$ and $\hat{H}_{weak}$, as discussed above. The different symmetry 'isolates' the effect of parity violation. We shall in Sec. 3 report on the development of the quantitative theory of parity violation and in Sec. 4 on the current development of experiments.

Before addressing theory and experiment in more detail it is necessary to discuss the quantum dynamics of chiral molecules taking also the order of magnitude of the symmetry violations into account. Figure 7 presents a simple one dimensional quantum mechanical model for the case of the symmetrical electromagnetic Hamiltonian $\hat{H}_{em}$, represented by a symmetrical potential $V(q)$ for the interconversion of the

Friedrich Hund (ML): Discovered Tunnel Effect 1927
with quantum dynamics of chiral molecules
Die Begreifbarkeit der Natur: Understanding Nature (1957)

Geist	Mind spirit
Seele	Soul
Leben	Life
Physikalisch-Chemische Prozesse	Physical-Chemical Processes
Atom	Atom
Elementarteilchen	Elementary Particles
Elementare Materie	Elementary Matter ('SMPP')

Friedrich Hund (1957): Can this continued at the bottom?
Martin Quack (1990): Yes by Geist=Underlying Fundamental Laws=Mind

Fig. 8. Friedrich Hund discovered the tunnel effect in 1927 in his investigations of the quantum dynamics of chiral molecules [113, 114]. He also discussed the stability of biomolecular chirality. The quoted text is from his lecture on understanding nature in 1957, where he considers physical-chemical processes as mediators between atomic and elementary particle physics and life (modified after Ref. 44 with translation from there, see also Ref. 40).

enantiomers (with the 'physical' notation 'left' enantiomer located at the potential minimum of the potential at q_L and 'right' at the minimum at q_R). Friedrich Hund (Fig. 8) treated the quantum mechanics of chiral molecules under the assumption of a symmetrical Hamiltonian as represented on the left hand side of Fig. 7.

He demonstrated, that the quantum mechanical behaviour of chiral molecules differed considerably from what one expects on the basis of the classical mechanical structural models [113, 114]. Let us consider in Fig. 9 as an example hydrogen-peroxide HOOH which in its equilibrium structure shows axial chirality with the two enantiomers shown (also considered as helical in analogy to the two snails shown in Fig. 9, we note that the two natural snails are not enantiomers at the microscopic level, both having proteins with the L-amino acids and DNA with D-sugars. According to Brunner, the D-sugar DNA leads to a preference of the structure on the left with respect to the right one of 20'000:1 [115, 116]). It turns out that this description with quasi-rigid localized structures is quite inadequate for HOOH in quantum mechanics. By continuously changing the torsional angle indicated in Fig. 9, one can transform one enantiomer into the other: Such a transformation from one enantiomer into the other is also possible in chiral molecules with a pyramidal, near tetrahedral structure as shown in Fig. 10 for the example of the ammonia isotopomer NHDT, where such a 'stereomutation' can be achieved by changing the angle with respect to a planar geometry of the four atoms. Such an 'inversion' in ammonia (NH_3) was actually discussed by Hund and the simple one dimensional model as discussed by him entered the textbooks.

The transformation via planar transition structures in both examples is associated with a potential energy change shown in Fig. 7 schematically.

One may represent the stereomutation reaction from L to R (using the 'physical' notation left to right) by the motion of an effective mass in a potential energy $V(q)$ as a function of the generalized coordinate q. In the time independent Schrödinger

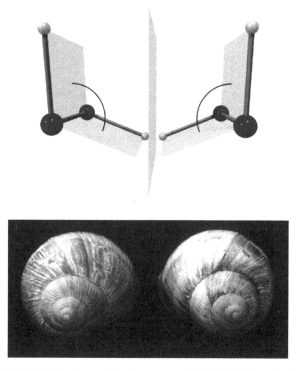

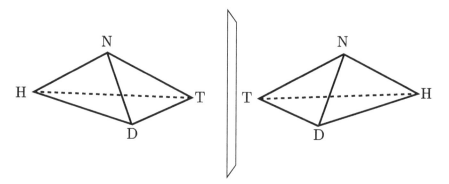

Fig. 9. The classical chiral structures of hydrogen-peroxide H-O-O-H and of snails (as example of axial or helical chirality, after [48]).

Fig. 10. The classical structures of the chiral enantiomers of the ammonia isotopomer NHDT (after [48]).

equation, Eq. (8), with an exactly symmetrical potential as shown on the left hand side of Fig. 7, the eigenfunctions must be either symmetrical (with 'positive parity' χ_+) or antisymmetrical (with 'negative parity', say $-\chi_-$) with respect to reflection at the 'mirror plane' situated at the center position q_c corresponding to a maximum in the potential energy ($V(q)$). Instead of a well defined structure one has in quantum mechanics quite generally a probability density $p_k(q)$ of finding a structure 'q' in the eigenfunction $\varphi_k(q)$ (with the square of the absolute magnitude $|\varphi_k|^2$, Eq. (10)).

It is seen that for the lowest energy levels (χ_+ at E_+ and χ_- at E_-) shown in the diagram this probability density has two equal maxima near the two enantiomeric structures, thus equal probability for finding each of the two enantiomers. In order to represent the experimental observation of localized enantiomers Hund considered the time dependent Schrödinger equation, Eq. (5), with the solutions for the time dependent wave functions (with time independent coefficients c_i) retaining just the two lowest terms in the sum (Eq. (7), corresponding to $\chi_+ = \varphi_1$ and $-\chi_- = \varphi_2$ and energies $E_+ = E_1$ and $E_- = E_2$). One obtains a time dependent probability density ($\Delta E_{1,2} = E_2 - E_1 = E_- - E_+$)

$$|\Psi(q,t)|^2 = \frac{1}{2}|\varphi_1 + \varphi_2 \exp(-2\pi i \Delta E_{1,2}t/h)|^2 \tag{22}$$

when one takes equal coefficients $c_1 = c_2 = 1/\sqrt{2}$. This is an oscillatory function with a period

$$\tau' = h/\Delta E_\pm = h/(E_2 - E_1) \tag{23}$$

Inspection of the graphical representation of the wave functions on the left hand side in Fig. 7 shows that this moves from a function λ located near the left minimum at $t = 0$

$$\lambda = \frac{1}{\sqrt{2}}(\chi_1 - \chi_-) \tag{24}$$

to a function ϱ situated near the right hand minimum of the potential

$$\varrho = \frac{1}{\sqrt{2}}(\chi_+ + \chi_-) \tag{25}$$

in a time equal half the period

$$t_{\lambda \to \varrho} = \tau'/2 = \frac{h}{2\Delta E_\pm} \tag{26}$$

With these considerations Hund made two important observations: Firstly the 'reaction' from one enantiomer to the other can occur, although neither of the energies E_+ and E_- is sufficient to overcome the 'barrier maximum' $V(q_c)$, a process which would be completely impossible in classical mechanics. This was the discovery of the quantum mechanical 'tunnel effect', with numerous later applications in chemistry and physics (see [46, 117] for the history and reviews). Secondly, when estimating parameters for the effective masses and potentials for the 'inversion' (from R to S) in the case of the chiral substituted methane derivatives, Hund found millions to billions of years for the stereomutation times, thus explaining the apparent stability of CHFClBr, amino acids or sugars for example (see [118] for a critical discussion of the estimates, which does, however, not invalidate the conclusions).

The situation is quite different for the chiral molecules hydrogen-peroxide HOOH and the ammonia isotopomer, NHDT, which we have chosen here on purpose for illustration as Hund's description is effectively correct for these examples. While the simple one-dimensional model for stereomutation used by Hund and discussed by us above for illustration of the concept has become textbook material, only recently full-dimensional 'exact' solutions of the time independent and time dependent Schrödinger equation have become possible, as this requires the solution in a space of 6 internal (vibrational) coordinates $(q_1, q_2, \ldots, q_6)$ for the four atoms (nuclei) in these molecules and a potential hypersurface $V(q_1, q_2, \ldots, q_6)$ [94–96, 119–122] (and further dimensions if external rotation is included [96]). This results in a 6-dimensional wave function $\Psi(q_1, q_2, \ldots, q_6, t)$ and probability density $(|\Psi|^2)$. As a visualisation in a high dimensional space is not possible, one integrates over 5 coordinates $(q_1, \ldots, q_5)$ and represents the probability density as a function of one 'reaction coordinate' corresponding to the torsion angle τ in HOOH (still exact). Thereby one obtains a time dependent 'wave packet' as shown in Fig. 11 for HOOH

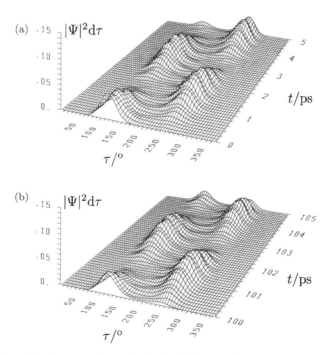

Fig. 11.   Six-dimensional wave packet evolution for H_2O_2 in its lowest quantum states. $|\Psi|^2$ shows the time-dependent probability as a function of the torsional coordinate, where the probability density is integrated over all other coordinates. a) shows the time interval from 0 to 5 ps and b) the time interval from 100 to 105 ps with identical initial conditions at $t = 0$ as in a) [94, 95]. The migration of the wave packet from the left to the right corresponds to a change from one enantiomer of HOOH to the other in Fig. 9 with a transfer time according to Eq. (26) of about 1.5 ps (after [46]).

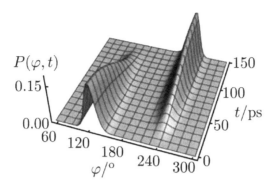

$$P(\varphi, t)$$

Fig. 12. Reduced probability density as a function of the inversion coordinate φ and time t for NHDT. The two enantiomeric structures correspond to $\varphi \cong 120°$ and $\varphi \cong 240°$, respectively, with $\varphi = 180°$ corresponding to the planar geometry. The graphic shows reduced probability densities (probability densities integrated over all other coordinates) for the field-free isolated molecule (after [96]).

which moves from one enantiomeric structure at the left to the other enantiomeric at the right in about 1.5 ps [94, 95].

The corresponding result for the chiral ammonia isotopomer is shown in Fig. 12 [96]. Because of the short lifetime of the chiral, enantiomeric structures of HOOH or NHDT one may call such molecules 'transiently chiral'. The time dependence could be easily followed by observing time dependent optical activity or circular dichroism (also vibrational circular dichroism VCD or Raman optical activity, ROA), which will follow the wave packet motion in a straightforward manner.

The recent exact treatment for hydrogen-peroxide and ammonia isotopomers has also been extended to show various interesting effects such as mode selective tunneling stereomutation [95] or tunneling enhancement and inhibition by coherent radiative excitation [96]. We shall not pursue this in detail here but rather turn to the consequences of the asymmmetry arising from parity violation, illustrated on the right hand side of Fig. 7. Here one can distinguish two dynamical limiting cases. If the asymmetry due to parity violation ($\Delta_{pv} E_{el}$ in the scheme of Fig. 7) is small compared to the tunneling splitting $\Delta E_{\pm}$ in the symmetrical case

$$|\Delta_{pv} E_{el}| \ll |\Delta E_{\pm}| \qquad (27)$$

then Hund's description as discussed above remains effectively valid. This is in fact the case for the examples HOOH and NHDT we have chosen here for illustration. The localization of the wave function is effectively due to the initial conditions only and corresponds to a symmetry breaking *de facto* ('fact like'). If on the other hand the asymmetry is large compared to the tunneling splitting

$$|\Delta_{pv} E_{el}| \gg |\Delta E_{\pm}| \qquad (28)$$

then the symmetry is broken *de lege* ('law like') and the stationary states of the time independent Schrödinger equation are in fact localized, in the example of Fig. 7

226

the ground state is located left λ at q_L and the first excited state on the right hand side ϱ at q_R. λ with E_λ and ϱ with E_ϱ are the eigenfunctions of 'infinite' lifetime, tunneling being almost completely suppressed. For a more detailed discussion of the concepts, the dynamics and including also a discussion of the classical concept 'spontaneous symmetry breaking' we refer to [23, 38–40, 45]. We anticipate here the results discussed below that in spite of the small magnitude of parity violation ($\Delta_\mathrm{pv}E_\mathrm{el} \approx 100\,\mathrm{aeV}$ in Fig. 7) for ordinary stable chiral molecules the case of *de lege* symmetry breaking applies, parity violation dominates completely over tunneling, very different from Hund's treatment of chiral molecules.

In this case the low energy eigenfunctions are localized (λ and ϱ) and do not have a well defined parity. The key idea for an experiment to study the extremely small values of $\Delta_\mathrm{pv}E$ follows the scheme of Fig. 4. One first prepares a state of well defined parity and then follows the slow time evolution of parity which arises from parity violation. Figure 13 shows how this can be done, in principle, following the proposal of Ref. 100. In the sequence of laser pulses shown in the figure one passes via an excited achiral state, which has spectroscopic energy eigenstates of essentially pure parity either $+$ or $-$ ($-$ in the figure). With a second pulse using the spectroscopic selection rule for strong electric dipole transitions ($- \rightarrow +$) one prepares a state of well defined parity ($+$ in the figure). This state is not an eigenstate of energy and evolves in time following the equation

$$\Psi(t) = \frac{1}{\sqrt{2}} \exp\left(-2\pi \mathrm{i} E_\lambda t/h\right) [\lambda + \varrho \cdot \exp\left(-2\pi \mathrm{i} \Delta_\mathrm{pv} E t/h\right)] \tag{29}$$

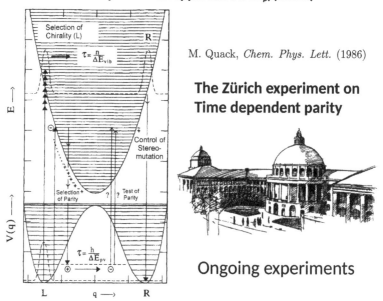

Fig. 13. The Zurich experiment on time dependent parity (after Ref. 100).

The probabilities of parities as a function of time are given by Eq. (30)

$$1 - p_+(t) = p_-(t) = \sin^2\left(\pi\Delta_{\mathrm{pv}}Et/h\right) \tag{30}$$

If one measures this time dependence one can extract $\Delta_{\mathrm{pv}}E$. Details of spectroscopic experiments will be discussed in Sec. 4. At the time the proposal was made [100] it seemed almost impossible to actually carry out such experiments as most of the spectroscopic ground work was missing. Also it turned out that existing theories for predicting the effect were inadequate quantitatively. We shall in the following sections describe two steps in the long road towards such experiments. In Sec. 3 we discuss the development of quantitative theory followed by a description of current experiments in Sec. 4.

3. The quantitative theory of parity violation in chiral molecules

From 1925 onwards quantum mechanics provided a detailed theoretical understanding of how molecular structure and bonding arises [56–59, 71, 123–125] and modern electronic structure theory as summarised in reviews and text books would claim that this understanding is essentially complete [126–130], particularly when relativistic effects are included as well [131, 132]. Indeed, Dirac already in 1929 made the statement [124]:

'*The underlying physical laws for the mathematical theory of a large part of physics and the whole of chemistry are thus completely known and the difficulty is only that the exact application of these laws leads to equations much too complicated to be soluble. It therefore becomes desirable that approximate practical methods of applying quantum mechanics should be developed, which can lead to an explanation of complex atomic systems without too much computation.*'

This is one of the most frequently cited quotations in the quantum chemistry literature, even though often severely abbreviated [133]. However, the statement was incorrect, as the weak nuclear force introducing parity violation was completely unknown by that time, but turns out to have important consequences in stereochemistry, as we shall see.

The weak nuclear force was introduced by Enrico Fermi in his early theory of radioactive β-decay in 1934 [134]. Initially, there was no suspicion of a violation of space inversion symmetry by this force, until the suggestion by Lee and Yang [25] that assuming parity violation one could explain the so called $\Theta - \tau$ puzzle of elementary particle physics, which led to the dramatically fast discovery of parity violation in high energy physics [26–30]. This story is told in numerous places including historical articles and textbooks (see Refs. 31, 73, 135–137 for example). Figure 14 shows a pictorial representation of the current view of fundamental forces in the standard model of particle physics (SMPP) as shown on the website of a large accelerator [138]. We have modified this picture by adding the importance of the weak nuclear force for the stereochemistry of chiral molecules. We have also added the Feynman diagram for the example of the electromagnetic interaction of two electrons to illustrate, how action at a distance is viewed in the SMPP [139–141]

The Forces in Nature: Theory of electroweak quantum chemistry
Stereochemistry meets high energy physics

Type	Intensity of Forces (Decreasing Order)	Particle Binding (Field Quantum)	Important in
Strong Nuclear Force	~ 1	Gluons (no mass)	Atomic Nucleus
Electro-Magnetic Force	$\sim 10^{-3}$	Photons (no mass)	Atoms and Molecules
Weak Nuclear Force	$\sim 10^{-5}$	Bosons Z, W^+, W^- (heavy)	Radioactive β-Decay, Neutrino induced dec. **Chiral Molecules**
Gravitation	$\sim 10^{-38}$	Gravitons (?)	Sun and Planets etc.

The exchange of particles is responsible for the force

Piazza Armerina   CERN AC_Z04_V25/B/1992

Fig. 14. Forces in the standard model of particle physics (SMPP) and important effects. This is taken from the CERN website Ref. [138], but the importance of the weak interaction for chiral molecules has been added here from our work (modified after [40], in turn, adapted from Ref. [138], Public Domain. We note that, while not referred to in Ref. [138], the motif of the lightly dressed ladies throwing balls can be found in a mosaic at Piazza Armerina. Sicily, from the 4th century AD). We also added the Feynman diagram to the left representing the electromagnetic interaction between two electrons (e^-) via the photon (γ) as field quantum (see also [139]).

(the diagrams are also sometimes called Stueckelberg diagrams, e.g. by Gell-Mann, or Feynman-Stueckelberg diagrams due to earlier contributions of Stueckelberg to the developments [142–147]). According to this view, the electromagnetic force, which is included in the 'Schrödinger-Dirac' like ordinary quantum chemistry, leads to the Coulomb repulsion, say, between two electrons by means of photons as field particles. In the picture the two electrons are compared to the ladies on two boats throwing a ball. If we do not see the exchange of the ball, we will only observe the accelerated motion of the boats resulting from the transfer of momentum in throwing the ball. We could interpret this motion as resulting from a repulsive 'force' between the two ladies on the boats. Similarly, we interpret the motion of the electrons resulting from 'throwing photons as field particles' as arising from the Coulomb law which forms the basis of the Hamiltonian in ordinary quantum chemistry. Different from the simple quasiclassical analogy, the Feynman diagram, which almost looks like representing such a picture, can be translated into quantitative mathematical equations, which then form the basis of a quantitative theoretical treatment of the interactions and dynamics [148]. The Coulomb force with the $1/r$ potential energy law is of long range. The other fundamental forces arise similarly, but with other field particles. The strong force with very short range (0.1 to 1 fm) mediated by the gluons is important in nuclear physics but has only indirect influence in chemistry by providing the structures of the nuclei, which enter as parameters in chemistry, but there is otherwise normally no need to retain the strong force explicitly in chemistry. The weak force, on the other hand, is mediated by the $W^\pm$ and Z^0 bosons of very high mass (86.316 and 97.894 Dalton, of the order of the mass of a Rb to Mo nucleus) and short lifetime (0.26 yoctoseconds = $0.26 \cdot 10^{-24}$ s, Ref. 47). This force is weak and of short range (< 0.1 fm) and one might think that, similar to the even weaker gravitational force (mediated by the still hypothetical graviton of spin 2), it should not contribute significantly to the forces between the particles in molecules (nuclei and electrons). Indeed, the weak force, because of its short range, becomes effective in atoms and molecules, when the electrons penetrate the nucleus and then it leads only to a very small perturbation on the molecular dynamics, which ordinarily might be neglected completely. However, in fact the weak force leads to a fundamental change as it has a different symmetry, violating space inversion symmetry, which is *exactly* valid for the strong and electromagnetic interactions (also gravitational interactions) according to present knowledge. When including the parity violating weak interaction, qualitative theory therefore tells us already that the ground state energies (and other properties) of enantiomers of chiral molecules will be different. The question remains then just how different the energies will be quantitatively. We shall not go into technical details here of the formulation of the quantitative theory of electroweak quantum chemistry (a term coined by us in [101, 149]). However, some brief remarks may be in order concerning the development and current status of the theory of parity violation in chiral molecules, which has been reviewed in detail elsewhere [23, 40, 112, 150, 151] (see also Refs. 22, 42, 101, 149, 152). The history of this theory can be broadly summarized in three phases.

In a first phase, after the discovery of parity violation in nuclear and high energy physics [25–30] in 1957, qualitative suggestions were made concerning the role of parity violation in chiral molecules, with estimates which were often wrong by many orders of magnitude (in the period of about 1960–1980 [153–155], see also the reviews in [22, 23, 38, 40] for many further references). In a second phase, attempts towards a quantitative theory started in about 1980, based on earlier work on the theory for atoms [156, 157], extended approximately to molecules by Hegström, Rein and Sandars [158], Mason and Tranter [159] and others [160, 161] (see also the further citations in [23, 38, 40, 42]). It turned out, however, that these approaches were quite inadequate quantitatively.

A third phase started when in 1993–1995 we carefully reinvestigated in relation to our experimental project [38, 100] also the theory, rederiving it from its foundations in the standard model and critically analysing the steps towards 'electroweak quantum chemistry' [101, 102, 149]. Indeed, we found values for the parity violating energy differences $\Delta_{pv} E$ in typical chiral 'benchmark' molecules often by one to two orders of magnitude larger as compared to previous results. In spite of some initial scepticism, which some audiences expressed towards the two-order of magnitude increase reported by us also in our lectures at the time (also in [162], for example), our discovery of the new, much larger values triggered further work and were subsequently confirmed by quite a few other theory groups [163–166] (see [23, 40, 42, 112, 150] for further references). Today there seems to be general agreement on the new orders of magnitude from a number of quite different approaches (see for instance the recent summary in [152]), although the current results still scatter within about a factor of 2, which has various more technical reasons not to be discussed in detail here.

Figure 15 illustrates the big quantitative jump discovered in our theoretical work in the mid 1990s, which stimulated much further theoretical work (and also experimental efforts). We summarize here the main steps in the theoretical development, which is described 'from scratch' up to practically useful equations in Ref. [101] (see also the review in Ref. [150], in particular, containing also some historical remarks). The starting point are the fully relativistic equations from the SMPP describing electron-neutron, electron-proton and electron-electron interactions in the order of their importance. In principle, one starts with electron-quark interactions at a more fundamental level (see [150]). One takes the low energy semirelativistic limit and then also neglects the electron-electron parity violating interactions which can be considered to be a smaller contribution only. For not too heavy nuclei of interest in simple inorganic and organic molecules (and also biomolecules) appropriate for fundamental studies one can use to a good approximation the Breit-Pauli form of the semirelativistic one- and two-electron spin-orbit interaction. For molecules involving heavy elements one can use as an alternative the Dirac Fock theory given

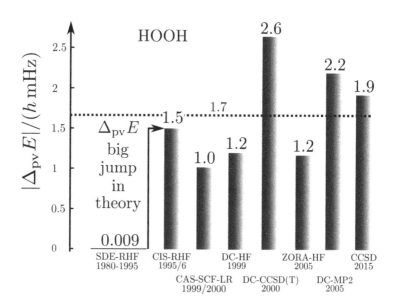

Fig. 15. Graphical representation of the 'big jump' in theory occurring in 1995 (after Ref. [42], values of the parity violating energy differences $\Delta_{\mathrm{pv}}E$ are shown for the benchmark molecule hydrogen-peroxide H_2O_2, note the extremely small unit for energy $\Delta_{\mathrm{pv}}E/h$ in millihertz).

by Laerdahl and Schwerdtfeger in 1999 [164] and related approaches [166, 167], see also the reviews [112, 150, 151, 168]. Finally, one can represent the nucleus by a point like object with the electric charge $-Z_A\,e$ (Z_A = number of protons) and an electroweak charge (neglecting also radiative corrections)

$$Q_A = [1 - 4\sin^2(\Theta_W)]Z_A - N_A \tag{31}$$

with the number of neutrons N_A. This last approximation can be rather easily improved upon by considering the actual shape of the nucleus with its distribution of neutrons and protons [169] from the theory of nuclear structure, where we note here the interesting sensitivity of parity violating effects upon the neutron distribution, indeed, with the parameter $[1 - 4\sin^2(\Theta_W)] = 0.0724$ [60]. We note that the value of the Weinberg parameter $\sin^2\Theta_W$ (or weak mixing parameter) depends on the scheme used and the momentum transfer. At every step in these approximations one can, in principle, check these by the inclusion of the neglected effects, and none of the approximations is fundamental, although considerable computational effort may be necessary to improve upon them. The convergence of current theories as described by Fig. 15 would indicate that at least no 'trivial' effects have been overlooked, although tests by experiment seem advisable after past experience. We can add here that precision experiments on molecular parity violation can contribute in general ways to fundamental aspects of physics as well. If the nuclear structure is well known, as is the case for the lighter nuclei up to Argon at least, then precision experiments can contribute to a better understanding of the weak mixing parameter according to Eq. (31). On the other hand, for heavier nuclei, where the structure is less well known, precision experiment on parity violation could provide significant information on the neutron distribution in the nucleus, whereas many other experiments are most sensitive to the proton distribution.

At this point it may also be of interest to refer to the extensive theoretical and experimental work available for atomic parity violation (see [156, 157, 170–176] for a small selection). While quite a number of successful spectroscopic observations of parity violation exist for atoms, these are restricted to very heavy atoms and this restriction remains probably valid for the foreseeable future. Because of uncertainties in both electronic and nuclear structure for heavy atoms, a quantitative theoretical analysis leads to relatively large uncertainties with respect to an analysis of fundamental parameters in the theory (such as $\sin^2(\Theta_W)$). A similar restriction is true also for recent efforts on studies of parity violation in diatomic molecules (so far unsuccessful, in contrast to atoms). We shall not discuss these further here and refer to the recent review by Berger and Stohner [168] which has a focus on atoms and diatomic molecules.

In contrast, as we shall see below, certain spectroscopic studies on parity violation in chiral molecules have the promise to be successful in molecules involving only the lighter elements. They are thus of interest also in terms of fundamental physics. Furthermore, studies of parity violation in chiral molecules provide a direct link to effects that may be important for biochemistry and the evolution of biomolecular homochirality, which provides a further fundamental motivation for such studies. The scatter in the current results for $\Delta_{pv} E$ of the benchmark molecule HOOH should not be taken as an indication that accurate calculations are not possible: Indeed, the scatter is in part due to the fact that a somewhat hypothetical value of $\Delta_{pv} E^{el}$ is computed and to secondary effects. For molecules, where $\Delta_{pv} E_0$ corresponds to a measurable quantity as a ground state energy difference, this can be accurately calculated, if also vibrational effects are taken into account (see discussion in [23, 55]; we use here the symbol $\Delta_{pv} E_0$, when we wish to refer explicitly to the overall ground state, rotational — vibrational — electronic, perhaps also hyperfine).

Without going into any detail of the theory for chiral molecules, we shall provide here a simplified summary of some of the main aspects and a few exemplary results. An instructive approximate form for the parity violating potential in chiral molecules is given by Eq. (32) from perturbation theory

$$E_{pv} = 2\,Re \left\{ \sum_m \frac{\langle\psi_0|\hat{H}_{pv}^{e-nucl.}|\psi_m\rangle\langle\psi_m|\hat{H}_{SO}|\psi_0\rangle}{E_m - E_0} \right\} \tag{32}$$

where ψ_0 would be the electronic wave function of the ground state, say, a singlet state and ψ_m is some excited (triplet) state. The sum extends over many excited electronic states for convergence. $\hat{H}_{SO}$ is the spin-orbit Hamiltonian. The parity violating Hamiltonian $\hat{H}_{pv}$ has a dominant term

$$\hat{H}_{pv,1}^{e-nucl.} = \frac{\pi G_F}{m_e hc\sqrt{2}} \sum_{j=1}^n \sum_{A=1}^N Q_A \left\{ \hat{p}_j \cdot \hat{s}_j, \delta^3(\vec{r}_j - \vec{r}_A) \right\}_+ \tag{33}$$

Here we use SI units with common symbols for the fundamental constants [60] and in particular the Fermi constant

$$G_F = 1.4358510(8) \cdot 10^{-62} \, \text{Jm}^3 \tag{34}$$

One notes the small value for G_F. The sums extend over all electrons (n) and all nuclei (N) in the molecule. $\{ , \}_+$ is the symbol for the anticommutator and $\hat{p}_j$ is the momentum and $\hat{s}_j$ the spin operator for the jth-electron. The Dirac-delta function $\delta(\vec{r}_j - \vec{r}_A)$ vanishes at all values except when $\vec{r}_j = \vec{r}_A$ in the point nucleus approximation (see above). This part of the parity violating Hamiltonian is sufficient when discussing properties independent of nuclear spin. When nuclear spin I_A is to be considered as for hyperfine structure or for NMR experiments one has to include a nuclear spin dependent term

$$\hat{H}_{pv,2}^{e-nucl.} = \frac{\pi G_F}{m_e hc\sqrt{2}} \sum_{j=1}^n \left[\sum_{A=1}^N (\lambda_A)(1 - 4\sin^2\Theta_W) \left\{ \hat{\mathbf{p}}_j\hat{\mathbf{I}}_A, \delta^3(\vec{r}_j - \vec{r}_A) \right\}_+ \right.$$
$$\left. + (2i\lambda_A)(1 - 4\sin^2\Theta_W)(\hat{\mathbf{s}}_j \times \hat{\mathbf{I}}_A) \left[\hat{\mathbf{p}}_j, \delta^3(\vec{r}_j - \vec{r}_A) \right]_- \right] \tag{35}$$

Here again $\{ , \}_+$ is the anticommutator and $[,]_-$ is the commutator. We note that because of Eq. (31) neutrons provide the dominant contribution to the weak charge of the nucleus (except for the case of the proton, of course). We can also note that while NMR spectroscopy is among the most powerful spectroscopic methods [177] and has been suggested for the study of parity violation in chiral molecules [178], conclusive results are to be expected only for such experiments, if carried out in molecular beams as discussed in Refs. 23, 39, as they were done at the very start of NMR spectroscopy [179, 180]. We note furthermore, that as the neutron number is different for different isotopes, parity violation introduces a fundamentally new isotope effect arising from the weak nuclear charge Q_A: Isotopic chirality is important for parity violation as pointed out in Refs. 23, 38 and calculated quantitatively in Ref. [181]. We should also mention, that Eq. (32), which we gave here, because it leads to some instructive insight on the structure of the potential, in actual calculations this has slow convergence of the sum, when carried out numerically in this way [101, 149]. In practice one uses response theory equivalent to Eq. (32) as established through propagator methods (see Refs. 23, 40, 46, 102 for discussions). In contrast to atoms, the parity violating potential E_{pv} in polyatomic (chiral) molecules depends on $S = 3N - 6$ internal coordinates describing the structure of the molecule. Similar to the ordinary electronic potential energy of the molecule it is a potential hypersurface in an S-dimensional space. As chiral molecules necessarily have at least 4 atoms, this is an at least 6-dimensional space (for the examples H_2O_2 or NHDT the space would be just 6-dimensional). We can thus write with some general internal coordinates more explicitly

$$E_{pv} = E_{pv}(q_1, q_2, q_3, ..., q_s) \qquad (36)$$

This potential is antisymmetric with respect to the operation of space inversion $\hat{P}$ or $\hat{E}^*$, we can thus write with a symbolic notation for the inverted coordinates $\bar{q}_i$:

$$\hat{E}^* E_{pv\,R}(q_1, q_2, q_3, \ldots, q_s) = E_{pv\,S}(\bar{q}_1, \bar{q}_2, \bar{q}_3, \ldots, \bar{q}_s) = -E_{pv\,R}(q_1, q_2, q_3, \ldots, q_s) \qquad (37)$$

We can define a parity violating energy difference

$$\Delta_{pv} E(q_1, q_2, q_3, \ldots, q_s) = E_{pv\,R}(q_1, q_2, q_3, \ldots, q_s) - E_{pv\,S}(\bar{q}_1, \bar{q}_2, \bar{q}_3, \ldots, \bar{q}_s) \qquad (38)$$

Here we omit the extra index "el" for simplicity. The absolute values of these parity violating energy differences for given structures are very small, on the order of sub-feV typically for molecules composed of lighter atoms only. The sign of $\Delta_{pv} E$ depends on the structure and we note that for some well-defined convention (say R and S in the CIP convention or P and M for axially chiral molecules or D and L in another convention), $\Delta_{pv} E$ may change sign even within a given domain (say all R). We denote by $E_{pv\,R}$ or $E_{pv\,S}$ that the potential refers to an R or S structure in the given convention.

This is illustrated for the example ClSSCl in Fig. 16, where the potentials are shown as a function of just one coordinate, the torsional angle τ as indicated. The parity violating potentials shown in color are antisymmetric with respect to inversion at the point $\tau = 180°$, where they are zero by symmetry. On the other hand one sees a sign change also in the range of some chiral geometries with an 'accidental' value $E_{pv} = 0$ at chiral geometries of about $\tau \approx 80°$ and $\tau \approx 280°$. Nevertheless

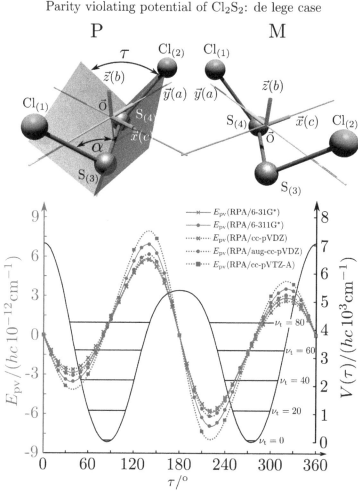

Fig. 16. Calculated torsional potential (full line, right ordinate scale) and parity-violating potential (left ordinate scale, lines with various symbols for various approximations in the electroweak quantum chemistry [97]) for ClSSCl. The equilibrium structure and the definition of the torsional angle τ are shown in the upper part of the figure [99] (after Ref. [97]).

it is true that two mirror image structures have exactly the same absolute value but opposite sign for E_{pv} (but possibly being zero), as the potential is strictly anti-symmetric. On the other hand, the Born Oppenheimer electronic potentials shown by the full black line in Fig. 16 are strictly symmetric with respect to inversion at $\tau = 180°$. This result is true also going beyond the Born-Oppenheimer approximation at higher levels of approximation and in fact the symmetry remains true for the exact electromagnetic Hamiltonian at all orders of precision: The difference of energies between mirror image (space inverted) structures is exactly zero by symmetry as long as only strong and electromagnetic interactions are included. We note that in [182] higher dimensional parity violating potential hypersurfaces are discussed and some graphical representations for 2-dimensional surfaces are shown. As discussed in Sec. 2 with Eqs. (27) and (28) one must discuss the relative magnitude of the parity violating potentials in relation to tunneling processes connecting the two enantiomers. Only when parity violation dominates over tunneling, Eq. (28), one will have localized wave functions and a measurable ground state energy difference $\Delta_{pv}E_0$ between the enantiomers. We have systematically studied this for the series of hydrogen isotopomers XYYX, i.e. for the chalcogenic Y and X=H,D,T including also mixed compounds (i.e. with X,X′ different and Y,Y′ different) and further compounds ZYYZ with other elements Z and further compounds and this has been reviewed recently [46]. It turns out that for the example HOOH the tunneling splittings $|\Delta E_{\pm}|$ for the symmetrical potentials are much larger than the $|\Delta_{pv}E|$. Thus in HOOH one does not find a measurable parity violating ground state difference $|\Delta_{pv}E_0|$, the ground state is delocalized and has almost pure parity, as also excited rovibrational-tunneling states, as discussed in Sec. 2. For the higher elements in the series the tunneling splittings become comparable to $|\Delta_{pv}E_{el}|$. TSeSeT being the first example where parity violation dominates, and for the tellurium and polonium compounds parity violation dominates already for the deuterated compounds, thus these molecules might, in principle, be useful for measurements of $\Delta_{pv}E_0$ disregarding other problems arising from chemical properties and radioactivity.

$\Delta_{pv}E$ roughly scales with the 5th power of the heaviest nuclei in the molecule as suggested by an equation originally given by Zel'dovich [183, 184] and complemented in [23] by a geometry dependent factor f_{geo} (for Zel'dovich $f_{geo} = 1$)

$$\frac{\Delta_{pv}E}{h} = f_{geo}(q_1, q_2, q_3, ...q_s) \cdot 10^4 \cdot \left(\frac{Z_{eff}}{100}\right)^5 \text{Hz} \qquad (39)$$

This would provide a simple estimate, if one takes some weighted average over the charge numbers of the nuclei in the molecule (in the simplest case the weighted average of the two heaviest nuclei [23]). The simple scaling with Z^5 is suggested by the form of the operators, the effect of spin orbit coupling and of the weak nuclear charges Q_A taking the simple approximation that the number of neutrons is roughly proportional to Z as well. The problem in such scaling formulae arises from the complicated form of $f_{geo}(q_1, q_2, q_3, ...q_s)$ and a critical discussion can be found in Ref. [182]. An obvious way to find molecules with smaller tunneling splittings is to use compounds ZYYZ (or ZYYX etc.) with heavier Z-atoms and interesting examples are ClOOCl [103, 185, 186] and ClSSCl [97, 99]. Figure 16 shows the torsional potential and the parity violating potential as a function of the torsional angle (with scales on the ordinate axes being different by 15 orders of magnitude as shown).

ClSSCl is an example where the tunneling splitting was shown to be at least 40 orders of magnitude smaller then the parity violating potentials [97] as recently reconfirmed [187]. Thus in this molecule, the ground states of each enantiomer are clearly localized at the chiral structures and one can significantly measure a ground state energy difference $\Delta_{pv}E_0$ between the enantiomers of this molecule. This measurable energy difference for the ground state (or excited rovibrational states possibly including hyperfine substructures) can be calculated by the expectation values for the corresponding rovibronic (possibly nuclear spin hyperfine) state $\varphi_{evr}^{(k)}$:

$$\Delta_{pv}E^{(k)} = \langle\varphi_{evr}^{(k)}|E_{pv\,R}(q_1,\ldots,q_s)|\varphi_{evr}^{(k)}\rangle - \langle\varphi_{evr}^{(k)}|E_{pv\,S}(\bar{q}_1,\ldots,\bar{q}_s)|\varphi_{evr}^{(k)}\rangle \quad (40)$$

The ground state energy difference $\Delta_{pv}E_0$ is a special case with $k = 0$. Sometimes the $\Delta_{pv}E_0$ can be approximated by taking the values of the parity violating potentials at the equilibrium geometries $\Delta_{pv}E_{el}(q_e)$. For ClSSCl, for example, one finds $\Delta_{pv}E_0 = hc \cdot 1.35 \cdot 10^{-12}\,\mathrm{cm}^{-1}$ and $\Delta_{pv}E_{el}(q_e) = hc \cdot 1.29 \cdot 10^{-12}\,\mathrm{cm}^{-1}$, Ref. 97, where the vibrational averaging was restricted to the torsional coordinate to obtain an estimate for the uncertainties introduced by vibrational averaging (see also Ref. 55).

4. Towards spectroscopic experiments

4.1. Developing experimental concepts and spectroscopic techniques for the study of parity violation in chiral molecules

Successful spectroscopic experiments promise great progress in fundamental aspects of molecular and biomolecular stereochemistry and possibly also concerning precision experiments on parameters of the standard model of particle physics (SMPP). At the same time such experiments also present a major challenge because of the small magnitude of the predicted effects, and there are no 'low hanging fruits' to be harvested, even though the high hanging fruits may be of particularly good taste. As of today there seem to be only two reasonably advanced efforts world wide, one in Zürich which started with first publications from 1986 [100] onwards and one in Paris, starting with first publications in 1999 [188]. The two projects have been reviewed for example in Refs. [23, 39, 45, 46, 112] and [189, 190]. Other groups have expressed interest in developing experimental projects (e.g. [191]), but there does not seem to be any concrete published record on progress in any of these. There have been proposals on a variety of experimental concepts on how to address parity violation in chiral molecules and we have summarized these before [23, 39]. None of these other schemes seems to be particularly promising or easier than the schemes used in current projects, but one should certainly also consider these and further possibilities [277]. Here we shall discuss only the two different concepts for the currently ongoing projects. Figure 17 shows a graphical survey of the historical development, from which one can get an overview over these really long term efforts.

The different concepts can be understood with the energy level scheme shown in Fig. 3. In a scheme originally proposed by Letokhov in 1975 [192, 200] one attempts to measure a difference in the high resolution spectra of the separate enantiomers S and R, which may for a particular transition have the frequency ν_S and ν_R in the different enantiomers. As one can see from the scheme in Fig. 3 this corresponds

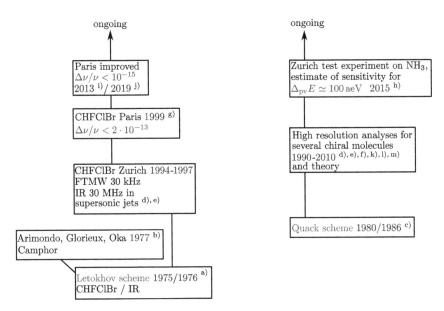

Fig. 17. Graphical survey of the historical development of the two currently pursued experimental concepts (with a) Ref. 192, b) Ref. 193, c) Ref. 100, d) Ref. 194, e) Ref. 195, f) Ref. 196, g) Ref. 188, h) Ref. 104, i) Ref. 189, j) Ref. 190, k) Ref. 197, l) Ref. 198, m) Ref. 199.

to a difference of parity violating energy differences $\Delta_{\mathrm{pv}}E$ in two corresponding molecular levels

$$\Delta_{\mathrm{pv}}E^* - \Delta_{\mathrm{pv}}E = h(\nu_{\mathrm{R}} - \nu_{\mathrm{S}}) \tag{41}$$

The group of Letokhov has already tried to see a corresponding splitting of lines in infrared spectra of CHFClBr in a racemic mixture [200] measuring at sub-Doppler resolution of about $\Delta\nu/\nu \simeq 10^{-8}$, which is almost ten orders of magnitude away from the effect as calculated later [54]. Shortly thereafter Arimondo, Glorieux and Oka did sub-Doppler Lamb-Dip spectroscopy on camphor with separated enantiomers, with about a similar precision [193]. Again, the effect was calculated later to be much smaller [201] for camphor as well. We have in Zurich followed Letokhov's scheme in parallel to our scheme, achieving high resolution analyses with hyperfine structure analysis of microwave and rovibrational infrared of CHFClBr [194, 195] spectra of supersonic jets. This work identified coincidences with CO_2 laser lines and came to the conclusion, that with ultrahigh resolution sub-Doppler spectroscopy using these or related coincidences one might approach resolutions sufficient to identify parity violating effects [195]. Such experiments were subsequently actually carried out in Paris [188] achieving $\Delta\nu/\nu \simeq 2 \cdot 10^{-13}$ still several orders of magnitude away from the theoretically predicted effects [54]. With subsequent improvements reviewed in [189, 190], the currently ongoing experiments appear to be promising for molecules involving very heavy elements, where the parity violating effects are relatively large. So far, no experiment along these lines has been successful, however.

In the other concept proposed by us in 1986 [100], actually informally reported earlier on a few occasions between 1977 and 1986 already (see [47] and references cited therein), one makes use of an 'achiral' excited molecular energy level of well defined parity which has radiative electric dipole transition moments connecting to both enantiomers (see Figs. 3 and 18). Using such a level, one can prepare a coherent superposition of well defined parity in the ground state and follow the time evolution of parity due to parity violation (see Sec. 2). From this, one obtains $\Delta_{pv}E$ directly and separately (one could also measure some $\Delta_{pv}E^*$ separately, of course [23]). This scheme can also be carried out in frequency resolved experiments, when $\Delta_{pv}E$ can be isolated as a spectroscopic 'combination difference' [38]. We shall discuss now properties for the time dependent scheme in more detail, as this has some advantages [42]:

(i) the requirements for spectral resolution are less severe, being sufficient for state selection,

(ii) there is no need to prepare pure enantiomers, the experiment works for racemic mixtures as well as for separate enantiomers,

(iii) one measures the parity violating energy difference between the corresponding levels of the enantiomers directly and separately for every pair of levels instead of only a 'difference of differences', as given by Eq. (41),

(iv) the technique has been demonstrated to be able to measure smaller absolute values of the parity violating energy difference than by the other scheme, thus allowing one to study molecules involving only light nuclei.

4.2. *Towards a measurement of $\Delta_{pv}E$ and a test of sensitivity with an achiral molecule*

Our approach in Zurich [100] can be illustrated with the scheme outlined in Fig. 18. It uses the idea that either one may have a planar excited electronic state, where the rovibronic levels have essentially well defined parity or else one has a modest barrier for stereomutation in the ground state, where one can thus reach by vibrational excitation with infrared lasers levels near to or above the barrier, where they can have large tunneling splittings and thus satisfy the condition in Eq. (27), guaranteeing that they have essentially well defined parity, a typical tunneling switching situation [186]. Then one can carry out an experiment following the schemes in Fig. 4 and Fig. 19. One first prepares with a sequence of two laser pulses a state of well defined parity at low energy, where the inequality Eq. (28) applies. One can use rapid adiabatic passage (RAP) in a molecular beam experiment [104] or chirped laser pulses [103] or also stimulated Raman adiabatic passage (STIRAP) [105]. This prepared parity state is time dependent and evolves under parity violation according to Eqs. (5), (29) and (30) (in the two state approximation).

This is a periodic time evolution with a period on the order of a second, depending on the value of $\Delta_{pv}E$. In the molecular beam setup shown in Fig. 19 the evolution time after preparation will be on the order of milliseconds with typical

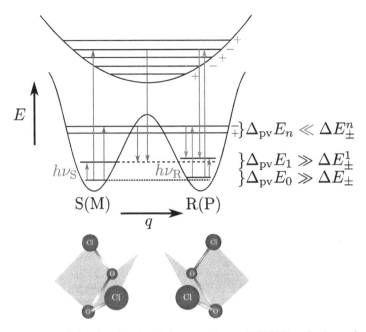

$$\begin{aligned} &+\}\Delta_{\mathrm{pv}}E_n \ll \Delta E^n_\pm \\ &\}\Delta_{\mathrm{pv}}E_1 \gg \Delta E^1_\pm \\ &\}\Delta_{\mathrm{pv}}E_0 \gg \Delta E_\pm \end{aligned}$$

Fig. 18. Enantiomers of chiral molecules of the general type X-Y-Y-X and scheme for the experiment to measure $\Delta_{\mathrm{pv}}E$ (red arrows for time-dependent experiment) and $\Delta_{\mathrm{pv}}\nu := \nu_{\mathrm{S}} - \nu_{\mathrm{R}}$ (blue arrows). The combination of selected red and green arrows in the scheme leads to a measurement of $\Delta_{\mathrm{pv}}E$ in the frequency domain. The excited state of well defined parity (plus or minus signs) can be in an electronically excited state or in an excited vibrational-tunneling state as shown in Ref. [186]. The red arrow in absorption corresponds to a transition between an S and a negative parity level (energy $h\nu_{\mathrm{S}-}$). The green arrow in absorption corresponds to a transition between an R and a negative parity level (energy $h\nu_{\mathrm{R}-}$). The difference is $h\nu_{\mathrm{R}-} - h\nu_{\mathrm{S}-} = \Delta_{\mathrm{pv}}E_0$. The energy difference between the two blue absorption lines within R and S, $h\nu_{\mathrm{S}} - h\nu_{\mathrm{R}} = \Delta_{\mathrm{pv}}E_0 - \Delta_{\mathrm{pv}}E_1$, is the difference of parity violating energy differences (modified after [48, 112] and [186]).

molecular speeds and a flight path on the order of $1\,\mathrm{m}$. One can use the approximation $\sin^2 x \approx x^2$ for small values of the argument resulting in

$$1 - p_+(t) = p_-(t) \simeq \pi^2 \Delta_{\mathrm{pv}}E^2 t^2/h^2 \qquad (42)$$

This change in the populations can be observed spectroscopically because the rovibronic spectra of the parity isomers differ, as shown in Fig. 20. When one has a full assignment of the rovibronic spectrum of the R or S enantiomer (or a racemic mixture) one can label each line by a parity symbol '+' or '-', depending upon whether it connects to a positive or negative parity of the upper level in the transition, all lines being allowed in the spectrum for the ordinary chiral molecule (R or S or racemic mixture) of the upper part in Fig. 20. After preparation of a 'negative parity isomer' in the preparation steps only the blue lines marked with a + will appear due to electric dipole selection rules. As the character of the prepared state changes parity in time following Eqs. (30) and (42), this can be probed by detecting transitions at frequencies corresponding to the red lines marked '-'. This

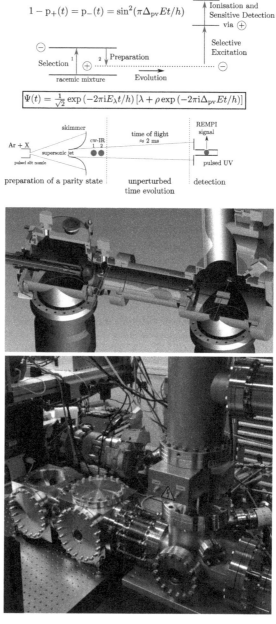

Fig. 19. Experimental molecular beam setup showing the three laser beams of the three steps (top part, scheme, middle part, modified after [104] and below a photograph of the time-of-flight part of the actual setup). We note that the laser systems used in the experiment need actually much more space than this 'core' part of the experiment. In the back one can also see a part of the setup for comb based high-resolution cavity-ring down spectroscopy of molecules in a supersonic jet expansion [202].

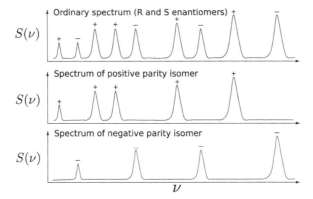

Fig. 20. Schematic illustration of the spectral changes in the superposition experiment (modified after [45, 112]).

detection can be carried out sensitively, say, by a UV laser multiphoton excitation-ionization technique. The sensitivity of this detection, i.e. how small a population in Eq. (42) is detectable, determines in essence the magnitude of $\Delta_{\mathrm{pv}}E$ which can be detected without the signal disappearing in the noise. We note that in a real experiment neither the parity selection will be perfect nor will be the time evolution be completely free from external influences (collisions, fields, thermal radiation, etc.), these effects determining the 'noise' background limiting the measurement. On the other hand, the frequency resolution need only be sufficient to detect separate rovibrational levels, and for this a resolution of 1 MHz (often only 10 or 100 MHz, depending on the spectrum) will be sufficient and is readily available by current high resolution laser spectroscopic techniques for instance with the frequency comb based optical parametric oscillators (OPO's) used by us in [104], but also with other laser techniques.

We have tested the method and its sensitivity with the setup shown in Fig. 19 and with an achiral molecule NH_3, where the spectrum is extremely well known and assigned [104]. These experiments, which also demonstrated new hyperfine structure resolution on excited vibrational states of NH_3, show, of course, effective parity conservation in the time evolution, because NH_3 is achiral with levels of well defined parity, which would also be the case for the chiral isotopomer NHDT having a large tunneling splitting in the ground state [96]. However, the test experiment can be used for an estimate of the sensitivity of the current setup and therefore of the values $\Delta_{\mathrm{pv}}E$, which would be detectable. It was concluded that values of $\Delta_{\mathrm{pv}}E \simeq 100\,\mathrm{aeV}$ should be measurable with the current experimental setup, and systematic improvements would allow the measurement of even smaller values of $\Delta_{\mathrm{pv}}E$. Such values are predicted for chiral molecules involving only atoms not heavier than chlorine. Indeed, for the molecule ClOOCl parity violation is calculated to completely dominate over tunneling, making the molecule suitable, in principle, for a measurement of $\Delta_{\mathrm{pv}}E$. In Ref. [103] we have reported a complete simulation of the experiment with preparation and detection steps. The current progress in the experiment depends on obtaining adequate assignments of spectra for suitable chiral molecules identifying positive and negative parity levels as indicated schematically in Fig. 20. This turns out to be a non-trivial task with progress to be discussed in Sec. 4.3. We shall, however, first discuss an interesting conceptual aspect. The prepared parity isomers in the experiment have the character of being at the same time R and S enantiomers corresponding to the wave functions χ_+ and χ_- indicated in Fig. 7, which carry equal weight for λ and ϱ which would be the chiral enantiomers. Such states are classically impossible and sometimes are called 'Schrödinger's cat' after the famous discussion of Schrödinger, which identified quantum mechanical states of a cat which is prepared in a thought experiment in a state where it is at the same time dead and alive (with some probability [203]), a situation which is classically absurd (see also a cartoon in [45]). We have noted occasionally that Schrödinger's 'quantum cat' analogy corresponds to a cruel experiment on an animal, which frequently comes out dead in the final step. We therefore have proposed (for the analogous 'parity isomer') as a better analogy a quantum chameleon which stays alive and only changes color in the course of time, which would be a natural thing for a chameleon to do any way (the change of a spectrum as in Fig. 20 is, of course, the analogue of a change of color [47, 106, 204]).

4.3. Development of spectroscopic techniques and high resolution analyses for chiral molecules and the spectroscopic realization of a quantum chameleon

Identifying suitable molecules and their high resolution spectra as suitable for 'ultra high' resolution in view of a spectroscopic detection of parity violation in chiral molecules is important for all experimental concepts. Particularly for our approach as proposed in 1986 [100], it is essential to have adequately resolved and analysed spectra in order to identify lines connecting to states of well defined parity in the excited state as discussed in Sec. 4.2.

However, until 1986 there existed not a single example of a chiral molecule, where an analysis of an optical spectrum (IR, VIS, UV) had been achieved with full rovibrational resolution as needed for this approach. Such analyses seemed very difficult although not impossible at this time for molecules of the necessary minimum complexity related to chirality. In the meantime there has been much progress in developments of spectroscopic techniques which is reviewed elsewhere [76, 85, 205–208] (see also many further articles contained in [205]).

Our progress in developing advanced techniques of high resolution laser and also Fourier transform infrared (FTIR) spectroscopy has enabled the analysis of high resolution optical spectra for chiral and achiral molecules with full rotational and vibrational (and sometimes hyperfine-structure) assignments. This occurred over several generations of new developments in techniques. As an example we show in Fig. 21 the currently worldwide highest resolution FTIR spectrometer setup at the

Our group's unique high resolution FTIR setup currently still with
the worlds highest resolution
(synchrotron-based at the *Swiss Light Source - SLS*)

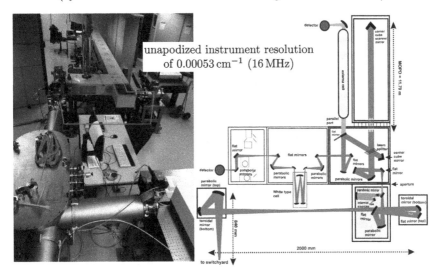

unapodized instrument resolution
of $0.00053\,\text{cm}^{-1}$ ($16\,\text{MHz}$)

Fig. 21. Photograph and schematic diagram of the high resolution FTIR prototype (Bruker) spectrometer built at the Swiss Light Source (SLS), using synchrotron radiation in the infrared (after Ref. 209, see also Refs. 206, 208, 210).

Swiss Light Source (synchrotron) built as prototype for our group. Other important developments in our group relate to high resolution diode laser supersonic jet and frequency-comb based spectroscopy with optical parametric oscillators (OPO).

Specifically, progress has been made also on the analysis of high resolution spectra of chiral molecules and Tab. 2 provides a survey of chiral molecules for which parity violation has been studied theoretically and preliminary high resolution analyses of optical spectra are available by now. The table retains only molecules, for which tunneling splittings in the ground state are negligible such that $\Delta_{\text{pv}}E_0$ is an actually measurable ground state energy difference of the stable enantiomers according to the scheme on the right hand side of Fig. 7. These molecules are thus in principle suitable for studies of parity violation by one of the techniques discussed in Sec. 4.2, although not all are really very favourable for such studies, if $|\Delta_{\text{pv}}E|$ is small. Here, we shall discuss one relatively favourable case: 1,2-Dithiine ($C_4H_4S_2$) shown in Fig. 22 [211, 212].

For this molecule theory predicts $\Delta_{\text{pv}}E/(hc) \simeq 1.1 \cdot 10^{-11}\,\text{cm}^{-1}$ (corresponding to $\Delta_{\text{pv}}E \simeq 1\,\text{feV}$) which is well in the range accessible to our current experimental technique. Also the tunneling splitting in the ground state has been estimated to be well below $10^{-20}\,\text{cm}^{-1}$, which guarantees that $\Delta_{\text{pv}}E_0$ is, in principle, measurable [211]. At the same time the barrier for stereomutation is calculated to be around $2500\,\text{cm}^{-1}$, which makes large tunneling splittings at this and higher levels possible,

Table 2.   Summary of chiral molecules for which theoretical studies of $\Delta_{\mathrm{pv}}E$ and high resolution spectroscopic analyses exist.

Molecule	$\lvert\Delta_{\mathrm{pv}}E\rvert/(hc\,10^{-14}\,\mathrm{cm}^{-1})$	references
ClOOCl	57.5	103, 152, 186
ClSSCl	130.0	97, 99
$PF^{35}Cl^{37}Cl$	2.8	181, 208
HSSSH	160	213, 214
$CHF^{35}Cl^{37}Cl$	a)	206, 207, 215–217
CHFClBr	190	54, 55, 194, 195
CDFClBr	190	55, 218
CHFClI	b)	219, 220
CHFBrI	b)	219, 221
CHDTOH	37	222
CHF=C=CHF	14	152, 223, 224
CHF=C=CHCl	70	152, 224
CHCl=C=CHCl	110	152, 224
D-oxirane (CH_2CHDO)	(0.02)c)	225
D_2 oxirane (trans CHDCHDO)	c)	226
F-oxirane (CH_2CHFO)	17	196, 227, 228
cyclo-CH_2CD_2SO	d)	197, 229
cyclo-CH_2CHDSO	d)	197, 229
cyano-oxirane CH_2CHCNO	10	230, 231
cyano-aziridine $CH_2CHCNNH$	10	230, 232
1,2-Dithiine $C_4H_4S_2$	1100	211, 212
Alanine (CH_3CHNH_2COOH)	5e)	101, 149, 233

Note: a) a preliminary value is small, b) here theory calculated frequency shift values for the C-F stretching fundamental transition, c) small preliminary value, d) preliminary estimate is small, e) strongly conformer dependent and in any case small.

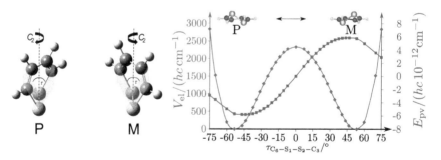

Fig. 22.   The enantiomers of 1,2-Dithiine are shown on the left together with the symmetric parity conserving (blue line, left ordinate scale) and antisymmetric parity violating potentials (red line, right ordinate scale) on the right (after [211, 212]).

in a range accessible to our comb-based high resolution laser (OPO) supersonic jet techniques [104, 202] (in addition to high resolution FTIR spectroscopy [206, 212]). It is clear, however, that the analysis of rotation-vibration-tunneling spectra in a molecule of this complexity still presents a major challenge, particularly so at high energy in a very dense spectrum. Clearly progress on similar molecules, also newly designed ones, remains desirable in the future.

At present we shall mention here an example of an achiral molecule of almost comparable complexity, where we have successfully achieved such an analysis of rotation-vibration-tunneling states in a tunneling switching situation [234–236]. This has allowed us to demonstrate non-classical 'quantum chameleon states' involving ground state levels which are separated by energies of corresponding to about $hc \cdot 0.8\,\mathrm{cm}^{-1}$. The example is m-D-phenol which exists in the two planar isomeric forms as syn- and anti- isomer. While the electronic Born-Oppenheimer potential hypersurface has minima of identical energy for these isomers, small zero point energy effects lead to a small ground state energy difference of $\Delta E(\mathrm{syn - anti})$, which was predicted by theory to be much larger than the ground state tunneling splitting in the symmetric tunneling potential for ordinary phenol, which is about $hc \cdot 0.001\,\mathrm{cm}^{-1}$ [234].

Thus in the lowest vibrational states of m-D-phenol, one has energy eigenstates separated by about $0.8\,\mathrm{cm}^{-1}$ with wave functions corresponding to localized syn- and anti- isomers. The situation is analogous to the asymmetric tunneling situation with parity violation in the right hand side of Fig. 7 with localized wave functions as well. However, the asymmetry in m-D-phenol has a different physical origin and is much larger ($0.8\,\mathrm{cm}^{-1}$ corresponding to about $0.1\,\mathrm{meV}$ instead of $100\,\mathrm{aeV}$ for parity violation, 12 orders of magnitude smaller). Nevertheless also for the m-D-phenol we have a tunneling switching situation given the modest barrier for isomerization of only about $1000\,\mathrm{cm}^{-1}$. Thus for the torsional level $\mathrm{v_T} = 2$ at $600\,\mathrm{cm}^{-1}$ and higher the eigenfunctions are delocalized and show an approximate symmetry similar to the symmetric case of ordinary phenol, a typical tunneling switching situation. Therefore this molecule has been analysed successfully to demonstrate the 'quantum chameleon states' which might be used for molecular quantum switches [234–236] going far beyond possibilities of 'classical' molecular switches [15–17]. For a detailed discussion of these achiral quantum switches we refer to Ref. 47 (and references cited therein). We conclude here with a brief note on parity violation in isotopically chiral molecules such as PF $^{35}\mathrm{Cl}\,^{37}\mathrm{Cl}$ and CHF $^{35}\mathrm{Cl}\,^{37}\mathrm{Cl}$ or D-oxirane ($\mathrm{CH_2CHDO}$) and related ones. As pointed out in Refs. 23, 38, 42, 181, 207, in such molecules the parity violating weak interaction introduces a fundamentally new isotope effect, which arises from the difference in the weak nuclear charge Q_A (Eq. 33) of the different isotopes and not from the mass difference or the difference in nuclear spin as for the ordinary, well- known isotope effects. Because of the nature of the nuclear structure of the isotopes and their values for Q_A, the parity violating energy differences for isotopically chiral molecules are particularly interesting for theory, but they are also typically smaller than in molecules, which are chiral by 'ordinary chemical substitution' of symmetric molecules, say CHFClBr derived from methane or PFClBr derived from $\mathrm{PCl_3}$, see also table 2, and thus they are more difficult for experiments in this respect. This does not exclude, however, larger effects of parity violation in specially selected isotopically chiral molecules. For instance for $\mathrm{CH_3Re^{16}O}\,^{17}\mathrm{O}\,^{18}\mathrm{O}$ suggested by us as an interesting candidate at an early time [38], our preliminary calculations indicate much larger values (even by order of magnitude [279]) for the parity violating energy difference than, say, in PF ^{35}Cl ^{37}Cl. We also draw attention here to the possibility of measuring parity violation in achiral spherical top molecules by a level crossing experiment proposed by us in 1995, as reviewed in [23, 47]. While the effects are minute in CH_4 they can be much larger in CF_4, for which recent high resolution spectroscopy might be a first step towards such experiments (see [277, 278] and references cited therein).

5. Parity violation and biomolecular homochirality

In his Nobel prize lecture 1975 on 'Chirality in chemistry' [10], V. Prelog concludes with a brief statement on the question of biomolecular homochirality as shown in Fig. 23. Prelog refers here to the observation that in the biopolymers of life (with chiral amino acids in the proteins and chiral sugars in DNA) only the L-amino acids are used and only the D-sugars. Strictly speaking in other circumstances (not in the biopolymers) the other enantiomeric forms do occur also naturally in biological processes. Also the uniqueness of selection, say, of the whole series of natural L-amino acids depends upon the convention used. For instance with the R, S nomenclature there are not only S-amino acids selected. For example L-cystein is selected as D/L nomenclature identifies it as 'L' enantiomer by convention, but when using the CIP convention it will be R-cystein as this corresponds to L-cystein according to the rule in this case. But the basic fact remains true, as stated by Prelog, that for each particular amino acid and each sugar only one enantiomer occurs in the biopolymers.

Why this is so has been an enigma for more than a century, the phenomenon has been noted (with limited knowledge) already by Pasteur [4–7] and by Fischer

Chirality and Biomolecular Homochirality
'The time at my disposition also does not permit me to deal with the manifold biochemical and biological aspects of molecular chirality. Two of these must be mentioned, however, briefly. The first is the fact that although most compounds involved in fundamental life processes, such as sugars and amino acids, are chiral and **although the energy of both enantiomers and the probability of their formation in an achiral environment are equal, only one enantiomer occurs in Nature;** the enantiomers involved in life processes are the same in men, animals, plants and microorganisms, independent on their place and time on Earth. Many hyptheses have been conceived about this subject, which can be regarded as one of the first problems of **molecular theology**. One possible explanation is that the creation of living matter was an extremely improbable event. which occured only once.' **- Vladimir Prelog (ML), Nobel Lecture, 12 Dec 1975**

Fig. 23.   Citation on chirality and biomolecular homochirality from Ref. 10 (emphasis added here).

[237] for instance, identifying this 'homochirality' as a simple chemical signature of life. While the statement of this long standing enigma by Prelog is thus essentially correct, we have noted in many friendly private discussions with him in Zurich in the early 1980ies and in the publication dedicated to him at his 80th birthday meeting [100], that the premisses are incorrect: Because of parity violation the energies of the enantiomers in an achiral environment are in fact not equal (that one might consider free space as chiral is a separate story, see [44, 48] for some of the history).

The question is then, how important this small asymmetry arising from parity violation will be for the origin of biomolecular homochirality [39]. We anticipate here the short answer: We do not know, but at least one can say that the discovery in 1995 [149] of an increase by a factor of 10 to 100 of parity violating energy differences $\Delta_{\text{pv}}E$ calculated from theory has made the possibility of an important effect of parity violation in the evolution of biomolecular homochirality more likely than it seemed before, but in any case the question remains open. There is an obvious relation to the other open question concerning the origin of life. How did life arise from 'non-living matter'? This question, which refers to the frontier between the non-living and living refers to what is the very nature of life. The question 'What is life?' also has a long history with many debates in the scientific and the non-scientific literature. Schrödinger asked the question from a physicist's perspective in his famous little book arising from lectures in 1943 [238]. We shall take here the pragmatic point of view that we can distinguish the 'living' from the 'non-living' by inspection in a fairly straight-forward way [45]. On these three open questions there exists, indeed, a huge literature, which we cannot refer to in any meaningful way here. In [39, 45, 48] we have provided more extended discussion related to various questions on life in a 'decision diagram' or flow diagram reproduced here in Fig. 24. Without going into details we summarize here that at every step the related question is really open. However, there are for each step 'opinions' from what we have called communities of belief [38, 41, 45, 48]. For instance, whether life is rare in the universe, possibly singular, only existing on our planet, as surmised by Monod [239] and perhaps Prelog [10], or else whether it is frequent, as seems to be the current majority opinion, according to some informal, non-representative 'polls' in lectures [45, 48], is in reality completely open [41, 43, 48]. This corresponds to the first step on top in the diagram of Fig. 24 (see also the reviews [45, 48]).

Here we shall focus on the question on homochirality addressed in the lower part of the diagram. One has essentially two large communities of belief (with numerous different 'denominations' within each community). The first one assumes that homochirality arises simply by a *chance selection* in the early stages of the evolution, either pre-biological (non-living matter) or a little later in early living species. This we have called the '*de facto*' selection (by chance). The other community assumes that the '*de lege*' asymmetry arising from parity violation has been important and the selection occurred *by necessity*, preferring one form over the other in analogy to the diagram in Fig. 7. The question could, in principle be answered by observation of possibly many different forms of life on exoplanets [240–242], which might be

248

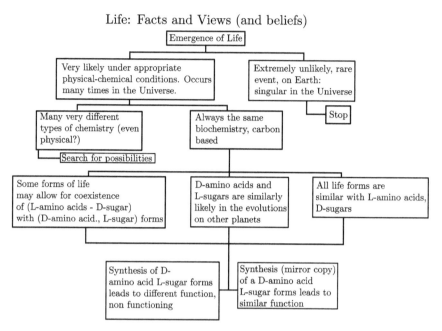

Life: Facts and Views (and beliefs)

Fig. 24. Summary as 'flow diagram' for opinions and beliefs on the emergence of life and homochi-rality (modified after Refs. 45, 48).

possible by a spectroscopic search for a systematic homochirality on these planets [42, 45] or perhaps within our solar system by space flight. The *de facto* hypotheses would predict a statistical distribution of 'L' or 'D' homochirality (if any). The *de lege* hypotheses would predict a preponderance of one form (perhaps even a unique selection of always the same form). Numerous mechanisms have been proposed to explain the origin of biomolecular homochirality. Our lack of understanding does not arise from an absence of such explanations. Indeed there are many different more or less plausible explanations, but we do not know which one is correct.

The long list of existing proposals, can clearly be grouped into broad classes of '*de facto*' selection (by chance) and '*de lege*' selection (by necessity). We shall not enter into any further detailed discussion here, but we note that some of these have had firm believers with heated debates among the different communities of belief and sometimes false statements of evidence (see the discussions in [39, 41, 45] for further information). The question arises on how to obtain conclusive evidence. One approach might be to repeat total synthesis of life and evolution in the laboratory, clearly a difficult enterprise so far without much success (see e.g. [243, 244]). We note here, that so far not even the total synthesis of an enantiomeric copy of a simple existing bacterium has been successful, although already proposed in 1990 [245].

Another approach would be to investigate the conditions for certain mechanisms by systematic experiments complemented by theory and thereby approach an answer in a stepwise way. In this context we can refer to the lectures by Blackmond and Soai at this meeting [272, 273]. There has been a most interesting series of experiments by Blackmond and coworkers [246–252] in relation to the famous auto-catalytic Soai mechanism [253–258] (see also [259]). Using minor perturbations by substituting 'heavy' (non-hydrogen) isotopes such as $^{12}C/^{13}C$ or $^{14}N/^{15}N$ an esti-mate was made about the threshold energy differences (e.g. in terms of $\Delta G^{\neq}$ in transition states) that can induce an amplification to homochirality in a Soai-type mechanism. This threshold was found to be about 10^{-5} Jmol^{-1}. While such small values are, indeed, interesting, they are obviously much larger than anything that might be expected to arise from parity violation, perhaps 10^{-10} Jmol^{-1} or at most a little larger. Sometimes this gap on the order of 4 to 5 orders of magnitude has already been interpreted as proving the unimportance of parity violation in the evolution of biomolecular homochirality. However, such a conclusion is quite unjus-tified as it refers only to relatively simple mechanisms such as the Soai reaction. It is possible and even likely, that real-life mechanisms in early prebiotic or early biotic chemistry would be much more complex and more sensitive to small energy effects. In that sense the result, that energy differences on the order of 10^{-5} Jmol^{-1} are estimated to suffice for a selection of homochirality on short time scales in simple autocatalytic mechanisms would lead to rather optimistic expectations to bridge the remaining gap of a few orders of magnitude with more sensitive and more complex mechanisms on longer time scales.

It has sometimes also been argued that a chemical evolution of homochirality is impossible on long time scales because chiral molecules in reactive environments would always racemize on long time scales. This argument is obviously invalid, as we have 'living proof' in the long lived existing homochirality of life, which, indeed, has used for millions of years chiral molecules in a permanently reactive chemical environment and has successfully avoided racemization by a 'complex autocatalytic mechanism', if one may say so.

Without entering into any further discussion of the many other suggestions and hypotheses, we conclude here with the statement that at present there is simply no proof or even compelling argument to exclude either a *de lege* mechanism (involving electroweak parity violation) or a *de facto* mechanism (by chance). Both are per-fectly possible origins of today's biomolecular homochirality and it remains our task to find out, which of the two applies, possibly both under different circumstances.

6. Conclusion and outlook

Chiral molecules are of greatest importance for many aspects of chemistry and they are the building blocks of biological matter. The relation of chirality with space inversion symmetry establishes a close connection with fundamental physics. Parity violation changes our concepts for the stereochemistry, structure and dynam-ics of chiral molecules. While tunneling according to Hund's treatment dominates over parity violation in transiently chiral molecules such as hydrogen-peroxide (HOOH) [94, 95], the chiral ammonia isotopomer (NHDT) or the aniline isotopomer (C_6H_5NHD), for example [96, 274], we know now that for ordinary chiral molecules, which are stable for days or years, at least, such as CHFClBr, alanine or other amino acids, sugars etc., parity violation dominates completely over tunneling and is the decisive factor for a conceptually correct understanding of their structure and energetics [23, 100].

In contrast to the classical picture of van't Hoff with two symmetrically equivalent enantiomers with exactly equal energies, there is a small but physically significant and in principle measurable energy difference $\Delta_{pv}E$ between their ground states, resulting in a non-zero reaction enthalpy and free energy for stereomutation. The new and much larger orders of magnitude discovered for this energy difference in our theoretical work from 1995, has in the meantime been confirmed independently by theoretical work in other groups and can in this sense be considered to be well established (see Sec. 3). This has also improved the outlook for spectroscopic experiments, which are ongoing (see Sec. 4), although still without conclusive results, so far.

However, much of the spectroscopic ground work for such experiments has been completed by now and one can expect most significant results for all of the possible outcomes of such experiments:

(1) If experiments confirm the theoretically predicted values for $\Delta_{pv}E$, then one can analyse the results of the precision experiments in terms of the standard model of particle physics (SMPP) in a range of quantum systems not yet tested in previous experiments, for example in molecules involving only 'light' nuclei up to, say, Chlorine.
(2) If one finds, on the other hand, values of $\Delta_{pv}E$ different from the theoretical predictions, this will lead to a fundamental revision of current theories for $\Delta_{pv}E$ with even the potential for 'new physics'. Surprises are possible!
(3) Finally, a profound experimental and theoretical understanding of $\Delta_{pv}E$ in chiral molecules can be a first step towards a possible understanding of the implications of parity violation for the evolution of biomolecular homochirality.

We have obtained also further results concerning fundamental concepts of physical-chemical stereochemistry. The possible preparation of 'parity isomers' of chiral molecules introduces the structural concept of 'bistructural' molecules (being 'quantum chameleons', at the same time R and S enantiomers but very different from a 'racemic mixture'). The time evolution of the spectra of these bistructural molecules corresponds to the new intramolecular primary process of parity change with time arising from parity violation by the weak nuclear force. The concept of such bistructural isomers can be extended to other types of isomers and has potential for quantum technology in the more distant future. Based on high resolution spectroscopic results, we have already demonstrated such exotic chameleon states for bistructural syn- and anti- m-D-phenol [234–236] and also for different nuclear spin symmetry isomers (ortho- and para-) in ClSSCl [99].

Main Ideas and Observations

We live in a world		Symmetry
1.	of matter (not antimatter)	C (CP, CPT)
2.	of L-amino acids and D-sugars (and not D-amino acids and L-sugars) in ordinary life (proteins, DNA, RNA)	P
3.	where time runs forward (and not backwards) (also combinations CP, CPT,...)	T

What is the origin of these observations? (*de facto* vs. *de lege?*)
Are they basic 'quasi-fossils' of the evolution of matter and life?

Fig. 25. Three fundamental observations in our world, with so far unexplained relation to fundamental symmetries and their violation (after Ref. 45).

We conclude here with a brief discussion going beyond these now well established results. Charles Darwin is occasionally quoted with a statement 'It is as absurd to think about the origin of life as it is absurd to think about the origin of matter'. However, if we do start this 'absurd thinking' we note three basic 'asymmetries' in our world (Fig. 25). We might consider these as 'quasi-fossils' carrying information about the evolution of matter and life [45]. It turns out that the asymmetries C and CP between matter and antimatter, while experimentally and theoretically known in the SMPP [260], do not seem to be able so far to explain quantitatively the current preponderance of matter over antimatter in the universe [275, 276]. There remain open questions on this quasi-fossil from the origin of matter. Similarly, the observation of the current 'homochirality of life' (with L-amino acids in proteins and D-sugars in DNA) has contradictory explanations (*de facto*, by chance, or *de lege* - involving parity violation, see Sec. 5) and we do not know, which of the explanations is correct. The interpretation of this quasi-fossil from the origin of life [45] remains completely open in spite of much work, which exists already.

We can finally go beyond these two asymmetries by considering the general scheme for chiral molecules and their antimatter counter parts, in Fig. 26, which can be considered 'antimatter stereochemistry' (where we use the 'physical nomenclature' L and R for the stereoisomers and L* and R* for their antimatter counterparts) [62]. As we have discussed, CPT symmetry, which so far has never been found violated and is considered to be exact in the current SMPP, requires the pair of molecules L and R* to be exactly equivalent energetically (and similarly for R and L*). We have discussed that this can be used for a most sensitive spectroscopic test

Stereochemistry and CPT - Symmetry (Violation)

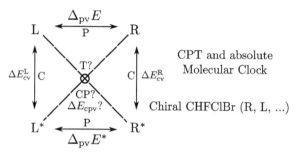

Fig. 26. Stereochemistry with antimatter (using the 'physical' notation L and R for enantiomers of ordinary matter, and L* and R* for antimatter) as a possible test for CPT symmetry violation (after Refs. [23, 62, 107]).

for CPT symmetry [23, 107], much more sensitive than other tests that have been made or proposed so far [261–266] (see also Ref. 47). CPT symmetry can be related to a 'non-observable' which can be considered to concern the observation of a generalized direction of time when including antimatter with an absolute chiral molecular clock for example [41–43]. Whether a violation of CPT symmetry will ever be found remains open, but even at the level of a direction of time in ordinary matter there remains an open question whether the second law of thermodynamics concerning the increase of entropy with time is due to a *de lege* or a *de facto* symmetry violation [23, 41–43], and there are further interesting alternatives to be mentioned [267]. We might refer in this context to various results and searches on the violation of time reversal symmetry and also its relation to the observation of the permanent electric dipole moment of the electron [261, 268–270] expected to be extremely small in the current SMPP (see also the review in Ref. [23]). Thus there remain numerous open questions related to 'chiral matter' in general and chiral molecules, where we do not know the answers yet. But we can conclude by quoting with David Hilbert: 'Wir müssen wissen! Wir werden wissen!' (We must know. We shall know.)

Acknowledgments

We are grateful to Mats Larsson and his colleagues for their hospitality in Stockholm and also for their patience in waiting for our manuscript. Our work has received financial support from an ERC Advanced Grant, from the Swiss National Science Foundation and ETH Zürich, in particular the laboratory for physical chemistry. We acknowledge gratefully recent support from and many fruitful discussions with Frédéric Merkt. Over the years numerous coworkers have contributed to our projects, too numerous to be mentioned in detail and we refer to the list of references and the lists given in Refs. [87, 271]. This publication as already the lecture, on which it is based, is dedicated to the memory of our friend and colleague for many years and decades Richard R. Ernst.

References

1. R. Noyori, Asymmetric Catalysis: Science and Opportunities (Nobel Lecture), *Angew. Chem. Int. Ed.* **41** 2008–2022 (2002).
2. W. T. Kelvin, Baltimore Lectures on Molecular Dynamics and the Wave Theory of Light. Founded on Mr. A. S. Hathaway's Stenographic Report of Twenty Lectures Delivered in Johns Hopkins University, Baltimore, in October (1884): Followed by Twelve Appendices on Allied Subjects. (C. J. Clay, London, 1904).
3. W. T. Kelvin, The Molecular Tactics of a Crystal (Clarendon Press, Oxford, 1894).
4. C. Bourgois (ed.), Louis Pasteur, J. H. van't Hoff, A. Werner, Sur la dissymétrie moléculaire (Cox and Wyman Ltd, P.F.C. Dole, 1986), ISBN 2-267-00454-2.
5. L. Pasteur, Sur les relations qui peuvent exister entre la forme cristalline, la composition chimique et le sens de la polarisation rotatoire, *Annal. Chim. Phys.* **24** 442–459 (1848).
6. L. Pasteur, Mémoire sur la relation qui peut exister entre la forme cristalline et la composition chimique, et sur la cause de la polarisation rotatoire., *C. R. Hebd. Séances Acad. Sci.* **26** 535–538 (1848).
7. L. Pasteur, Recherches sur les propriétés spécifiques des deux acides qui composent l'acide racémique, *Ann. Chim. Phys.* **28** 56–99 (1850).
8. J. H. van't Hoff, Vorlesungen über theoretische und physikalische Chemie: Die chemische Statik (Vieweg, Braunschweig, 1899), Heft 2.
9. A. Werner, Sur les composés métalliques à dissymétrie moléculaire, *(conférence faite devant la société chimique de France, 24 mai 1912)* (1912), reprinted in Ref. 4.
10. V. Prelog, Chirality in Chemistry, in Les prix Nobel en 1975, (Nobel lecture, The Nobel foundation Imprimerie Royale P.A. Norstedt & Söhner, Stockholm, 1976).
11. W. S. Knowles, Asymmetric Hydrogenations (Nobel Lecture), *Angew. Chem. Int. Ed.* **41** 1998–2007 (2002).
12. K. B. Sharpless, Searching for New Reactivity (Nobel Lecture), *Angew. Chem. Int. Ed.* **41** 2024–2032 (2002).
13. B. List, R. A. Lerner and C. F. Barbas, Proline-Catalyzed Direct Asymmetric Aldol Reactions, *J. Am. Chem. Soc.* **122** 2395–2396 (2000), and The Nobel Prize in Chemistry 2021. https://www.nobelprize.org/prizes/chemistry/2021/press-release/.
14. K. A. Ahrendt, C. J. Borths and D. W. C. MacMillan, New Strategies for Organic Catalysis: The First Highly Enantioselective Organocatalytic Diels-Alder Reaction, *J. Am. Chem. Soc.* **122** 4243–4244 (2000), and The Nobel Prize in Chemistry 2021. https://www.nobelprize.org/prizes/chemistry/2021/press-release/.
15. B. L. Feringa, The Art of Building Small: From Molecular Switches to Motors (Nobel Lecture), *Angew. Chem. Int. Ed.* **56** 11060–11078 (2017).
16. J.-P. Sauvage, From Chemical Topology to Molecular Machines (Nobel Lecture), *Angew. Chem. Int. Ed.* **56** 11080–11093 (2017).
17. J. F. Stoddart, Mechanically Interlocked Molecules (MIMs)—Molecular Shuttles, Switches, and Machines (Nobel Lecture), *Angew. Chem. Int. Ed.* **56** 11094–11125 (2017).
18. K. Mislow, Introduction to Stereochemistry (Benjamin, New York, 1965).
19. V. Prelog and G. Helmchen, Basic Principles of the CIP-System and Proposals for a Revision, *Angew. Chem. Int. Ed.* **21** 567–583 (1982).
20. H.-U. Blaser and H.-J. Federsel (eds.), Asymmetric catalysis on industrial scale: Challenges, approaches and solutions, Second edn. (John Wiley & Sons, Ltd, Weinheim, 2010).
21. M. Beller and H.-U. Blaser (eds.), Organometallics as catalysts in the fine chemical industry (Springer, Heidelberg, 2012).

22. M. Quack and J. Hacker (eds.), Symmetrie und Asymmetrie in Wissenschaft und Kunst, in Nova Acta Leopoldina NF 127, Nr. 412, (Wissenschaftliche Verlagsgesellschaft, Stuttgart, 2016). (Book, 275 pages, with contributions in German and English from several authors).

23. M. Quack, Fundamental Symmetries and Symmetry Violations from High Resolution Spectroscopy, in Handbook of High Resolution Spectroscopy, eds. M. Quack and F. Merkt ch. 18, pp. 659–722, (Wiley, Chichester, New York, 2011).

24. H. Fritzsch, Symmetrien der Physik, in Symmetrie und Asymmetrie in Wissenschaft und Kunst, Nova Acta Leopoldina NF 127 Nr. 412, eds. M. Quack and J. Hacker pp. 75–90, (Wissenschaftliche Verlagsgesellschaft, Stuttgart, 2016).

25. T. D. Lee and C. N. Yang, Question of Parity Conservation in Weak Interactions, *Phys. Rev.* **104** 254–258 (1956).

26. C. S. Wu, E. Ambler, R. W. Hayward, D. D. Hoppes and R. P. Hudson, Experimental Test of Parity Conservation in Beta Decay, *Phys. Rev.* **105** 1413–1415 (1957).

27. H. Schopper, Circular polarization of gamma-rays - Further proof for parity failure in beta-decay, *Phil. Mag.* **2** 710–713 (1957).

28. H. Schopper, Die elastische Streuung von Gamma-Strahlen bei kleinen Streuwinkeln, *Z. Physik* **147** 253–260 (1957).

29. R. L. Garwin, L. M. Lederman and M. Weinrich, Observation of the failure of conservation of parity and charge conjugation in meson decays - the magnetic moment of the free muon, *Phys. Rev.* **105** 1415–1417 (1957).

30. J. I. Friedman and V. L. Telegdi, Nuclear emulsion evidence for parity nonconservation in the decay chain π^+-μ^+-ϵ^+, *Phys. Rev.* **105** 1681–1682 (1957).

31. L. Hoddeson, L. Brown, M. Riordan and M. Dresden, The Rise of the Standard Model (Cambridge Uni. Press, Cambridge, 1999).

32. S. L. Glashow, Partial-symmetries of weak interactions, *Nucl. Phys.* **22** 579–588 (1961).

33. S. Weinberg, A Model of Leptons, *Phys. Rev. Lett.* **19** 1264–1266 (1967).

34. A. Salam, Weak and electromagnetic interactions, in Elementary Particle Theory: Relativistic Groups and Analyticity, Proc. of the 8th Nobel Symposium held May 19-25, (1968) at Aspenäs-gården, Lerum, in the county of Älvsborg, Sweden, pp. 367–377, (Almkvist & Wiksell, Stockholm, 1968).

35. M. J. G. Veltman, Nobel Lecture: From weak interactions to gravitation, *Rev. Mod. Phys.* **72** 341–349 (2000).

36. G. 't Hooft, Nobel Lecture: A confrontation with infinity, *Rev. Mod. Phys.* **72** 333–339 (2000).

37. P. Jenni, The Long Journey to the Higgs Boson and Beyond at the Large Hadron Collider (LHC), in Symmetrie und Asymmetrie in Wissenschaft und Kunst, Nova Acta Leopoldina NF 127 Nr. 412, eds. M. Quack and J. Hacker pp. 99–117, (Wissenschaftliche Verlagsgesellschaft, Stuttgart, 2016).

38. M. Quack, Structure and dynamics of chiral molecules, *Angew. Chem. Int. Ed.* **28** 571–586 (1989), *Angew. Chem.*, **1989**, *101*, 588-604.

39. M. Quack, How important is parity violation for molecular and biomolecular chirality?, *Angew. Chem., Int. Ed.* **41** 4618–4630 (2002), *Angew. Chem.*, **2002**, *114*, 4812-4825.

40. M. Quack, Electroweak quantum chemistry and the dynamics of parity violation in chiral molecules, in Modelling Molecular Structure and Reactivity in Biological Systems, Proc. 7th WATOC Congress, Cape Town January 2005, eds. K. J. Naidoo, J. Brady, M. J. Field, J. Gao and M. Hann pp. 3–38, (Royal Soc. Chem., Cambridge, 2006).

41. M. Quack, Intramolekulare Dynamik: Irreversibilität, Zeitumkehrsymmetrie und eine absolute Moleküluhr, *Nova Acta Leopoldina* **81** 137–173 (1999).

42. M. Quack, Frontiers in Spectroscopy (Concluding paper to Faraday Discussion 150, *Farad. Disc.* **150** 533–565 (2011), see also pages 126–127 therein.

43. M. Quack, Time and Time Reversal Symmetry in Quantum Chemical Kinetics, in Fundamental World of Quantum Chemistry. A Tribute to the Memory of Per-Olov Löwdin, eds. E. J. Brändas and E. S. Kryachko pp. 423–474, (Kluwer Adacemic Publishers, Dordrecht, 2004).

44. M. Quack, The Concept of Law and Models in Chemistry, *Europ. Rev.* **22** S50–S86 (2014).

45. M. Quack, On Biomolecular Homochirality as a Quasi-Fossil of the Evolution of Life, *Adv. Chem. Phys.* **157** 249–290 (2014).

46. M. Quack and G. Seyfang, Tunnelling and Parity Violation in Chiral and Achiral Molecules :Theory and High Resolution Spectroscopy, in Tunnelling in Molecules: Nuclear Quantum Effects from Bio to Physical Chemistry, eds. J. Kästner and S. Kozuch pp. 192–244, (Royal Society of Chemistry, Cambridge, England, 2020).

47. M. Quack, G. Seyfang and G. Wichmann, Fundamental and approximate symmetries, parity violation and tunneling in chiral and achiral molecules, *Adv. Quant. Chem.* **81** 51–104 (2020), and *Chem. Science* **13**, 10598–10643 (2022).

48. M. Quack, Die Spiegelsymmetrie des Raumes und die Chiralität in Chemie, Physik, und in der biologischen Evolution, in Symmetrie und Asymmetrie in Wissenschaft und Kunst, Nova Acta Leopoldina NF 127, Nr. 412, eds. M. Quack and J. Hacker pp. 119–166, (Wissenschaftliche Verlagsgesellschaft, Stuttgart, 2016).

49. M. Quack, The symmetries of time and space and their violation in chiral molecules and molecular processes, in Conceptual Tools for Understanding Nature. Proc. 2nd Int. Symp. of Science and Epistemology Seminar, Trieste April 1993, eds. G. Costa, G. Calucci and M. Giorgi pp. 172–208, (World Scientific Publ., Singapore, 1995).

50. F. Merkt and M. Quack, Molecular Quantum Mechanics and Molecular Spectra, Molecular Symmetry, and Interaction of Matter with Radiation, in Handbook of High-Resolution Spectroscopy, eds. M. Quack and F. Merkt ch. 1, pp. 1–55, (Wiley, Chichester, New York, 2011).

51. R. Marquardt and M. Quack, Chapter 1 - Foundations of Time Dependent Quantum Dynamics of Molecules Under Isolation and in Coherent Electromagnetic Fields, in Molecular Spectroscopy and Quantum Dynamics, eds. R. Marquardt and M. Quack pp. 1–41, (Elsevier, 2021).

52. J. H. van't Hoff, La chimie dans l'espace, Rotterdam (1887), reprinted in Sur la dissymétrie moléculaire, Collection Epistème, Paris, (1986) (see Ref. 4).

53. M. Quack and J. Stohner, Molecular chirality and the fundamental symmetries of physics: Influence of parity violation on rovibrational frequencies and thermodynamic properties, *Chirality* **13** 745–753 (2001), (Erratum: Chirality, 15, 375-376 (2003)).

54. M. Quack and J. Stohner, Influence of parity violating weak nuclear potentials on vibrational and rotational frequencies in chiral molecules, *Phys. Rev. Lett.* **84** 3807–3810 (2000).

55. M. Quack and J. Stohner, Combined multidimensional anharmonic and parity violating effects in CDBrClF, *J. Chem. Phys.* **119** 11228–11240 (2003).

56. E. Schrödinger, Quantisierung als Eigenwertproblem I, *Ann. d. Physik* **79** 361–376 (1926).

57. E. Schrödinger, Quantisierung als Eigenwertproblem II, *Ann. d. Physik* **79** 489–527 (1926).

58. E. Schrödinger, Quantisierung als Eigenwertproblem III, *Ann. d. Physik* **80** 437–490 (1926).

59. E. Schrödinger, Quantisierung als Eigenwertproblem IV, *Ann. d. Physik* **81** 109–139 (1926).

60. R. Marquardt and M. Quack (eds.), Molecular Spectroscopy and Quantum Dynamics, 1st edn. (Elsevier, Amsterdam, 2020), ISBN 978-012-817234-6.

61. M. Quack, Reaction dynamics and statistical mechanics of the preparation of highly excited states by intense infrared radiation, *Adv. Chem. Phys.* Vol. 50 pp. 395–473, (John Wiley & Sons, Chichester and New York, 1982).

62. M. Quack, Molecular Quantum Dynamics from High-Resolution Spectroscopy and Laser Chemistry, *J. Mol. Struct.* **292** 171–195 (1993).

63. W. Heisenberg, Die physikalischen Prinzipien der Quantentheorie (Hirzel Verlag, Leipzig, 1980).

64. P. A. M. Dirac, The Principles of Quantum Mechanics, 4th edn. (Clarendon Press, Oxford, 1958).

65. M. Quack, Comments on the role of Symmetries in intramolecular quantum dynamics, *Farad. Disc.* **102** 90–93, 358–360 (1995).

66. M. Quack, Molecular femtosecond quantum dynamics between less than yoctoseconds and more than days: Experiment and theory, in Femtosecond Chemistry, Proc. Berlin Conf. Femtosecond Chemistry, Berlin (March 1993), eds. J. Manz and L. Wöste ch. 27, pp. 781–818, (Verlag Chemie, Weinheim, 1995).

67. M. Quack, Detailed symmetry selection-rules for reactive collisions, *Mol. Phys.* **34** 477–504 (1977).

68. M. Quack, Detailed symmetry selection rules for chemical reactions, in Symmetries and properties of non-rigid molecules: A comprehensive survey., eds. J. Maruani and J. Serre, Studies in Physical and Theoretical Chemistry, Vol. 23, Vol. 23 pp. 355–378, (Elsevier Publishing Co., Amsterdam, 1983).

69. A. Beil, D. Luckhaus, M. Quack and J. Stohner, Intramolecular vibrational redistribution and unimolecular reaction: Concepts and new results on the femtosecond dynamics and statistics in CHBrClF, *Ber. Bunsenges. Phys. Chem.* **101** 311–328 (1997).

70. D. Luckhaus and M. Quack, The role of potential anisotropy in the dynamics of the CH-chromophore in CHX_3 (C_{3v}) symmetrical tops, *Chem. Phys. Lett.* **205** 277–284 (1993).

71. M. Born, W. Heisenberg and P. Jordan, Zur Quantenmechanik. II., *Z. Physik* **35** 557–615 (1926).

72. E. Wigner, Über die Erhaltungssätze in der Quantenmechanik, *Nachr. Ges. Wiss. Gött.* 375–381 (1927).

73. W. Pauli, Die Verletzung von Spiegelungs-Symmetrien in den Gesetzen der Atomphysik, *Experientia* **14** 1–5 (1958).

74. E. R. Cohen, T. Cvitas, J. G. Frey, B. Holmström, K. Kuchitsu, R. Marquardt, I. Mills, F. Pavese, M. Quack, J. Stohner, H. L. Strauss, M. Takami and A. J. Thor, Quantities, Units and Symbols in Physical Chemistry, 3rd edn. (IUPAC and Royal Society of Chemistry, RSC Publishing, Cambridge, 2007).

75. M. Quack, Molecules in Motion, *Chimia* **55** 753–758 (2001).

76. M. Quack, Spectra and dynamics of coupled vibrations in polyatomic molecules, *Ann. Rev. Phys. Chem.* **41** 839–874 (1990).

77. H. R. Dübal and M. Quack, Tridiagonal Fermi Resonance Structure in the IR-Spectrum of the Excited CH Chromophore in CF_3H, *J. Chem. Phys.* **81** 3779–3791 (1984).

78. R. Marquardt, M. Quack, J. Stohner and E. Sutcliffe, Quantum-Mechanical Wavepacket Dynamics of the CH Group in Symmetric Top X_3CH Compounds Using Effective-Hamiltonians from High-Resolution Spectroscopy, *J. Chem. Soc. Farad. Trans. 2* **82** 1173–1187 (1986).

79. R. Marquardt and M. Quack, The Wave Packet Motion and Intramolecular Vibrational Redistribution in CHX_3 Molecules under Infrared Multiphoton Excitation, *J. Chem. Phys.* **95** 4854–4876 (1991).

80. M. Quack and J. Stohner, Femtosecond quantum dynamics of functional-groups under coherent infrared multiphoton excitation as derived from the analysis of high-resolution spectra, *J. Phys. Chem.* **97** 12574–12590 (1993).

81. A. Beil, D. Luckhaus and M. Quack, Fermi resonance structure and femtosecond quantum dynamics of a chiral molecule from the analysis of vibrational overtone spectra of CHBrClF, *Ber. Bunsenges. Phys. Chem.* **100** 1853–1875 (1996).

82. A. Beil, H. Hollenstein, O. L. A. Monti, M. Quack and J. Stohner, Vibrational spectra and intramolecular vibrational redistribution in highly excited deuterobromochlorofluoromethane CDBrClF: Experiment and theory, *J. Chem. Phys.* **113** 2701–2718 (2000).

83. J. Pochert, M. Quack, J. Stohner and M. Willeke, Ab initio calculation and spectroscopic analysis of the intramolecular vibrational redistribution in 1,1,1,2-tetrafluoroiodoethane CF_3CHFI, *J. Chem. Phys.* **113** 2719–2735 (2000).

84. Y. B. He, H. Hollenstein, M. Quack, E. Richard, M. Snels and H. Bürger, High resolution analysis of the complex symmetric CF_3 stretching chromophore absorption in CF_3I, *J. Chem. Phys.* **116** 974–983 (2002).

85. S. Albert, K. Keppler Albert, H. Hollenstein, C. Manca Tanner and M. Quack, Fundamentals of Rotation-Vibration Spectra, in Handbook of High-Resolution Spectroscopy; Fundamentals of Rotation-Vibration Spectra, eds. M. Quack and F. Merkt ch. 3, pp. 117–173, (Wiley, Chichester, New York, 2011).

86. S. Albert, E. Bekhtereva, I. Bolotova, Z. Chen, C. Fábri, H. Hollenstein, M. Quack and O. Ulenikov, Isotope effects on the resonance interactions and vibrational quantum dynamics of fluoroform $^{12,13}CHF_3$, *Phys. Chem. Chem. Phys.* **19** 26527–26534 (2017).

87. M. Quack, Molecular spectra, reaction dynamics, symmetries and life, *Chimia* **57** 147–160 (2003).

88. K. von Puttkamer, H. R. Dübal and M. Quack, Time-Dependent Processes in Polyatomic-Molecules During and after Intense Infrared Irradiation, *Farad. Disc.* **75** 197–210 (1983).

89. M. Quack and M. A. Suhm, Potential Energy Surfaces, Quasi-Adiabatic Channels, Rovibrational Spectra, and Intramolecular Dynamics of $(HF)_2$ and Its Isotopomers from Quantum Monte Carlo Calculations, *J. Chem. Phys.* **95** 28–59 (1991).

90. K. von Puttkamer and M. Quack, Vibrational spectra of $(HF)_2$, $(HF)_n$ and their D-isotopomers - mode selective rearrangements and nonstatistical unimolecular decay, *Chem. Phys.* **139** 31–53 (1989).

91. M. Hippler, L. Oeltjen and M. Quack, High-resolution continuous-wave-diode laser cavity ring-down spectroscopy of the hydrogen fluoride dimer in a pulsed slit jet expansion: Two components of the N = 2 triad near 1.3 μm, *J. Phys. Chem. A* **111** 12659–12668 (2007).

92. A. Kushnarenko, E. Miloglyadov, M. Quack and G. Seyfang, Intramolecular vibrational energy redistribution in $HCCCH_2X$ (X = Cl, Br, I) measured by femtosecond pump-probe experiments in a hollow waveguide, *Phys. Chem. Chem. Phys.* **20** 10949–10959 (2018).

93. G. Seyfang and M. Quack, Atomare und molekulare Tunnelprozesse, *Nachrichten aus der Chemie,* **66** 307–315 (2018).

94. B. Fehrensen, D. Luckhaus and M. Quack, Mode selective stereomutation tunnelling in hydrogen peroxide isotopomers, *Chem. Phys. Lett.* **300** 312–320 (1999).

95. B. Fehrensen, D. Luckhaus and M. Quack, Stereomutation dynamics in hydrogen peroxide, *Chem. Phys.* **338** 90–105 (2007).

96. C. Fábri, R. Marquardt, A. Császár and M. Quack, Controlling tunneling in ammonia isotopomers, *J. Chem. Phys.* **150**, p. 014102 (2019).

97. R. Berger, M. Gottselig, M. Quack and M. Willeke, Parity violation dominates the dynamics of chirality in dichlorodisulfane, *Angew. Chem. Int. Ed.* **40** 4195–4198 (2001), *Angew. Chem.* **113** 4342–4345 (2001).

98. P. L. Chapovsky and L. J. F. Hermans, Nuclear Spin Conversion in Polyatomic Molecules, *Ann. Rev. Phys. Chem.* **50** 315–345 (1999).

99. G. Wichmann, G. Seyfang and M. Quack, Time-dependent dynamics of nuclear spin symmetry and parity violation in dichlorodisulfane (ClSSCl) during and after coherent radiative excitation, *Mol. Phys.* **119**, p. e1959073 (29 pages) (2021), with supplementary material (44 pages).

100. M. Quack, On the measurement of the parity violating energy difference between enantiomers, *Chem. Phys. Lett.* **132** 147–153 (1986).

101. A. Bakasov, T. K. Ha and M. Quack, Ab initio calculation of molecular energies including parity violating interactions, *J. Chem. Phys.* **109** 7263–7285 (1998).

102. R. Berger and M. Quack, Multiconfiguration linear response approach to the calculation of parity violating potentials in polyatomic molecules, *J. Chem. Phys.* **112** 3148–3158 (2000).

103. R. Prentner, M. Quack, J. Stohner and M. Willeke, Wavepacket Dynamics of the Axially Chiral Molecule Cl-O-O-Cl under Coherent Radiative Excitation and Including Electroweak Parity Violation, *J. Phys. Chem. A* **119** 12805–12822 (2015).

104. P. Dietiker, E. Miloglyadov, M. Quack, A. Schneider and G. Seyfang, Infrared laser induced population transfer and parity selection in $^{14}NH_3$: A proof of principle experiment towards detecting parity violation in chiral molecules, *J. Chem. Phys.* **143**, p. 244305 (2015).

105. E. Miloglyadov, M. Quack, G. Seyfang and G. Wichmann, Precision experiments for parity violation in chiral molecules: the role of STIRAP, *J. Phys. B: At. Mol. Opt. Phys.* **52** ch. A2.3, 11–13, (51–52) (2019), in 'Roadmap on STIRAP applications' by Bergmann, K., Nägerl, H. C., Panda, C., Gabrielse, G., Miloglyadov, E., Quack, M., Seyfang, G., Wichmann, G., Ospelkaus, S., Kuhn, A., Longhi, S., Szameit, A., Pirro, P., Hillebrands, B., Zhu, X. F., Zhu, J., Drewsen, M., Hensinger, W. K., Weidt, S., Halfmann, T., Wang, H. L., Paraoanu, G. S., Vitanov, N. V., Mompart, J., Busch, T., Barnum, T. J., Grimes, D. D., Field, R. W., Raizen, M. G., Narevicius, E., Auzinsh, M., Budker, D., Pálffy, A. and Keitel, C. H., *J. Phys. B: At. Mol. Opt. Phys.* **52** (2019), 202001 (55 pages).

106. M. Quack, Telluride public lecture (1995), unpublished, in part contained in Ref. 43.

107. M. Quack, On the measurement of CP-violating energy differences in matter-antimatter enantiomers, *Chem. Phys. Lett.* **231** 4–6 (1994).

108. M. Quack, Comments on intramolecular dynamics and femtosecond kinetics, Proc. 20th Solvay Conference "Chemical reactions and their control on the femtosecond time scale", *Adv. Chem. Phys.* **101** 83–84, 92–93, 202, 277–278, 282, 373–388, 443, 453–456, 459, 586–591, 595 (1997).

109. D. Luckhaus, M. Quack and J. Stohner, Femtosecond quantum structure, equilibration and time-reversal for the CH-chromophore dynamics in CHD_2F, *Chem. Phys. Lett.* **212** 434–443 (1993).

110. A. Einstein, Grundzüge der Relativitätsteherie (Vieweg, Wiesbaden, 1922).
111. M. Gardner and J. Mackey, The Ambidextrous Universe: Mirror Asymmetry and Time-Reversed Worlds (Basic Books, New York, 1964).
112. M. Quack, J. Stohner and M. Willeke, High-resolution spectroscopic studies and theory of parity violation in chiral molecules, *Ann. Rev. Phys. Chem.* **59** 741–769 (2008).
113. F. Hund, Zur Deutung der Molekelspektren II., *Z. Physik* **42** 93–120 (1927).
114. F. Hund, Zur Deutung der Molekelspektren III. Bemerkungen über das Schwingungs- und Rotationsspektrum bei Molekeln mit mehr als zwei Kernen., *Z. Physik* **43** 805–826 (1927).
115. H. Brunner, Rechts oder links in der Natur und anderswo (Wiley-VCH, Weinheim, 1999).
116. H. Brunner, Bild und Spiegelbild (GNT Verlag, Berlin, 2021).
117. M. Quack and G. Seyfang, Atomic and Molecular Tunneling Processes in Chemistry, in Molecular Spectroscopy and Quantum Dynamics, eds. R. Marquardt and M. Quack ch. 7, pp. 231–282, (Elsevier, Amsterdam, 1st edition, 2020). ISBN 978-012-817234-6.
118. R. Janoschek, Theories on the origin of biomolecular homochirality. In: Janoschek, R. (Ed.): Chirality - From Weak Bosons to the α-Helix. Chapt. 2, pp. 18-33. Berlin: Springer (1991).
119. R. Marquardt and M. Quack, Global Analytical Potential Energy Surfaces for High Resolution Molecular Spectroscopy and Reaction Dynamics, in Handbook of High-Resolution Spectroscopy, eds. M. Quack and F. Merkt ch. 12, pp. 511–549, (Wiley, Chichester, New York, 2011).
120. B. Kuhn, T. R. Rizzo, D. Luckhaus, M. Quack and M. A. Suhm, A new six-dimensional analytical potential up to chemically significant energies for the electronic ground state of hydrogen peroxide, *J.Chem. Phys.* **111** 2565–2587 (1999), 135 pages of supplementary material published as AIP Document No PAPS JCPS A6-111-302905 by American Institute of Physics, Physics Auxiliary Publication Service, 500 Sunnyside, Blvd., Woodbury, N.Y. 1179-29999.
121. R. Marquardt, K. Sagui, W. Klopper and M. Quack, Global analytical potential energy surface for large amplitude nuclear motions in ammonia, *J. Phys. Chem. B* **109** 8439–8451 (2005).
122. R. Marquardt, K. Sagui, J. Zheng, W. Thiel, D. Luckhaus, S. Yurchenko, F. Mariotti and M. Quack, A Global Analytical Potential Energy Surface for the Electronic Ground State of NH_3 from High Level Ab Initio Calculations, *J. Phys. Chem. A* **117** 7502–7522 (2013).
123. W. Heisenberg, Über quantentheoretische Umdeutung kinematischer und mechanischer Beziehungen, *Z. Physik* **33** 879–893 (1925).
124. P. A. M. Dirac, Quantum mechanics of many-electron systems, *Proc. Roy. Soc. London Series A* **123** 714–733 (1929).
125. W. Heitler and F. London, Wechselwirkung neutraler Atome und homöopolare Bindung nach der Quantenmechanik, *Z. Physik* **44** 455–472 (1927).
126. K. Ruedenberg, The Physical Nature of the Chemical Bond, *Rev. Mod. Phys.* **34** 326–376 (1962).
127. T. Helgaker, P. Jorgensen and J. Olsen, Molecular Electronic-Structure Theory (Wiley, 2014).
128. Schaefer, H. F. III., Quantum Chemistry, The development of ab initio methods in molecular electronic structure theory (Oxford University Press, Oxford, 1984).
129. L. Pauling, The Nature of the Chemical Bond; An Introduction to Modern Structural Chemistry, Third edn. (Cornell University Press Ithaca, New York, 1960).

130. G. Frenking and S. Shaik, The Chemical Bond: Fundamental Aspects of Chemical Bonding (Wiley-VCH, Weinheim, 2014).

131. M. Reiher and A. Wolf, Relativistic Quantum Chemistry, The Fundamental Theory of Molecular Science, 1st edn. (Wiley VCH Weinheim, 2009).

132. P. Pyykkö, The Physics behind Chemistry and the Periodic Table, *Chem. Rev.* **112** 371–384 (2012).

133. F. Neese, High-Level Spectroscopy, Quantum Chemistry, and Catalysis: Not just a Passing Fad, *Angew. Chem. Int. Ed.* **56** 11003–11010 (2017).

134. E. Fermi, Versuch einer Theorie der Beta-Strahlen, *Z. Physik* **88** 161–177 (1934).

135. M. Quack, Error and Discovery: Why Repeating Can Be New, *Angew. Chem. Int. Edit.* **52** 9362–9370 (2013), Irrtum und Erkenntnis: Wenn Wiederholen neu ist, Angew. Chem. 2013, 125, 9530-9538.

136. T. D. Lee, Weak Interactions and Nonconservation of Parity, *Science* **127** 569–573 (1958).

137. C. N. Yang, Law of Parity Conservation and Other Symmetry Laws, *Science* **127** 565–569 (1958).

138. CERN, AC_Z04_V25/B/1992, *Website, https://cds.cern.ch/record/39722.*

139. H. Fritzsch, The Fundamental Constants: A Mystery of Physics (World Scientific, 2005).

140. R. P. Feynman, The Theory of Positrons, *Phys. Rev.* **76** 749–759 (1949).

141. R. P. Feynman, Space-Time Approach to Quantum Electrodynamics, *Phys. Rev.* **76** 769–789 (1949).

142. E. C. G. Stueckelberg, Relativistisch invariante Störungstheorie des Diracschen Elektrons I. Teil: Streustrahlung und Bremsstrahlung, *Ann. d. Physik* **413** 367–389 (1934).

143. E. Stueckelberg, Die Wechselwirkungskräfte in der Elektrodynamik und in der Feldtheorie der Kernkräfte. Teil II und III, *Helv. Phys. Acta* **11** 51–80 (1938).

144. E. Stueckelberg, Un nouveau modèle de l'électron ponctuel en théorie classique, *Helv. Phys. Acta* **14** 51–80 (1941).

145. E. Stueckelberg, La signification du temps propre en mécanique ondulatoire, *Helv. Phys. Acta* **14** 322–323 (1941).

146. E. Stueckelberg, La mécanique du point matériel en théorie de relativité et en théorie des quanta, *Helv. Phys. Acta* **15** 23–37 (1942).

147. E. Stueckelberg and A. Petermann, La normalisation des constantes dans la théorie des quanta, *Helv. Phys. Acta* **26** 499–520 (1953).

148. M. Veltman, Diagrammatica: The Path to Feynman Rules (Cambridge University Press, Cambridge, 1994), (reprinted 1995).

149. A. Bakasov, T. K. Ha and M. Quack, Ab initio calculation of molecular energies including parity violating interactions, in Chemical Evolution, Physics of the Origin and Evolution of Life, Proc. of the 4th Trieste Conference (1995), eds. J. Chela-Flores and F. Raulin pp. 287–296, (Kluwer Academic Publishers, Dordrecht, 1996).

150. M. Quack and J. Stohner, Parity violation in chiral molecules, *Chimia* **59** 530–538 (2005), (Erratum for printer's errors: Chimia, **59**, 712-712 (2005)).

151. R. Berger, Parity-violation effects in molecules, in Relativistic Electronic Structure Theory. Part 2 (Applications), ed. P. Schwerdtfeger (Elsevier, Amsterdam, 2004).

152. L. Horný and M. Quack, Computation of molecular parity violation using the coupled-cluster linear response approach, *Mol. Phys.* **113** 1768–1779 (2015).

153. Y. Yamagata, A hypothesis for the asymmetric appearance of biomolecules on earth, *J. Theor. Bio.* **11** 495–498 (1966).

154. A. S. Garay and P. Hraskó, Neutral currents in weak interactions and molecular asymmetry, *J. Mol. Evo.* **6** 77–89 (1975).

155. R. A. Harris and L. Stodolsky, Quantum beats in optical activity and weak interactions, *Phys. Lett. B* **78** 313–317 (1978).
156. M. A. Bouchiat and C. Bouchiat, Parity violation induced by weak neutral currents in atomic physics 1., *J. Physique* **35** 899–927 (1974).
157. M. A. Bouchiat and C. Bouchiat, Parity violation induced by weak neutral currents in atomic physics 2., *J. Physique* **36** 493–509 (1975).
158. R. Hegström, D. W. Rein and P. G. H. Sandars, Calculation of parity non-conserving energy difference between mirror-image molecules, *J. Chem. Phys.* **73** 2329–2341 (1980).
159. S. F. Mason and G. E. Tranter, The parity-violating energy difference between enantiomeric molecules, *Mol. Phys.* **53** 1091–1111 (1984).
160. O. Kikuchi and H. Wang, Parity-violating energy shift of glycine, alanine, and serine in the zwitterionic forms, *Bull. Chem. Soc. Jpn.* **63** 2751–2754 (1990).
161. L. Wiesenfeld, Effect of atomic number on parity-violating energy differences between enantiomers, *Mol. Phys.* **64** 739–745 (1988).
162. M. Quack, Fundamental Symmetry Principles in Relation to the Physical-Chemical Foundations of Molecular Chirality and Possible Biological Consequences, *Lecture Notes in 'Seventh College on Biophysics' (Structure and Function of Biopolymers, Experimental and Theoretical Techniques) 4-29 March 1996, UNESCO and International Centre for Theoretical Physics* (1996), preprint H4.SMR/916-27, distributed and printed by ICTCP Trieste, Italy, March 1996.
163. P. Lazzeretti and R. Zanasi, On the calculation of parity-violating energies in hydrogen peroxide and hydrogen disulphide molecules within the random-phase approximation, *Chem. Phys. Lett.* **279** 349–354 (1997).
164. J. K. Laerdahl and P. Schwerdtfeger, Fully relativistic ab initio calculations of the energies of chiral molecules including parity-violating weak interactions, *Phys. Rev. A* **60** 4439–4453 (1999).
165. A. C. Hennum, T. Helgaker and W. Klopper, Parity-violating interaction in H_2O_2 calculated from density-functional theory, *Chem. Phys. Lett.* **354** 274–282 (2002).
166. P. Schwerdtfeger, T. Saue, J. N. P. van Stralen and L. Visscher, Relativistic second-order many-body and density-functional theory for the parity-violation contribution to the C-F stretching mode in CHFClBr, *Phys. Rev. A* **71**, p. 012103 (2005).
167. J. K. Laerdahl, P. Schwerdtfeger and H. M. Quiney, Theoretical Analysis of Parity-Violating Energy Differences between the Enantiomers of Chiral Molecules, *Phys. Rev. Lett.* **84** 3811–3814 (2000).
168. R. Berger and J. Stohner, Parity violation, *WIREs Comp. Mol. Sci.* **9**, p. e1396 (2019).
169. D. Andrae, M. Reiher and J. Hinze, A comparative study of finite nucleus models for low-lying states of few-electron high-Z atoms, *Chem. Phys. Lett.* **320** 457–468 (2000).
170. R. Conti, P. Bucksbaum, S. Chu, E. Commins and L. Hunter, Preliminary Observation of Parity Nonconservation in Atomic Thallium, *Phys. Rev. Lett.* **42** 343–346 (1979).
171. S. C. Bennett and C. E. Wieman, Measurement of the $6S \rightarrow 7S$ Transition Polarizability in Atomic Cesium and an Improved Test of the Standard Model, *Phys. Rev. Lett.* **82** 2484–2487 (1999).
172. V. M. Shabaev, K. Pachucki, I. I. Tupitsyn and V. A. Yerokhin, QED Corrections to the Parity-Nonconserving $6s-7s$ Amplitude in ^{133}Cs, *Phys. Rev. Lett.* **94**, p. 213002 (2005).

173. K. Tsigutkin, D. Dounas-Frazer, A. Family, J. E. Stalnaker, V. V. Yashchuk and D. Budker, Observation of a Large Atomic Parity Violation Effect in Ytterbium, *Phys. Rev. Lett.* **103**, p. 071601 (2009).

174. M. Bouchiat, J. Guena, L. Pottier and L. Hunter, New observation of a parity violation in cesium, *Phys. Lett. B* **134** 463–468 (1984).

175. D. N. Stacey, Experiments on the Electro-Weak Interaction in Atoms, *Physica Scripta* **T40** 15–22 (1992).

176. I. B. Khriplovich, Parity nonconservation in atomic phenomena (Gordon and Breach Science Publishers, Philadelphia, 1991), (translated from the Russian by L. Ya. Yuzina.).

177. R. R. Ernst, G. Bodenhausen and A. Wokaun, Principles of nuclear magnetic resonance in one and two dimensions (Clarendon Press, Oxford, 1987).

178. A. L. Barra, J. B. Robert and L. Wiesenfeld, Parity non-conservation and NMR observables. calculation of Tl resonance frequency differences in enantiomers, *Phys. Lett. A* **115** 443–447 (1986).

179. I. I. Rabi, Space Quantization in a Gyrating Magnetic Field, *Phys. Rev.* **51** 652–654 (1937).

180. I. I. Rabi, J. R. Zacharias, S. Millman and P. Kusch, A New Method of Measuring Nuclear Magnetic Moment, *Phys. Rev.* **53** 318–318 (1938).

181. R. Berger, G. Laubender, M. Quack, A. Sieben, J. Stohner and M. Willeke, Isotopic chirality and molecular parity violation, *Angew. Chem. Int. Ed.* **44** 3623–3626 (2005), Angew. Chem., 117, 3689-3693 (2005).

182. A. Bakasov, R. Berger, T. K. Ha and M. Quack, Ab initio calculation of parity-violating potential energy hypersurfaces of chiral molecules, *Int. J. Quant. Chem.* **99** 393–407 (2004).

183. Y. B. Zel'dovich, Parity nonconservation in the first order in the weak-interaction constant in electron scattering and other effects, *Soviet physics, JETP* **9** 682–683 (1959).

184. Y. B. Zel'dovich, D. B. Saakyan and I. I. Sobel'man, Energy difference between right-hand and left-hand molecules, due to parity nonconservation in weak interactions of electrons with nuclei, *Soviet physics, JETP Lett.* **25** 94–97 (1977), (Pis'ma Zh. Eksp. Teor. Fiz. **25** No. 2 106-109 (1977)).

185. R. Prentner, M. Quack, J. Stohner and M. Willeke, On tunneling and parity violation in ClOOCl, *Farad. Discuss. Chem. Soc.* **150** 130–132 (2011).

186. M. Quack and M. Willeke, Stereomutation tunneling switching dynamics and parity violation in chlorineperoxide Cl-O-O-Cl, *J. Phys. Chem. A* **110** 3338–3348 (2006).

187. N. Sahu, J. O. Richardson and R. Berger, Instanton calculations of tunneling splittings in chiral molecules, *J. Comput. Chem.* **42** 210–221 (2021).

188. C. Daussy, T. Marrel, A. Amy-Klein, C. T. Nguyen, C. J. Bordé and C. Chardonnet, Limit on the Parity Nonconserving Energy Difference between the Enantiomers of a Chiral Molecule by Laser Spectroscopy, *Phys. Rev. Lett.* **83** 1554–1557 (1999).

189. S. K. Tokunaga, C. Stoeffler, F. Auguste, A. Shelkovnikov, C. Daussy, A. Amy-Klein, C. Chardonnet and B. Darquié, Probing weak force-induced parity violation by high-resolution mid-infrared molecular spectroscopy, *Mol. Phys.* **111** 2363–2373 (2013).

190. A. Cournol, M. Manceau, M. Pierens, L. Lecordier, D. B. A. Tran, R. Santagata, B. Argence, A. Goncharov, O. Lopez, M. Abgrall, Y. Le Coq, R. Le Targat, H. Alvarez Martinez, W. K. Lee, D. Xu, P. E. Pottie, R. J. Hendricks, T. E. Wall, J. M. Bieniewska, B. E. Sauer, M. R. Tarbutt, A. Amy-Klein, S. K. Tokunaga and B. Darquié, A new experiment to test parity symmetry in cold chiral molecules using vibrational spectroscopy, *Quant. Electr.* **49** 288–292 (2019).

191. M. Schnell and J. Küpper, Tailored molecular samples for precision spectroscopy experiments, *Farad. Disc.* **150** 33–49 (2011).

192. V. S. Letokhov, On difference of energy levels of left and right molecules due to weak interactions, *Phys. Lett. A* **53** 275–276 (1975).

193. E. Arimondo, P. Glorieux and T. Oka, Observation of inverted infrared Lamb dips in separated optical isomers, *Opt. Commun.* **23** 369–372 (1977).

194. A. Beil, D. Luckhaus, R. Marquardt and M. Quack, Intramolecular energy-transfer and vibrational redistribution in chiral molecules - experiment and theory, *Farad. Disc.* **99** 49–76 (1994).

195. A. Bauder, A. Beil, D. Luckhaus, F. Müller and M. Quack, Combined high resolution infrared and microwave study of bromochlorofluoromethane, *J. Chem. Phys.* **106** 7558–7570 (1997).

196. H. Hollenstein, D. Luckhaus, J. Pochert, M. Quack and G. Seyfang, Synthesis, structure, high-resolution spectroscopy, and laser chemistry of fluorooxirane and 2,2-[^{2}H$_2$] fluorooxirane, *Angew. Chem. Int. Ed.* **36** 140–143 (1997), Angew. Chem. 109, 136-40 (1997).

197. H. Gross, G. Grassi and M. Quack, The synthesis of [2-^{2}H$_1$]thiirane-1-oxide and [2,2-^{2}H$_2$]thiirane-1-oxide and the diastereoselective infrared laser chemistry of [2-^{2}H$_1$]thiirane-1-oxide, *Chem. Europ. J.* **4** 441–448 (1998).

198. S. Albert, K. K. Albert and M. Quack, Very-high-resolution studies of chiral molecules with a Bruker IFS 120 HR: The rovibrational spectrum of CDBrClF in the range 600 - 2300 cm^{-1} (Optical Society of America, Washington DC, 2003).

199. S. Albert and M. Quack, High resolution rovibrational spectroscopy of chiral and aromatic compounds, *ChemPhysChem* **8** 1271–1281 (2007).

200. O. N. Kompanets, A. R. Kukudzhanov, V. S. Letokhov and L. L. Gervits, Narrow resonances of saturated absorption of asymmetrical molecule CHFClBr and possibility of weak current detection in molecular physics, *Opt. Commun.* **19** 414–416 (1976).

201. J. Stohner and M. Quack, Recent results on parity violation in chiral molecules: Camphor and the influence of molecular parity violation, in Proceedings, 15th Symposium on Atomic and Surface Physics and Related Topics, Obergurgl, Österreich, 4. - 9.2.2006, eds. V. Grill and T. D. Märk pp. 196–199, (Innsbruck University Press, Innsbruck, 2006).

202. G. Wichmann, E. Miloglyadov, G. Seyfang and M. Quack, Nuclear spin symmetry conservation studied by cavity ring-down spectroscopy of ammonia in a seeded supersonic jet from a pulsed slit nozzle, *Mol. Phys.* **118**, p. e1752946 (2020).

203. E. Schrödinger, Die gegenwärtige Situation in der Quantenmechanik, *Naturwissenschaften* **23** 807–812, 823–828, 844–849 (1935).

204. M. Quack, On the fundamental and approximate symmetries in molecular quantum dynamics and the preparation of exotic superposition isomers, *Proc. Symp. MOLIM 2018 - Molecules in Motion Athens*, Merkt, F. and Quack, M. and Thanopulos, I. and Vayenas, C. G. eds., (2018), p.36.

205. M. Quack and F. Merkt, Handbook of High Resolution Spectroscopy (Wiley, Chichester, New York, 2011), (with preface of the editors and eight review articles from our group, as well as numerous articles by groups worldwide).

206. S. Albert, K. Keppler Albert and M. Quack, High Resolution Fourier Transform Infrared Spectroscopy, in Handbook of High Resolution Spectroscopy, eds. M. Quack and F. Merkt ch. 26, pp. 965–1019, (Wiley, Chichester, New York, 2011).

207. M. Hippler, E. Miloglyadov, M. Quack and G. Seyfang, Mass and Isotope Selective Infrared Spectroscopy, in Handbook of High Resolution Spectroscopy, eds. M. Quack and F. Merkt ch. 28, pp. 1069–1118, (Wiley, Chichester; New York, 2011).

208. M. Snels, V. Horká-Zelenková, H. Hollenstein and M. Quack, High Resolution FTIR and Diode Laser Spectroscopy of Supersonic Jets, in Handbook of High Resolution Spectroscopy, eds. M. Quack and F. Merkt ch. 27, pp. 1021–1067, (Wiley, Chichester, New York, 2011).

209. S. Albert, K. Keppler Albert, P. Lerch and M. Quack, Synchrotron-based highest resolution Fourier transform infrared spectroscopy of naphthalene ($C_{10}H_8$) and indole (C_8H_7N) and application to astrophysical problems, *Farad. Disc.* **150** 71–99 (2011).

210. I. B. Bolotova, O. N. Ulenikov, E. S. Bekhtereva, S. Albert, Z. Chen, H. Hollenstein, D. Zindel and M. Quack, High resolution Fourier transform infrared spectroscopy of the ground state, ν_3, 2 ν_3 and ν_4 levels of $^{13}CHF_3$, *J. Mol. Spectr.* **337** 96–104 (2017).

211. S. Albert, I. Bolotova, Z. Chen, C. Fábri, L. Horný, M. Quack, G. Seyfang and D. Zindel, High resolution GHz and THz (FTIR) spectroscopy and theory of parity violation and tunneling for 1,2-dithiine ($C_4H_4S_2$) as a candidate for measuring the parity violating energy difference between enantiomers of chiral molecules, *Phys. Chem. Chem. Phys.* **18** 21976–21993 (2016).

212. S. Albert, F. Arn, I. Bolotova, Z. Chen, C. Fábri, G. Grassi, P. Lerch, M. Quack, G. Seyfang, A. Wokaun and D. Zindel, Synchrotron-Based Highest Resolution Terahertz Spectroscopy of the ν_{24} Band System of 1,2-Dithiine ($C_4H_4S_2$): A Candidate for Measuring the Parity Violating Energy Difference between Enantiomers of Chiral Molecules, *J. Phys. Chem. Lett.* **7** 3847–3853 (2016).

213. S. Albert, I. Bolotova, Z. Chen, C. Fabri, M. Quack, G. Seyfang and D. Zindel, High-resolution FTIR spectroscopy of trisulfane HSSSH: a candidate for detecting parity violation in chiral molecules, *Phys. Chem. Chem. Phys.* **19** 11738–11743 (2017).

214. C. Fábri, L. Horný and M. Quack, Tunneling and Parity Violation in Trisulfane (HSSSH): An Almost Ideal Molecule for Detecting Parity Violation in Chiral Molecules, *ChemPhysChem* **16** 3584–3589 (2015).

215. M. Snels and M. Quack, High-resolution Fourier-transform infrared-spectroscopy of $CHCl_2F$ in supersonic jets - Analysis of ν_3, ν_7, and ν_8, *J. Chem. Phys.* **95** 6355–6361 (1991).

216. S. Albert, K. K. Albert and M. Quack, Rovibrational analysis of the ν_4 and $\nu_5 + \nu_9$ bands of $CHCl_2F$, *J. Mol. Struct.* **695** 385–394 (2004).

217. S. Albert, S. Bauerecker, M. Quack and A. Steinlin, Rovibrational analysis of the $2\nu_3$, $3\nu_3$ and ν_1 bands of $CHCl_2F$ measured at 170 and 298 K by high-resolution FTIR spectroscopy, *Mol. Phys.* **105** 541 – 558 (2007).

218. S. Albert, K. Keppler, V. Boudon, P. Lerch and M. Quack, Combined Synchrotron-based high resolution FTIR and IR - diode laser supersonic jet spectroscopy of the chiral molecule CDBrClF, *J. Mol. Spectr.* **337** 105–123 (2017).

219. P. Schwerdtfeger, J. K. Laerdahl and C. Chardonnet, Calculation of parity-violation effects for the C-F stretching mode of chiral methyl fluorides, *Phys. Rev. A* **65**, p. 042508 (2002).

220. P. Soulard, P. Asselin, A. Cuisset, J. R. Aviles Moreno, T. R. Huet, D. Petitprez, J. Demaison, T. B. Freedman, X. Cao, L. A. Nafie and J. Crassous, Chlorofluoroiodomethane as a potential candidate for parity violation measurements, *Phys. Chem. Chem. Phys.* **8** 79–92 (2006).

221. S. Albert, K. K. Albert, S. Bauerecker and M. Quack, CHBrIF and molecular parity violation: First high resolution rovibrational analysis of the CF-stretching mode, in Proc. 16th SASP 2008, eds. R. D. Beck, M. Drabbels and T. R. Rizzo pp. 79–82, (Innsbruck University Press (IUP), Innsbruck, 2008).

222. R. Berger, M. Quack, A. Sieben and M. Willeke, Parity-violating potentials for the torsional motion of methanol (CH_3OH) and its isotopomers CD_3OH, $^{13}CH_3OH$, CH_3OD, CH_3OT, CHD_2OH, and CHDTOH, *Helv. Chim. Acta* **86** 4048–4060 (2003).

223. M. Gottselig and M. Quack, Steps towards molecular parity violation in axially chiral molecules. I.Theory for allene and 1,3-difluoroallene, *J. Chem. Phys.* **123** 84305-1 – 84305-11 (2005).

224. L. Horný and M. Quack, On coupled cluster calculations of parity violating potentials in chiral molecules (Discussion contribution), *Farad. Disc.* **150** 152–154 (2011).

225. S. Albert, Z. Chen, K. Keppler, M. Quack, V. Schurig and O. Trapp, The Gigahertz and Terahertz spectrum of monodeuterooxirane (c-C_2H_3DO), *Phys. Chem .Chem. Phys.* **21** 3669–3675 (2019).

226. Z. Chen, S. Albert, K. Keppler, P. Lerch, M. Quack, V. Schurig and O. Trapp, High resolution Gigahertz- and Terahertz Spectroscopy of the isotopically chiral molecule trans 2,3 di deutero oxirane C-CHDCHDO, *Proc. International Symp. on Molecular Spectroscopy*, Urbana, Ill. 21-25 June (2021) and *Proc. MOLIM 2018 Molecules in Motion Int. Workshop on Molecular Quantum Dynamics and Kinetics*, Athens, Greece 8-10 October (2018).

227. F. Hobi, R. Berger and J. Stohner, Investigation of parity violation in nuclear spin-rotation interaction of fluorooxirane, *Mol. Phys.* **111** 2345–2362 (2013).

228. R. Berger, M. Quack and J. Stohner, Parity violation in fluorooxirane, *Angew. Chem. Int. Edit.* **40** 1667–1670 (2001).

229. H. Gross, Y. He, C. Jeitziner, M. Quack and G. Seyfang, Vibrational IR-multiphoton excitation of thiirane-1-oxide (C_2H_4SO) and d-thiirane-1-oxide (C_2H_3DSO), *Ber. Bunsenges. Phys. Chem.* **99** 358–365 (1995).

230. R. Berger, M. Quack and G. S. Tschumper, Electroweak quantum chemistry for possible precursor molecules in the evolution of biomolecular homochirality, *Helv. Chim. Acta* **83** 1919–1950 (2000).

231. S. Albert, P. Lerch, K. Keppler and M. Quack, THz Spectroscopy of cyano-oxirane (c-C_2H_3OCN) and methyl oxirane (c-$C_2H_3OCH_3$) with synchrotron light, J. Stohner and C. Yeretzian (eds.) (Innsbruck University Press (IUP), Innsbruck, 2016).

232. K. Keppler, S. Albert, C. Manca Tanner, M. Quack and J. Stohner, Paper F04 in *Proc. 27th Coll. High. Res. Mol. Spectr.*, Cologne, Germany 29 August-3 September 2021 paper and Fall meeting of the swiss Chemical Society September 2021, *Chimia* **75(7/8)**, p. 670 (2021), (2022) Suppl. Paper PC-118 (ISSN 009-4293).

233. R. Berger and M. Quack, Electroweak quantum chemistry of alanine: Parity violation in gas and condensed phases, *ChemPhysChem* **1** 57–60 (2000).

234. S. Albert, P. Lerch, R. Prentner and M. Quack, Tunneling and Tunneling Switching Dynamics in Phenol and Its Isotopomers from High-Resolution FTIR Spectroscopy with Synchrotron Radiation, *Angew. Chem. Int. Edit.* **52** 346–349 (2013).

235. S. Albert, Z. Chen, C. Fábri, P. Lerch, R. Prentner and M. Quack, A combined Gigahertz and Terahertz (FTIR) spectroscopic investigation of meta-D-phenol: observation of tunnelling switching, *Mol. Phys.* **114** 2751–2768 (2016).

236. C. Fábri, S. Albert, Z. Chen, R. Prentner and M. Quack, A Molecular Quantum Switch Based on Tunneling in Meta-D-phenol C_6H_4DOH, *Phys. Chem. Chem. Phys.* **20** 7387–7394 (2018).

237. E. Fischer, Einfluss der Configuration auf die Wirkung der Enzyme. Berichte der Deutschen Chemischen Gesellschaft *27*, 2985-2993, **1894**.

238. E. Schrödinger, What is life?: the physical aspect of the living cell (Univ. Press, Cambridge, 1951), (Based on Lectures delivered under the auspices of the Institute at Trinity College, Dublin, in February 1943).

239. J. Monod, Le Hasard et la Nécessité - Essai sur la philosophie naturelle de la biologie moderne (Editions du Seuil, Paris, 1970).

240. X. Bonfils, X. Delfosse, S. Udry, T. Forveille, M. Mayor, C. Perrier, F. Bouchy, M. Gillon, C. Lovis, F. Pepe, D. Queloz, N. C. Santos, D. Ségransan and J.-L. Bertaux, The HARPS search for southern extra-solar planets - XXXI. The M-dwarf sample, *Astronomy & Astrophysics* **549**, p. A109 (2013).

241. M. Mayor, Nobel Lecture: Plurality of worlds in the cosmos: A dream of antiquity, a modern reality of astrophysics, *Rev. Mod. Phys.* **92**, p. 030502 (2020).

242. D. Queloz, Nobel Lecture: 51 Pegasi b and the exoplanet revolution, *Rev. Mod. Phys.* **92**, p. 030503 (2020).

243. P. Schwille, Biologische Selbstorganisation im Reagenzglas - ein Weg zur künstlichen Zelle?, in Was ist Leben?, Nova Acta Leopoldina, eds. J. Hacker and M. Hecker pp. 119–130, 2012.

244. P. Schwille, How Simple Could Life Be?, *Angew. Chem. Int. Ed.* **56** 10998–11002 (2017).

245. M. Quack, The Role of Quantum Intramolecular Dynamics in Unimolecular Reactions, *Philosoph. Trans. Royal Soc.* A **332** 203–220 (1990).

246. N. A. Hawbaker and D. G. Blackmond, Energy threshold for chiral symmetry breaking in molecular self-replication, *Nature Chem.* **11** 957–962 (2019).

247. D. G. Blackmond, C. R. McMillan, S. Ramdeehul, A. Schorm and J. M. Brown, Origins of Asymmetric Amplification in Autocatalytic Alkylzinc Additions, *J. Am. Chem. Soc.* **123** 10103–10104 (2001).

248. J. Yu, A. X. Jones, L. Legnani and D. G. Blackmond, Prebiotic access to enantioenriched glyceraldehyde mediated by peptides, *Chem. Sci.* **12** 6350–6354 (2021).

249. D. G. Blackmond, Asymmetric autocatalysis and its implications for the origin of homochirality, *Proc. Nat. Acad. Sci.* **101** 5732–5736 (2004).

250. J. I. Murray, J. N. Sanders, P. F. Richardson, K. N. Houk and D. G. Blackmond, Isotopically Directed Symmetry Breaking and Enantioenrichment in Attrition-Enhanced Deracemization, *J. Am. Chem. Soc.* **142** 3873–3879 (2020).

251. D. G. Blackmond, Autocatalytic Models for the Origin of Biological Homochirality, *Chem. Rev.* **120** 4831–4847 (2020).

252. L. Legnani, A. Darù, A. X. Jones and D. G. Blackmond, Mechanistic Insight into the Origin of Stereoselectivity in the Ribose-Mediated Strecker Synthesis of Alanine, *J. Am. Chem. Soc.* **143** 7852–7858 (2021).

253. K. Soai, T. Shibata, H. Morioka and K. Choji, Asymmetric autocatalysis and amplification of enantiomeric excess of a chiral molecule, *Nature* **378** 767–768 (1995).

254. T. Shibata, J. Yamamoto, N. Matsumoto, S. Yonekubo, S. Osanai and K. Soai, Amplification of a Slight Enantiomeric Imbalance in Molecules Based on Asymmetric Autocatalysis: The First Correlation between High Enantiomeric Enrichment in a Chiral Molecule and Circularly Polarized Light, *J. Am. Chem. Soc.* **120** 12157–12158 (1998).

255. K. Soai, S. Osanai, K. Kadowaki, S. Yonekubo, T. Shibata and I. Sato, d- and l-Quartz-Promoted Highly Enantioselective Synthesis of a Chiral Organic Compound, *J. Am. Chem. Soc.* **121** 11235–11236 (1999).

256. T. Kawasaki, Y. Matsumura, T. Tsutsumi, K. Suzuki, M. Ito and K. Soai, Asymmetric Autocatalysis Triggered by Carbon Isotope ($^{13}C^{12}C$) Chirality, *Science* **324** 492–495 (2009).

257. I. Sato, H. Urabe, S. Ishiguro, T. Shibata and K. Soai, Amplification of Chirality from Extremely Low to Greater than 99.5% ee by Asymmetric Autocatalysis, *Angew. Chem. Int. Ed.* **42** 315–317 (2003).

258. Y. Kaimori, Y. Hiyoshi, T. Kawasaki, A. Matsumoto and K. Soai, Formation of enantioenriched alkanol with stochastic distribution of enantiomers in the absolute asymmetric synthesis under heterogeneous solidvapor phase conditions, *Chem. Comm.* **55** 5223–5226 (2019).

259. S. V. Athavale, A. Simon, K. N. Houk and S. E. Denmark, Demystifying the asymmetry-amplifying, autocatalytic behaviour of the Soai reaction through structural, mechanistic and computational studies, *Nature Chem.* **12** 412–423 (2020).

260. M. Fidecaro and H.-J. Gerber, The fundamental symmetries in the neutral kaon system—a pedagogical choice, *Rep. Progr. Phys.* **69** 1713–1770 (2006).

261. G. Gabrielse, Probing nature's fundamental symmetries. One slow particle at a time, in Symmetrie und Asymmetrie in Wissenschaft und Kunst, Nova Acta Leopoldina NF 127 Nr. 412, eds. M. Quack and J. Hacker pp. 91–98, (Wissenschaftliche Verlagsgesellschaft, Stuttgart, 2016).

262. C. Zimmermann and T. W. Hänsch, Antiwasserstoff: Die Schlüsseltechniken zur Erzeugung und Spektroskopie sind vorhanden, *Physik. Blätter* **49** 193–196 (1993).

263. T. W. Hänsch, Passion for precision (The Nobel foundation, Stockholm, 2006).

264. B. Schwingenheuer, R. A. Briere, A. R. Barker, E. Cheu, L. K. Gibbons, D. A. Harris, G. Makoff, K. S. McFarland, A. Roodman, Y. W. Wah, B. Winstein, R. Winston, E. C. Swallow, G. J. Bock, R. Coleman, M. Crisler, J. Enagonio, R. Ford, Y. B. Hsiung, D. A. Jensen, E. Ramberg, R. Tschirhart, T. Yamanaka, E. M. Collins, G. D. Gollin, P. Gu, P. Haas, W. P. Hogan, S. K. Kim, J. N. Matthews, S. S. Myung, S. Schnetzer, S. V. Somalwar, G. B. Thomson and Y. Zou, *CPT* Tests in the Neutral Kaon System, *Phys. Rev. Lett.* **74** 4376–4379 (1995).

265. H. Dehmelt, R. Mittleman, R. S. Van Dyck and P. Schwinberg, Past Electron-Positron $g - 2$ Experiments Yielded Sharpest Bound on *CPT* Violation for Point Particles, *Phys. Rev. Lett.* **83** 4694–4696 (1999).

266. The ALPHA Collaboration, M. Ahmadi *et al.*, Investigation of the fine structure of antihydrogen, *Nature* **578** 375–380 (2020).

267. R. Riek, A Derivation of a Microscopic Entropy and Time Irreversibility From the Discreteness of Time, *Entropy* **16** 3149–3172 (2014).

268. J. J. Hudson, B. E. Sauer, M. R. Tarbutt and E. A. Hinds, Measurement of the Electron Electric Dipole Moment Using YbF Molecules, *Phys. Rev. Lett.* **89**, p. 023003 (2002).

269. K. Z. Rushchanskii, S. Kamba, V. Goian, P. Vaněk, M. Savinov, J. Prokleška, D. Nuzhnyy, K. Knížek, F. Laufek, S. Eckel, S. K. Lamoreaux, A. O. Sushkov, M. Ležaić and N. A. Spaldin, A multiferroic material to search for the permanent electric dipole moment of the electron, *Nature Materials* **9** 649–654 (2010).

270. J. Doyle, Lecture at this meeting (2021); J. M. Doyle, Z. D. Lasner, D. L. Augenbraun this volume p. 193.

271. R. R. Ernst, T. J. Carrington, G. Seyfang and F. Merkt, Editorial to the Special Issue of Molecular Physics, *Mol. Phys.* **111** 1939–1963 (2013).

272. K. Soai, Lecture at this meeting, (2021) and this volume p. 141.

273. D. Blackmond, Lecture at this meeting (2021)

274. M. Quack and M. Stockburger, Resonance fluorescence of aniline vapour, *J. Mol. Spectros.* **43**, 87–116 (1972).

275. M. Dine and A. Kusenko, Origin of the matter-antimatter asymmetry, *Rev. Mod. Phys.* **76**, 1–30 (2003).

276. M. Quack, Asymmetries as Quasi-Fossils of the Origin of Life and Matter Bunsen-Magazin **24**, 124–125 (2022) and Symmetry and Evolution: Molecules in Motion between less than Yoctoseconds and more than Days, Bunsen-Magazin **24**, 238–246 (2022).

277. M. Caviezel, V. Horka-Zelenkova, G. Seyfang and M. Quack, High-resolution FTIR and diode laser spectroscopy of trifluoromethylacetylene and tetrafluoromethane in a supersonic jet expansion, *Mol. Phys.* **120**, e2093285 (2022) DOI: 10.1080/00268976.2022.2093285

278. J. Agner, S. Albert, P. Allmendinger, U. Hollenstein, A. Hugi, P. Jouy, K. Keppler, M. Mangold, F. Merkt and M. Quack, High-resolution spectroscopic measurements of cold samples in a supersonic beam using a QCL dual-comb spectrometer *Mol. Phys.* **120**, e2094297 (2022) DOI: 10.1080/00268976.2022.2094297

279. L. Horny and M. Quack, Parity violation in $CH_3Re\ ^{16}O\ ^{17}O^{18}O$ (unpublished preliminary results).

CPSIA information can be obtained
at www.ICGtesting.com
Printed in the USA
LVHW050841020323
740311LV00003B/102

9 789811 26